PENGUIN BOOKS

COMPLEXITY

M. Mitchell Waldrop received his doctorate in elementary particle physics from the University of Wisconsin in 1975. The author of *Man-Made Minds*, a book about artificial intelligence, he spent ten years as a writer for *Science* magazine, for which he is now a contributing correspondent. He is currently at work on a book about computers and networking. He and his wife, Amy Friedlander, live in Washington, D.C.

M. MITCHELL WALDROP

COMPLEXITY

THE EMERGING SCIENCE AT THE EDGE
OF ORDER AND CHAOS

PENGUIN BOOKS

PENGUIN BOOKS

Published by the Penguin Group
Penguin Books Ltd, 27 Wrights Lane, London W8 5TZ, England
Penguin Putnam Inc., 375 Hudson Street, New York, New York 10014, USA
Penguin Books Australia Ltd, Ringwood, Victoria, Australia
Penguin Books Canada Ltd, 10 Alcorn Avenue, Toronto, Ontario, Canada M4V 3B2
Penguin Books India (P) Ltd, 11, Community Centre, Panchsheel Park, New Delhi – 110 017, India
Penguin Books (NZ) Ltd, Private Bag 102902, NSMC, Auckland, New Zealand
Penguin Books (South Africa) (Pty) Ltd, 5 Watkins Street, Denver Ext 4, Johannesburg 2094, South Africa

Penguin Books Ltd, Registered Offices: Harmondsworth, Middlesex, England

First published in the USA by Simon & Schuster 1992
First published in Great Britain by Viking 1993
Published in Penguin Books 1994
5

Copyright © M. Mitchell Waldrop, 1992
All rights reserved

The moral right of the author has been asserted

Printed in England by Clays Ltd, St Ives plc

To A. E. F.

Contents

VISIONS OF THE WHOLE   9

1. The Irish Idea of a Hero   15
2. The Revolt of the Old Turks   52
3. Secrets of the Old One   99
4. "You Guys Really *Believe* That?"   136
5. Master of the Game   144
6. Life at the Edge of Chaos   198
7. Peasants Under Glass   241
8. Waiting for Carnot   275
9. Work in Progress   324

BIBLIOGRAPHY   360
ACKNOWLEDGMENTS   365
INDEX   367

Visions of the Whole

This is a book about the science of *complexity*—a subject that's still so new and so wide-ranging that nobody knows quite how to define it, or even where its boundaries lie. But then, that's the whole point. If the field seems poorly defined at the moment, it's because complexity research is trying to grapple with questions that defy all the conventional categories. For example:

- Why did the Soviet Union's forty-year hegemony over eastern Europe collapse within a few months in 1989? And why did the Soviet Union itself come apart less that two years later? Why was the collapse of communism so fast and so complete? It surely had something to do with two men named Gorbachev and Yeltsin. And yet even they seemed to be swept up in events that were far beyond their control. Was there some global dynamic at work that transcends individual personalities?
- Why did the stock market crash more than 500 points on a single Monday in October 1987? A lot of the blame goes to computerized trading. But the computers had been around for years. Is there any reason why the crash came on that particular Monday?
- Why do ancient species and ecosystems often remain stable in the fossil record for millions of years—and then either die out or transform themselves into something new in a geological instant? Perhaps the dinosaurs got wiped out by an asteroid impact. But there weren't *that* many asteroids. What else was going on?
- Why do rural families in a nation such as Bangladesh still produce an

average of seven children apiece, even when birth control is made freely available—and even when the villagers seem perfectly well aware of how they're being hurt by the country's immense overpopulation and stagnant development? Why do they continue in a course of behavior that's so obviously disastrous?

- How did a primordial soup of amino acids and other simple molecules manage to turn itself into the first living cell some four billion years ago? There's no way the molecules could have just fallen together at random; as the creationists are fond of pointing out, the odds against that happening are ludicrous. So was the creation of life a miracle? Or was there something else going on in that primordial soup that we still don't understand?

- Why did individual cells begin to form alliances some 600 million years ago, thereby giving rise to multicellular organisms such as seaweed, jellyfish, insects, and eventually humans? For that matter, why do humans spend so much time and effort organizing themselves into families, tribes, communities, nations, and societies of all types? If evolution (or free-market capitalism) is really just a matter of the survival of the fittest, then why should it ever produce anything other than ruthless competition among individuals? In a world where nice guys all too often finish last, why should there be any such thing as trust or cooperation? And why, in spite of everything, do trust and cooperation not only exist but flourish?

- How can Darwinian natural selection account for such wonderfully intricate structures as the eye or the kidney? Is the incredibly precise organization that we find in living creatures really just the result of random evolutionary accidents? Or has something more been going on for the past four billion years, something that Darwin didn't know about?

- What *is* life, anyway? Is it nothing more than a particularly complicated kind of carbon chemistry? Or is it something more subtle? And what are we to make of creations such as computer viruses? Are they just pesky imitations of life—or in some fundamental sense are they really alive?

- What is a mind? How does a three-pound lump of ordinary matter, the brain, give rise to such ineffable qualities as feeling, thought, purpose, and awareness?

- And perhaps most fundamentally, why is there something rather than nothing? The universe started out from the formless miasma of the Big Bang. And ever since then it's been governed by an inexorable tendency toward disorder, dissolution, and decay, as described by the second law of thermodynamics. Yet the universe has also managed to bring forth structure on every scale: galaxies, stars, planets, bacteria, plants, animals, and brains. How? Is the cosmic compulsion for disorder matched by an equally powerful compulsion for order, structure, and organization? And if so, how can both processes be going on at once?

At first glance, about the only thing that these questions have in common is that they all have the same answer: "Nobody knows." Some of them don't even seem like scientific issues at all. And yet, when you look a little closer, they actually have quite a lot in common. For example, every one of these questions refers to a system that is *complex*, in the sense that a great many independent agents are interacting with each other in a great many ways. Think of the quadrillions of chemically reacting proteins, lipids, and nucleic acids that make up a living cell, or the billions of interconnected neurons that make up the brain, or the millions of mutually interdependent individuals who make up a human society.

In every case, moreover, the very richness of these interactions allows the system as a whole to undergo *spontaneous self-organization*. Thus, people trying to satisfy their material needs unconsciously organize themselves into an economy through myriad individual acts of buying and selling; it happens without anyone being in charge or consciously planning it. The genes in a developing embryo organize themselves in one way to make a liver cell and in another way to make a muscle cell. Flying birds adapt to the actions of their neighbors, unconsciously organizing themselves into a flock. Organisms constantly adapt to each other through evolution, thereby organizing themselves into an exquisitely tuned ecosystem. Atoms search for a minimum energy state by forming chemical bonds with each other, thereby organizing themselves into structures known as molecules. In every case, groups of agents seeking mutual accommodation and self-consistency somehow manage to transcend themselves, acquiring collective properties such as life, thought, and purpose that they might never have possessed individually.

Furthermore, these complex, self-organizing systems are *adaptive*, in that they don't just passively respond to events the way a rock might roll around in an earthquake. They actively try to turn whatever happens to their advantage. Thus, the human brain constantly organizes and reorganizes its billions of neural connections so as to learn from experience (sometimes, anyway). Species evolve for better survival in a changing environment—and so do corporations and industries. And the marketplace responds to changing tastes and lifestyles, immigration, technological developments, shifts in the price of raw materials, and a host of other factors.

Finally, every one of these complex, self-organizing, adaptive systems possesses a kind of dynamism that makes them qualitatively different from static objects such as computer chips or snowflakes, which are merely complicated. Complex systems are more spontaneous, more disorderly, more alive than that. At the same time, however, their peculiar dynamism is also a far cry from the weirdly unpredictable gyrations known as chaos. In the past two decades, chaos theory has shaken science to its foundations with the realization that very simple dynamical rules can give rise to extraordinarily intricate behavior; witness the endlessly detailed beauty of fractals, or the foaming turbulence of a river. And yet chaos by itself doesn't explain the structure, the coherence, the self-organizing cohesiveness of complex systems.

Instead, all these complex systems have somehow acquired the ability to bring order and chaos into a special kind of balance. This balance point—often called *the edge of chaos*—is were the components of a system never quite lock into place, and yet never quite dissolve into turbulence, either. The edge of chaos is where life has enough stability to sustain itself and enough creativity to deserve the name of life. The edge of chaos is where new ideas and innovative genotypes are forever nibbling away at the edges of the status quo, and where even the most entrenched old guard will eventually be overthrown. The edge of chaos is where centuries of slavery and segregation suddenly give way to the civil rights movement of the 1950s and 1960s; where seventy years of Soviet communism suddenly give way to political turmoil and ferment; where eons of evolutionary stability suddenly give way to wholesale species transformation. The edge of chaos is the constantly shifting battle zone between stagnation and anarchy, the one place where a complex system can be spontaneous, adaptive, and alive.

Complexity, adaptation, upheavals at the edge of chaos—these common themes are so striking that a growing number of scientists are convinced that there is more here than just a series of nice analogies. The movement's nerve center is a think tank known as the Santa Fe Institute, which was founded in the mid-1980s and which was originally housed in a rented convent in the midst of Santa Fe's art colony along Canyon Road. (Seminars were held in what used to be the chapel.) The researchers who gather there are an eclectic bunch, ranging from pony-tailed graduate students to Nobel laureates such as Murray Gell-Mann and Philip Anderson in physics and Kenneth Arrow in economics. But they all share the vision of an underlying unity, a common theoretical framework for complexity that would illuminate nature and humankind alike. They believe that they have in hand the mathematical tools to create such a framework, drawing from the past twenty years of intellectual ferment in such fields as neural networks, ecology, artificial intelligence, and chaos theory. They believe that their application of these ideas is allowing them to understand the spontaneous, self-organizing dynamics of the world in a way that no one ever has before—with the potential for immense impact on the conduct of economics, business, and even politics. They believe that they are forging the first rigorous alternative to the kind of linear, reductionist thinking that has dominated science since the time of Newton—and that has now gone about as far as it can go in addressing the problems of our modern world. They believe they are creating, in the words of Santa Fe Institute founder George Cowan, "the sciences of the twenty-first century."

This is their story.

1
The Irish Idea of a Hero

Sitting alone at his table by the bar, Brian Arthur stared out the front window of the tavern and did his best to ignore the young urban professionals drifting in to get an early start on Happy Hour. Outside, in the concrete canyons of the financial district, the typical San Francisco fog was turning into a typical San Francisco drizzle. That was fine by him. On this late afternoon of March 17, 1987, he wasn't in the mood to be impressed with brass fittings, ferns, and stained glass. He wasn't in a mood to celebrate Saint Patrick's Day. And he most definitely wasn't in a mood to carouse with ersatz Irishmen wearing bits of green on their pinstripes. He just wanted to silently sip his beer in frustrated rage. Stanford University Professor William Brian Arthur, native son of Belfast, Northern Ireland, was at rock bottom.

And the day had started so well.

That was the irony of it all. When he'd set out for Berkeley that morning, he'd actually been looking forward to the trip as a kind of triumphal reunion: local boy makes good. He'd really loved his years in Berkeley, back in the early 1970s. Perched on the hillsides north of Oakland, just across the bay from San Francisco, it was a pushy, vital, alive kind of place full of ethnics and street people and outrageous ideas. Berkeley was where he'd gotten his Ph.D. from the University of California, where he'd met and married a tall blonde doctoral student in statistics named Susan Peterson, where he'd spent his first "postdoc" year in the economics department. Berkeley, of all the places he'd lived and worked ever since, was the place he wanted to come home to.

Well now he *was* coming home, sort of. The event itself wouldn't be a big deal: just lunch with the chairman of the Berkeley economics department and one of his former professors there. But it was the first time he'd come back to his old department in years, and certainly the first time he'd ever done so feeling like an academic equal. He was coming back with twelve years of experience working all over the globe and a major reputation as a scholar of human fertility in the Third World. He was coming back as the occupant of an endowed chair of economics at Stanford—the sort of thing that rarely gets handed out to anyone under age fifty. At age forty-one, Arthur was coming back as someone who had made it in academia. And who knew? The folks at Berkeley might even start talking about a job offer.

Oh yes, he'd really been high on himself that morning. So why hadn't he, years ago, just stuck to the mainstream instead of trying to invent a whole new approach to economics? Why hadn't he played it safe instead of trying to get in step with some nebulous, half-imaginary scientific revolution?

Because he couldn't get it out of his head, that's why. Because he could see it almost everywhere he looked. The scientists barely seemed to recognize it themselves, most of the time. But after three hundred years of dissecting everything into molecules and atoms and nuclei and quarks, they finally seemed to be turning that process inside out. Instead of looking for the simplest pieces possible, they were starting to look at how those pieces go together into complex wholes.

He could see it happening in biology, where people had spent the past twenty years laying bare the molecular mechanisms of DNA, and proteins, and all the other components of the cell. Now they were also beginning to grapple with the essential mystery: how can several quadrillion such molecules organize themselves into an entity that moves, that responds, that reproduces, that is *alive*?

He could see it happening in the brain sciences, where neuroscientists, psychologists, computer scientists, and artificial intelligence researchers were struggling to comprehend the essence of mind: How do those billions of densely interconnected nerve cells inside our skulls give rise to feeling, thought, purpose, and awareness?

He could even see it happening in physics, where the physicists were still trying to come to terms with the mathematical theory of chaos, the intricate beauty of fractals, and the weird inner workings of solids and liquids. There was profound mystery here: Why is it that simple particles obeying simple rules will sometimes engage in the most astonishing, unpredictable behavior? And why is it that simple particles will spontaneously organize themselves into complex structures like stars, galaxies, snowflakes, and hurricanes—almost as if they were obeying a hidden yearning for organization and order?

The signs were everywhere. Arthur couldn't quite put the feeling into words. Nobody could, so far as he could tell. But somehow, he could sense that all these questions were really the *same* question. Somehow, the old categories of science were beginning to dissolve. Somehow, a new, unified science was out there waiting to be born. It would be a rigorous science, Arthur was convinced, just as "hard" as physics ever was, and just as thoroughly grounded in natural law. But instead of being a quest for the ultimate particles, it would be about flux, change, and the forming and dissolving of patterns. Instead of ignoring everything that wasn't uniform and predictable, it would have a place for individuality and the accidents of history. Instead of being about simplicity, it would be about—well, complexity.

And that was precisely where Arthur's new economics came in. Conventional economics, the kind he'd been taught in school, was about as far from this vision of complexity as you could imagine. Theoretical economists endlessly talked about the stability of the marketplace, and the balance of supply and demand. They transcribed the concept into mathematical equations and proved theorems about it. They accepted the gospel according to Adam Smith as the foundation for a kind of state religion. But when it came to instability and change in the economy—well, they seemed to find the very idea disturbing, something they'd just as soon not talk about.

But Arthur had embraced instability. Look out the window, he'd told his colleagues. Like it or not, the marketplace isn't stable. The *world* isn't stable. It's full of evolution, upheaval, and surprise. Economics had to take that ferment into account. And now he believed he'd found the way to do that, using a principle known as "increasing returns"—or in the King James translation, "To them that hath shall be given." Why had high-tech companies scrambled to locate in the Silicon Valley area around Stanford instead of in Ann Arbor or Berkeley? Because a lot of older high-tech companies were already there. Them that has gets. Why did the VHS video system run away with the market, even though Beta was technically a little bit better? Because a few more people happened to buy VHS systems early on, which led to more VHS movies in the video stores, which led to still more people buying VHS players, and so on. Them that has gets.

The examples could be multiplied endlessly. Arthur had convinced himself that increasing returns pointed the way to the future for economics, a future in which he and his colleagues would work alongside the physicists and the biologists to understand the messiness, the upheaval, and the spontaneous self-organization of the world. He'd convinced himself that increasing returns could be the foundation for a new and very different kind of economic science.

Unfortunately, however, he hadn't had much luck convincing anybody else. Outside of his immediate circle at Stanford, most economists thought his ideas were—strange. Journal editors were telling him that this increasing-returns stuff "wasn't economics." In seminars, a good fraction of the audience reacted with outrage: how *dare* he suggest that the economy was not in equilibrium! Arthur found the vehemence baffling. But clearly he needed allies, people who could open their minds and hear what he was trying to tell them. And that, as much as any desire for a homecoming, was the reason he'd gone to Berkeley.

So there they had all been, sitting down to sandwiches at the faculty club. Tom Rothenberg, one of his former professors, had asked the inevitable question: "So, Brian, what are you working on these days?" Arthur had given him the two-word answer just to get started: "Increasing returns." And the economics department chairman, Al Fishlow, had stared at him with a kind of deadpan look.

"But—we know increasing returns don't exist."

"Besides," jumped in Rothenberg with a grin, "if they did, we'd have to outlaw them!"

And then they'd laughed. Not unkindly. It was just an insider's joke. Arthur *knew* it was a joke. It was trivial. Yet that one sound had somehow shattered his whole bubble of anticipation. He'd sat there, struck speechless. Here were two of the economists he respected most, and they just—couldn't listen. Suddenly Arthur had felt naive. Stupid. Like someone who didn't know enough *not* to believe in increasing returns. Somehow, it had been the last straw.

He'd barely paid attention during the rest of the lunch. After it was over and everyone had said their polite good-byes, he'd climbed into his faded old Volvo and driven back over the Bay Bridge into San Francisco. He'd taken the first exit he could, onto the Embarcadero. He'd stopped at the first bar he found. And he'd come in here to sit amidst the ferns and to give some serious thought to getting out of economics entirely.

．　．　．

Somewhere around the bottom of his second beer, Arthur realized that the place was beginning to get seriously noisy. The yuppies were arriving in force to celebrate the patron saint of Ireland. Well, maybe it was time to go home. This certainly wasn't accomplishing anything. He got up and walked out to his car; the foggy drizzle was still coming down.

Home was in Palo Alto, thirty-five miles south of the city in the suburban flats around Stanford. It was sunset when he finally pulled into the driveway. He must have made some noise. His wife, Susan, opened the front door and watched him as he was walking across the lawn: a slim, prematurely gray man who doubtless looked about as fed up and bedraggled as he felt.

"Well," she said, standing there in the doorway, "how did it go in Berkeley? Did they like your ideas?"

"It was the pits," said Arthur. "Nobody there believes in increasing returns."

Susan Arthur had seen her husband returning from the academic wars before. "Well," she said, trying to find something comforting to say, "I guess it wouldn't be a revolution, would it, if everybody believed in it at the start?"

Arthur looked at her, struck speechless for the second time that day. And then he just couldn't help it. He started to laugh.

The Education of a Scientist

When you're growing up Catholic in Belfast, says Brian Arthur, speaking in the soft, high cadences of that city, a certain rebelliousness sets in naturally. It wasn't that he ever felt oppressed, exactly. His father was a bank manager and his family was solidly middle class. The only sectarian incident that ever involved him personally came one afternoon as he was walking home in his parochial school uniform: a bunch of Protestant boys started pelting him with bits of brick and stone, and one piece of brick hit him in the forehead. (He could hardly see for the blood pouring into his eyes—but he damn well threw that brick back.) Nor did he really feel that the Protestants were devils; his mother was a Protestant who converted to Catholicism when she married. He never even felt especially political. He tended to be much more interested in ideas and philosophy.

No, the rebelliousness is just something you pick up from the air. "The culture doesn't equip you to lead, but to undermine," he says. Look at whom the Irish admire: Wolfe Tone, Robert Emmet, Daniel O'Connell, Padraic Pearse. "All the Irish heroes were revolutionaries. The highest peak of heroism is to lead an absolutely *hopeless* revolution, and then give the greatest speech of your life from the dock—the night before you're hanged.

"In Ireland," he says, "an appeal to authority *never* works."

In an odd sort of way, Arthur adds, that streak of Irish rebelliousness is what got him started in his own academic career. Catholic Belfast tended to be rather contemptuous of intellectuals. So, of course, he became one. In fact, he can remember wanting to be a "scientist" as early as age four, long before he knew what a scientist was. The idea just seemed deliciously exotic and mysterious. And yet, having gotten that idea in his head, young Brian was nothing if not determined. At school he plunged into engineering and physics and hard-edged mathematics as soon as he could. And in 1966 he had taken first-class honors in electrical engineering at Queen's University in Belfast. "Oh, I suppose you'll end up a wee professor somewhere," said his mother, who was in fact very proud; no one in her generation of the family had ever even attended a university.

Later in 1966 that same determination had led him across the Irish Sea to England and the University of Lancaster, where he started graduate studies in a highly mathematical form of engineering known as operations research—basically, a set of techniques for calculating such things as how to organize a factory to get the most output for the least input, or how to keep a fighter jet under control when it is buffeted by unexpected forces. "At the time, British industry was in terrible shape," says Arthur. "I thought that maybe through science we could reorganize it and sort it out."

And in 1967, after the professors at Lancaster had proved insufferably stuffy and condescending—"Well," says Arthur, doing his best imitation of bored British snobbery, "it's nice to have an Irishman in the department; it adds a little colour"—he left for America and the University of Michigan in Ann Arbor. "From the moment I set foot here, I felt right at home," he says. "This was the sixties. The people were open, the culture was open, the scientific education was second to none. In the United States, anything seemed possible."

The one thing that wasn't possible in Ann Arbor, unfortunately, was ready access to the mountains and the sea, both of which Arthur loved. So he arranged to finish his Ph.D. work at Berkeley starting in the fall of 1969. And to support himself in the summer beforehand he applied for a job with McKinsey and Company, one of the top management consulting companies in the world.

That was a piece of incredibly good fortune. Arthur didn't realize until later just how lucky he was; people were clamoring to be hired at McKinsey. But it turned out that the company liked his operations research background and the fact that he knew German. They needed someone to work out of the Düsseldorf office. Was he interested?

Was he? Arthur had the time of his life. The last time he'd been in Germany he'd worked at a blue-collar summer job at 75 cents an hour. Now here he was, twenty-three years old, advising the board of directors of BASF on what to do with an oil and gas division or a fertilizer division worth hundreds of millions of dollars. "I learned that operating at the top was just as easy as operating at the bottom," he laughs.

But it was more than just an ego trip. Basically, McKinsey was selling modern American management techniques (a concept that didn't sound as funny in 1969 as it would have fifteen years later). "Companies in Europe at that time typically had hundreds of subdivisions," he says. "They didn't even know what they owned." Arthur discovered that he had a real taste for wading into messy problems like this and coming to grips with them firsthand. "McKinsey was genuinely first-rate," he says. "They weren't selling theories and they weren't selling fads. Their approach was to absolutely revel in the complexity, to live with it and breathe it. The McKinsey team would stay with a company for five or six months or more, studying a very complicated set of arrangements, until somehow certain patterns became clear. We'd all sit around on the edge of our desks and someone would say, 'This must be happening because of that,' and someone else would say, 'Then *that* must be so.' Then we'd go out and check it. And maybe the local executive would say, 'Well, you're almost right, but you forgot about such and such.' So we'd spend months clarifying and clarifying, until the issues were all worked out and the answer spoke for itself."

It didn't take very long for Arthur to realize that, when it came to real-world complexities, the elegant equations and the fancy mathematics he'd spent so much time on in school were no more than tools—and limited tools at that. The crucial skill was insight, the ability to see connections. And that fact, ironically, was what led him into economics. He remembers the occasion vividly. It was shortly before he was due to leave for Berkeley. He and his American boss, George Taucher, were driving one evening through West Germany's Ruhr Valley, the country's industrial heartland. And as they went, Taucher started talking about the history of each company they passed—who had owned what for a hundred years, and how the whole thing had built up in an absolutely organic, historical way. For Arthur it was a revelation. "I realized all of a sudden that this was economics." If he ever wanted to understand this messy world that fascinated him so much, if he ever wanted to make a real difference in people's lives, then he was going to have to learn economics.

So Arthur headed to Berkeley after that first summer on an intellectual high. And in total innocence, he announced that economics was what he would study.

Actually, he had no intention of completely shifting fields at this late date. He'd already finished most of his requirements for a Ph.D. in operations research at Michigan; the only remaining hurdle was to complete a dissertation, the large piece of original research with which a Ph.D. candidate supposedly demonstrates that he or she has mastered the craft. But he had more than enough time to do that: the University of California was insisting that he hang around Berkeley for another three years to fulfill its residency requirements. So Arthur was welcome to spend his extra time taking all the economics courses he could.

He did. "But after the McKinsey experience, I was very disappointed," he says. "This was nothing like the historical drama I'd been so fascinated with in the Ruhr Valley." In the lecture halls of Berkeley, economics seemed to be a branch of pure mathematics. "Neoclassical" economics, as the fundamental theory was known, had reduced the rich complexity of the world to a narrow set of abstract principles that could be written on a few pages. Whole textbooks were practically solid with equations. The brightest young economists seemed to be devoting their careers to proving theorem after theorem after theorem—whether or not those theorems had much to do with the world. "This extraordinary emphasis on mathematics surprised me," says Arthur. "To me, coming from applied mathematics, a theorem was a statement about an everlasting mathematical truth—not the dressing up of a trivial observation in a lot of formalism."

He couldn't help but feel that the theory was just too neat by half. It wasn't the mathematical rigor he objected to. He loved mathematics. After all those years of studying electrical engineering and operations research, moreover, he'd acquired considerably more background in mathematics than most of his economics classmates. No, what bothered him was the weird unreality of it all. The mathematical economists had been so successful at turning their discipline into ersatz physics that they had leached their theories clean of all human frailty and passion. Their theories described the human animal as a kind of elementary particle: "economic man," a godlike being whose reasoning is always perfect, and whose goals are always pursued with serenely predictable self-interest. And just as physicists could predict how a particle will respond to any given set of forces, economists could predict how economic man will respond to any given economic situation: he (or it) will just optimize his "utility function."

Neoclassical economics likewise described a society where the economy is poised forever in perfect equilibrium, where supply always exactly equals demand, where the stock market is never jolted by surges and crashes, where no company ever gets big enough to dominate the market, and where the magic of a perfectly free market makes everything turn out for the best. It was a vision that reminded Arthur of nothing so much as the eighteenth-century Enlightenment, when philosophers saw the cosmos as a kind of vast clockwork device kept in perfect running order by the laws of Sir Isaac Newton. The only difference was that the economists seemed to see human society as a perfectly oiled machine governed by the Invisible Hand of Adam Smith.

He just couldn't buy it. Granted, the free market was a wonderful thing, and Adam Smith had been a brilliant man. In fairness, moreover, neo-classical theorists had embroidered the basic model with all sorts of elab-orations to cover things like uncertainty about the future, or the transfer of property from one generation to the next. They had adapted it to fit taxation, monopolies, international trade, employment, finance, monetary policy—everything economists thought about. But none of that changed any of the fundamental assumptions. The theory still didn't describe the messiness and the irrationality of the human world that Arthur had seen in the valley of the Ruhr—or, for that matter, that he could see every day on the streets of Berkeley.

Arthur didn't exactly keep his opinions to himself. "I think I annoyed several of my professors by showing a great deal of impatience with theorems, and by wanting to know about the real economy," he says. He also knew he was hardly alone in those opinions: he could hear the grumbling in the hallways of any economics meeting he went to.

And yet, there was also a part of Arthur that found the neoclassical theory breathtakingly beautiful. As an intellectual tour de force it ranked right up there with the physics of Newton or Einstein. It had the kind of hard-edged clarity and precision that the mathematician in him couldn't help respond-ing to. Moreover, he could see why a previous generation of economists had welcomed it so enthusiastically. He'd heard horror stories about what economics was like when they were coming of age. Back in the 1930s, the English economist John Maynard Keynes had remarked that you could put five economists in a room and you'd get six different opinions. And from all reports, he was being kind. The economists of the 1930s and 1940s were long on insight, but they were often a trifle weak on logic. And even when they weren't, you'd still find that they came to very different con-clusions on the same problem: it turns out they were arguing from different, unstated assumptions. So these major wars would be fought out between different factions over government policies or theories of the business cycle. The generation of economists who crafted the mathematical theory in the 1940s and 1950s were the Young Turks of their day, a pack of brash upstarts determined to clean out the stables and make economics into a science as rigorous and as precise as physics. And they had come remarkably close; the Young Turks who had achieved it—Kenneth Arrow of Stanford, Paul Samuelson of MIT, Gerard Debreu of Berkeley, Tjalling Koopmans, and Lionel McKenzie of Rochester, among others—had deservedly gone on to become the Grand Old Men, the new establishment.

Besides, if you were going to do economics at all—and Arthur was still determined to do economics—what other theory were you going to use? Marxism? Well, this was Berkeley, and Karl Marx certainly had his followers. But Arthur wasn't one of them: so far as he was concerned, this business of class struggle proceeding in scientifically predictable stages was just plain silly. No, as the gambler once said, the game may be crooked, but it's the only game in town. So he kept on with his courses, determined to master the theoretical tools he couldn't quite believe in.

All this time, of course, Arthur had been working on his Ph.D. dissertation for operations research. And his adviser, mathematician Stuart Dreyfus, had proved to be both an excellent teacher and a kindred spirit. Arthur remembers stopping by Dreyfus's office to introduce himself shortly after he arrived at Berkeley in 1969. He met a long-haired bead-wearing graduate student coming out. "I'm looking for Professor Dreyfus," said Arthur. "Could you tell me when he's due back?"

"I'm Dreyfus," said the "student," who was in fact about forty.

Dreyfus reinforced all the lessons that Arthur had learned at McKinsey, and provided an ongoing antidote to the economics classes. "He believed in getting to the heart of a problem," says Arthur. "Instead of solving incredibly complicated equations, he taught me to keep simplifying the problem until you found something you could deal with. Look for what made a problem tick. Look for the key factor, the key ingredient, the key solution." Dreyfus would not let him get away with fancy mathematics for its own sake.

Arthur took Dreyfus's lessons to heart. "It was both good and bad," he says a bit sadly. Later on, his ideas on increasing returns might have gone down better with traditional economists if he'd hidden them in a thicket of mathematical formalism. In fact, colleagues urged him to do so. He wouldn't. "I wanted to say it as plainly and as simply as I could," he says.

In 1970 Arthur went back to Düsseldorf for a second summer with McKinsey and Company, and found it to be just as enthralling as the first. Sometimes he wonders if he should have kept up his contacts there and become a big-time international consultant after he graduated. He could have afforded a very luxurious lifestyle.

But he didn't. Instead he found himself being drawn to an economics specialty that focused on a problem even messier than industrial Europe: Third World population growth.

Of course, it didn't hurt that this specialty gave him the opportunity to go back and forth for study at the East-West Population Institute in Honolulu, where he could keep a surfboard ready for action on the beaches. But he was quite serious about it. This was the early 1970s, and the population problem was looming large. Stanford biologist Paul Ehrlich had just written his apocalyptic best-seller *The Population Bomb*. The Third World was full of newly independent former colonies struggling to achieve some kind of economic viability. And economists were full of theories about how to help them. The standard advice at the time tended to place a heavy reliance on economic determinism: to achieve its "optimum" population, all a country had to do was give its people the right economic incentives to control their reproduction, and they would automatically follow their own rational self-interest. In particular, many economists were arguing that when and if a country became a modern industrial state— organized along Western lines, of course—its citizenry would naturally undergo a "demographic transition," automatically lowering their birthrates to match those that prevailed in European countries.

Arthur, however, was convinced that he had a better approach, or at least a more sophisticated one: analyze population control in terms of "time-delayed" control theory, the subject of his Ph.D. dissertation. "The problem was one of timing," he says. "If a government manages to cut back on births today, it will affect school sizes in about 10 years, the labor force in 20 years, the size of the next generation in about 30 years, and the number of retirees in about 60 years. Mathematically, this is very much like trying to control a space probe far out in the solar system, where your commands take hours to reach it, or like trying to control the temperature of your shower when there's a half-minute delay between adjusting the tap and the hot water reaching you. If you don't take that delay into account properly, you can get scalded."

In 1973, Arthur included his population analysis as the final chapter in his dissertation: an equation-filled tome entitled *Dynamic Programming as Applied to Time-Delayed Control Theory*. "It was very much an engineering approach to the population problem," he says, looking back on it ruefully. "It was all just numbers." Despite all his experience with McKinsey and Dreyfus, and despite all his impatience with overmathematized economics, he was still feeling the same impulse that had led him into operations research in the first place: let's use science and mathematics to help run society rationally. "Most people in development economics have this kind of attitude," he says. "They're the missionaries of this century. But instead of bringing Christianity to the heathen, they're trying to bring economic development to the Third World."

What brought him back to reality with a jolt was going to work for a small New York think tank known as the Population Council. He arrived in 1974, after he had completed his doctorate and spent a year as a "postdoc" researcher in the Berkeley economics department. Physically, the Population Council was about as far from the Third World as you could get: it was set up in a Park Avenue skyscraper under the chairmanship of John D. Rockefeller III. But it did fund serious research into contraception, family planning, and economic development. And most important, from Arthur's point of view, it had a policy of getting its researchers away from their desks and out into the field as much as possible.

"Brian," the director would ask, "how much do you know about population and development in Bangladesh?"

"Very little."

"How would you like to find out?"

Bangladesh was a watershed for Arthur. He went there in 1975 with demographer Geoffrey McNicoll, an Australian who had been a fellow graduate student at Berkeley and who had been responsible for bringing Arthur to the Population Council in the first place. They arrived in the first plane permitted to land in the aftermath of a coup; they could still hear machine guns firing as they touched down. Then they proceeded into the countryside, where they acted like investigative reporters: "We talked to headmen in the villages, women in the villages, everyone. We interviewed and interviewed to understand how the rural society worked." In particular, they tried to find out why rural families were still producing an average of seven children apiece, even when modern birth control was made freely available—and even when the villagers seemed perfectly well aware of the country's immense overpopulation and stagnant development.

"What we found was that the terrible predicament of Bangladesh was the outcome of a network of individual and group interests at the village level," says Arthur. Since children could go to work at an early age, it was a net benefit to any individual family to have as many children as possible. Since a defenseless widow's relatives and neighbors might very well come in and take everything she possessed, it was in a young wife's interest to have as many sons as possible as quickly as possible, so that she would have grown sons to protect her in her old age. And so it went: "Patriarchs, women who were trying to hold onto their husbands, irrigation communities—all these interests combined to produce children and to stagnate development."

After six weeks in Bangladesh, Arthur and McNicoll returned to the United States to digest the information they had and to do further research in the anthropology and sociology journals. One of Arthur's first stops was Berkeley, where he dropped by the economics department in search of a reference. While he was there, he remembers, he happened to flip through a list of the latest course offerings. They were pretty much the same courses he had taken himself not so long ago. "But I had this very strange impression, as if I'd been off center a bit, that economics had changed in the year I'd been away. And then it dawned on me: *economics* hadn't changed. I had." After Bangladesh, all those neoclassical theorems that he'd worked so hard to learn seemed so—irrelevant. "Suddenly I felt 100 percent lighter, like a great weight had been lifted from me. I didn't have to believe this anymore! I felt it as a great freedom."

Arthur and McNicoll's eighty-page report, published in 1978, became something of a classic in social science—and was immediately banned in Bangladesh. (Much to the chagrin of the elite in Dacca, the capital, the authors had pointed out that the government had essentially no control of anything outside the capital; the countryside was essentially being run by local feudal godfathers.) But in any case, says Arthur, other missions for the Population Council in Syria and Kuwait only reinforced the lesson: the quantitative engineering approach—the idea that human beings will respond to abstract economic incentives like machines—was highly limited at best. Economics, as any historian or anthropologist could have told him instantly, was hopelessly intertwined with politics and culture. Perhaps the lesson was obvious, says Arthur, "But I had to learn it the hard way."

That insight likewise led him to abandon any hope of finding a general, deterministic theory of human fertility. Instead he began to conceive of fertility as part of a self-consistent pattern of folkways, myths, and social mores—a pattern, moreover, that was different for each culture. "You could measure something like income or childbearing in one country, and find that another country had the same levels of one, and totally different levels of the other. It would be a different pattern." Everything interlocked, and no piece of the puzzle could be considered in isolation from the others:

"The number of children interacted with the way their society was organized, and the way their society was organized had a lot to do with the number of children they had."

Patterns. Once he had made the leap, Arthur found that there was something about the concept that resonated. He had been fascinated by patterns all his life. Given a choice he would always take the window seat on airplanes, so he could look out on the ever-changing panorama below. He would generally see the same elements everywhere he went: rock, earth, ice, clouds, and so on. But these elements would be organized into characteristic patterns that might go on for half an hour. "So I asked myself the question, why does that geological pattern exist? Why is there a certain texture of rock formations and meandering rivers, and then half an hour later there's a totally different pattern?"

Now, however, he began to see patterns everywhere he went. In 1977, for example, he left the Population Council for a U.S.-Soviet think tank known as IIASA: the International Institute for Applied Systems Analysis. Created by Brezhnev and Nixon as a symbol of détente, it was housed in Maria Theresa's magnificent eighteenth-century "hunting lodge" in Laxenburg, a small village about ten miles outside of Vienna. It was also, as Arthur quickly determined, within ready driving distance from the ski slopes of the Tyrolean Alps.

"What struck me," he says, "was that if you went into one of these Alpine villages, it would have these ornate, Tyrolean roofs and balustrades and balconies, with characteristic pitches to the roofs, characteristic gables, and characteristic shutters on the windows. But rather than thinking that this was a nice jigsaw puzzle picture, I realized that there was not a single part of the village that wasn't there for a purpose, and interconnected with the other parts. The pitches of the roofs had to do with what would keep the right amount of snow on the roof for insulation in the winter. The degree of overhang of the gables beyond the balconies had to do with keeping snow from falling on the balconies. So I used to amuse myself looking at the villages, thinking that this part has this purpose, that part has that purpose, and they were all interconnected."

What also struck him, he says, was that just across the Italian border in the Dolomite Alps, the villages were suddenly not Tyrolean at all. It was no one thing that you could point to. It was just that myriad variant details added up to a totally different whole. And yet the Italian villagers and the Austrian villagers were coping with essentially the same problem of snowfall. "Over time," he says, "the two cultures had arrived at mutually self-consistent patterns that are different."

Epiphany on the Beach

Everyone has a research style, says Arthur. If you think of a research problem as being like a medieval walled city, then a lot of people will attack it head on, like a battering ram. They will storm the gates and try to smash through the defenses with sheer intellectual power and brilliance.

But Arthur has never felt that the battering ram approach was his strength. "I like to take my time as I think," he says. "So I just camp outside the city. I wait. And I think. Until one day—maybe after I've turned to a completely different problem—the drawbridge comes down and the defenders say, 'We surrender.' The answer to the problem comes all at once."

In the case of what he later came to call increasing returns economics, he had been camped for quite a long time. McKinsey. Bangladesh. His general disillusionment with standard economics. Patterns. None of it was quite the answer. But he can vividly remember when the drawbridge began to open.

It was in April 1979. His wife, Susan, was in a state of exhaustion after finishing her Ph.D. in statistics, and Arthur had arranged for an eight-week sabbatical from IIASA so that they could take a much-needed rest together in Honolulu. For himself, he made it a partial working vacation. From nine in the morning until three in the afternoon he would go over to the East-West Population Institute to work on a research paper while Susan continued to sleep—literally fifteen hours a day. Then in the late afternoon they would drive up to Hauula beach on the north side of Oahu: a tiny, almost deserted strip of sand where they could body-surf and lie around drinking beer, eating cheese, and reading. It was here, one lazy afternoon shortly after they arrived, that Arthur had opened up the book he had brought along for just such a moment: Horace Freeland Judson's *The Eighth Day of Creation*, a 600-page history of molecular biology.

"I was enthralled," he recalls. He read how James Watson and Francis Crick had discovered the double-helix structure of DNA in 1952. He read how the genetic code had been broken in the 1950s and 1960s. He read how scientists had slowly deciphered the intricately convoluted structures of proteins and enzymes. And as a lifetime laboratory klutz—"I've done miserably in every laboratory I've been in"—he read about the painstaking experiments that brought this science to life: the questions that made this or that experiment necessary, the months spent in planning each experiment and assembling the apparatus, and then the triumph or dejection when the answer was in hand. "Judson had the ability to bring the drama of science alive."

But what really galvanized him was the realization that here was a whole messy world—the interior of a living cell—that was at least as complicated as the messy human world. And yet it was a *science*. "I realized that I had been terribly unsophisticated about biology," he says. "When you're trained the way I was, in mathematics and engineering and economics, you tend to view science as something that only applies when you can use theorems and mathematics. But when it came to looking out the window at the domain of life, of organisms, of nature, I had this view that, somehow, science stops short." How do you write down a mathematical equation for a tree or a paramecium? You can't. "My vague notion was that biochemistry and molecular biology were just a bunch of classifications of this molecule or that. They didn't really help you understand anything."

Wrong. On every page, Judson was proving to him that biology was as much a science as physics had ever been—that this messy, organic, non-mechanistic world was in fact governed by a handful of principles that were as deep and profound as Newton's laws of motion. In every living cell there resides a long, helical DNA molecule: a chain of chemically encoded instructions, genes, that together constitute a blueprint for the cell. The genetic blueprints may be wildly different from one organism to the next. But in both, the genes will use essentially the same genetic code. That code will be deciphered by the same molecular code-breaking machinery. And that blueprint will be turned into proteins and membranes and other cellular structures in the same molecular workshops.

To Arthur, thinking of all the myriad forms of life on Earth, this was a revelation. At a molecular level, every living cell was astonishingly alike. The basic mechanisms were universal. And yet a tiny, almost undetectable mutation in the genetic blueprint might be enough to produce an enormous change in the organism as a whole. A few molecular shifts here and there might be enough to make the difference between brown eyes and blue, between a gymnast and a sumo wrestler, between good health and sickle-cell anemia. A few more molecular shifts, accumulating over millions of years through natural selection, might make the difference between a human and a chimpanzee, between a fig tree and a cactus, between an amoeba and a whale. In the biological world, Arthur realized, small chance events are magnified, exploited, built upon. One tiny accident can change everything. Life develops. It has a *history*. Maybe, he thought, maybe that's why this biological world seems so spontaneous, organic, and—well, alive.

Come to think of it, maybe that was also why the economists' imaginary world of perfect equilibrium had always struck him as static, machinelike, and dead. Nothing much could ever happen there; tiny chance imbalances in the market were supposed to die away as quickly as they occurred. Arthur couldn't imagine anything less like the real economy, where new products, technologies, and markets were constantly arising and old ones were constantly dying off. The real economy was not a machine but a kind of living system, with all the spontaneity and complexity that Judson was showing him in the world of molecular biology. Arthur had no idea yet how to use that insight. But it fired his imagination.

He kept reading: there was more. "Of all the drama in the book," says Arthur, "what appealed to me most was the work of Jacob and Monod." Working at the Institut Pasteur in Paris in the early 1960s, the French biologists Francois Jacob and Jacques Monod had discovered that a small fraction of the thousands of genes arrayed along the DNA molecule can function as tiny switches. Turn one of these switches on—by exposing the cell to a certain hormone, for example—and the newly activated gene will send out a chemical signal to its fellow genes. This signal will then travel up and down the length of the DNA molecule and trip other genetic switches, flipping some of them on and some of them off. These genes, in turn, start sending out chemical signals of their own (or stop sending them out). And as a result, still more genetic switches will be tripped in a mounting cascade, until the cell's collection of genes settles down into a new and stable pattern.

For biologists the implications of this discovery were enormous (so much so that Jacob and Monod later shared the Nobel Prize for it). It meant that the DNA residing in a cell's nucleus was not just a blueprint for the cell— a catalog of how to make this protein or that protein. DNA was actually the foreman in charge of construction. In effect, DNA was a kind of molecular-scale computer that directed how the cell was to build itself and repair itself and interact with the outside world. Furthermore, Jacob and Monod's discovery solved the long standing mystery of how one fertilized egg cell could divide and differentiate itself into muscle cells, brain cells, liver cells, and all the other kinds of cells that make up a newborn baby. Each different type of cell corresponded to a different pattern of activated genes.

To Arthur, the combination of déjà vu and excitement when he read this was overwhelming. Here it was again: patterns. An entire sprawling set of self-consistent patterns that formed and evolved and changed in response to the outside world. It reminded him of nothing so much as a kaleidoscope, where a handful of beads will lock in to one pattern and hold it—until a slow turn of the barrel causes them to suddenly cascade into a new configuration. A handful of pieces and an infinity of possible patterns.

Somehow, in a way he couldn't quite express, this seemed to be the essence of life.

When Arthur finished Judson's book he went prowling through the University of Hawaii bookstore, snatching up every book he could find on molecular biology. Back on the beach, he devoured them all. "I was captured," he says, "obsessed." By the time he returned to IIASA in June he was moving on pure intellectual adrenaline. He still had no clear idea how to apply all this to the economy. But he could feel that the essential clues were there. He continued to pour through biology texts all that summer. And in September, at the suggestion of a physicist colleague at IIASA, he started delving into the modern theories of condensed matter—the inner workings of liquids and solids.

He was as astonished as he had been at Hauula beach. He hadn't thought that physics was anything like biology. In fact, it *wasn't* like biology; the atoms and molecules that the physicists usually studied were much, much simpler than proteins and DNA. And yet, when you looked at those simple atoms and molecules interacting in massive numbers, you saw all the same phenomena: tiny initial differences producing enormously different effects. Simple dynamics producing astonishingly complex behaviors. A handful of pieces falling into a near-infinity of possible patterns. Somehow, at some very deep level that Arthur didn't know how to define, the phenomena of physics and biology were the same.

On the other hand, there was one very important difference at a practical level: the systems that physicists studied were simple enough that they could analyze them with rigorous mathematics. Suddenly, Arthur began to feel right at home. If he'd had any lingering doubts before, he knew now he was dealing with *science*. "These were not just fuzzy notions," he says.

He found that he was most impressed with the writings of the Belgian physicist Ilya Prigogine. Prigogine, as he later discovered, was considered by many other physicists to be an insufferable self-promoter who often exaggerated the significance of what he had accomplished. Nonetheless, he was an undeniably compelling writer. And perhaps not coincidentally, his work in the field of "nonequilibrium thermodynamics" had convinced the Swedish Academy of Sciences to award him the Nobel Prize in 1977.

Basically, Prigogine was addressing the question, Why is there order and structure in the world? Where does it come from?

This turns out to be a much tougher question than it might sound, especially when you consider the world's general tendency toward decay. Iron rusts. Fallen logs rot. Bathwater cools to the temperature of its surroundings. Nature seems to be less interested in creating structures than in tearing structures apart and mixing things up into a kind of average. Indeed, the process of disorder and decay seems inexorable—so much so that nineteenth-century physicists codified it as the second law of thermodynamics, which can be paraphrased as "You can't unscramble an egg." Left to themselves, says the second law, atoms will mix and randomize themselves as much as possible. That's why iron rusts: atoms in the iron are forever trying to mingle with oxygen in the air to form iron oxide. And that's why bathwater cools: fast-moving molecules on the surface of the water collide with slower-moving molecules in the air, and gradually transfer their energy.

Yet for all of that, we do see plenty of order and structure around. Fallen logs rot—but trees also grow. So how do you reconcile this growth of structure with the second law of thermodynamics?

The answer, as Prigogine and others realized back in the 1960s, lies in that innocuous-sounding phrase, "Left to themselves . . ." In the real world, atoms and molecules are almost never left to themselves, not completely; they are almost always exposed to a certain amount of energy and material flowing in from the outside. And if that flow of energy and material is strong enough, then the steady degradation demanded by the second law can be partially reversed. Over a limited region, in fact, a system can spontaneously organize itself into a whole series of complex structures.

The most familiar example is probably a pot of soup sitting on the stovetop. If the gas is off, then nothing happens. Just as the second law predicts, the soup will sit there at room temperature, in equilibrium with its surroundings. If the gas is turned on with a very tiny flame, then still nothing much happens. The system is no longer in equilibrium—heat energy is rising up through the soup from the bottom of the pot—but the difference isn't large enough to really disturb anything. But now turn the flame up just a little bit higher, moving the system just a little farther from equilibrium. Suddenly, the increased flux of heat energy turns the soup unstable. Tiny, random motions of the soup molecules no longer average out to zero; some of the motions start to grow. Portions of the fluid begin to rise. Other portions begin to fall. Very quickly, the soup begins to organize its motions on a large scale: looking down on the surface you can see a hexagonal pattern of convection cells, with fluid rising in the middle of each cell and falling along the sides. The soup has acquired order and structure. In a word, it has begun to simmer.

Such self-organizing structures are ubiquitous in nature, said Prigogine. A laser is a self-organizing system in which particles of light, photons, can spontaneously group themselves into a single powerful beam that has every photon moving in lockstep. A hurricane is a self-organizing system powered by the steady stream of energy coming in from the sun, which drives the winds and draws rainwater from the oceans. A living cell—although much too complicated to analyze mathematically—is a self-organizing system that survives by taking in energy in the form of food and excreting energy in the form of heat and waste.

In fact, wrote Prigogine in one article, it's conceivable that the economy is a self-organizing system, in which market structures are spontaneously organized by such things as the demand for labor and the demand for goods and services.

Arthur sat up immediately when he read those words. "The economy is a self-organizing system." That was it! That was precisely what he had been thinking ever since he'd read *The Eighth Day of Creation*, although he hadn't known how to articulate it. Prigogine's principle of self-organization, the spontaneous dynamics of living systems—now Arthur could finally see how to relate all of it to economic systems.

In hindsight it was all so obvious. In mathematical terms, Prigogine's central point was that self-organization depends upon self-reinforcement: a tendency for small effects to become magnified when conditions are right, instead of dying away. It was precisely the same message that had been implicit in Jacob and Monod's work on DNA. And suddenly, says Arthur, "I recognized it as what in engineering we would have called positive feedback." Tiny molecular motions grow into convection cells. Mild tropical winds grow into a hurricane. Seeds and embryos grow into fully developed living creatures. Positive feedback seemed to be the sine qua non of change, of surprise, of life itself.

And yet, positive feedback is precisely what conventional economics didn't have, Arthur realized. Quite the opposite. Neoclassical theory assumes that the economy is entirely dominated by *negative* feedback: the tendency of small effects to die away. In fact, he can remember listening with some puzzlement as his economics professors back in Berkeley had hammered away on the point. Of course, they didn't call it negative feedback. The dying-away tendency was implicit in the economic doctrine of "diminishing returns": the idea that the second candy bar doesn't taste nearly as good as the first one, that twice the fertilizer doesn't produce twice the yield, that the more you do of anything, the less useful, less profitable, or less enjoyble the last little bit becomes. But Arthur could see that the net effect was the same: just as negative feedback keeps small perturbations from running away and tearing things apart in physical systems, diminishing returns ensure that no one firm or product can ever grow big enough to dominate the marketplace. When people get tired of candy bars, they switch to apples or whatever. When all the best hydroelectric dam sites have been used, the utility companies start building coal-fired plants. When enough fertilizer is enough, farmers quit applying it. Indeed, negative feedback/ diminishing returns is what underlies the whole neoclassical vision of harmony, stability, and equilibrium in the economy.

But even back in Berkeley, Arthur the engineering student couldn't help but wonder: What happens if you have positive feedback in the economy? Or in the economics jargon, what happens if you have *increasing* returns?

"Don't worry about it," his teachers had reassured him. "Increasing-returns situations are extremely rare, and they don't last very long." And since Arthur didn't have any particular example in mind, he had shut up about it and gone on to other things.

But now, reading Prigogine, it all came flooding back to him. Positive feedback, increasing returns—maybe these things *did* happen in the real economy. Maybe they explained the liveliness, the complexity, the richness he saw in the real-world economy all around him.

Maybe so. The more he thought about it, in fact, the more Arthur came to realize what an immense difference increasing returns would make to economics. Take efficiency, for example. Neoclassical theory would have us believe that a free market will always winnow out the best and most efficient technologies. And, in fact, the market doesn't do too badly. But then, Arthur wondered, what are we to make of the standard QWERTY keyboard layout, the one used on virtually every typewriter and computer keyboard in the Western world? (The name QWERTY is spelled out by the first six letters along the top row.) Is this the most efficient way to arrange the keys on a typewriter keyboard? Not by a long shot. An engineer named Christopher Scholes designed the QWERTY layout in 1873 specifically to slow typists down; the typewriting machines of the day tended to jam if the typist went too fast. But then the Remington Sewing Machine Company mass-produced a typewriter using the QWERTY keyboard, which meant that lots of typists began to learn the system, which meant that other typewriter companies began to offer the QWERTY keyboard, which meant that still more typists began to learn it, et cetera, et cetera. To them that hath shall be given, thought Arthur—increasing returns. And now that QWERTY is a standard used by millions of people, it's essentially locked in forever.

Or consider the Beta versus VHS competition in the mid-1970s. Even in 1979 it was clear that the VHS videotape format was well on its way to cornering the market, despite the fact that many experts had originally rated it slightly inferior to Beta technologically. How could this have happened? Because the VHS vendors were lucky enough to gain a slightly bigger market share in the beginning, which gave them an enormous advantage in spite of the technological differences: the video stores hated having to stock everything in two different formats, and consumers hated the idea of being stuck with obsolete VCRs. So everyone had a big incentive to go with the market leader. That pushed up VHS's market share even more, and the small initial difference grew rapidly. Once again, increasing returns.

Or take this endlessly fascinating business of patterns. Pure neoclassical theory tells us that high-tech firms will tend to distribute themselves evenly across the landscape: there's no reason for any of them to prefer one location over another. But in real life, of course, they flock to places like California's Silicon Valley and Boston's Route 128 to be near other high-tech firms. Them that has gets—and the world acquires structure. In fact, Arthur suddenly realized, that's why you get patterns in any system: a rich mixture of positive and negative feedbacks can't *help* producing patterns. Imagine spilling a little water onto the surface of a highly polished tray, he says; it beads up into a complex pattern of droplets. And it does so because two countervailing forces are at work. There is gravity, which tries to spread out the water to make a very thin, flat film across the whole surface. That's negative feedback. And there is surface tension, the attraction of one water molecule to another, which tries to pull the liquid together into compact globules. That's positive feedback. It's the mix of the two forces that produces the complex pattern of beads. Moreover, that pattern is unique. Try the experiment again and you'll get a completely different arrangement of droplets. Tiny accidents of history—infinitesimal dust motes and invisible irregularities in the surface of the tray—get magnified by the positive feedback into major differences in the outcome.

Indeed, thought Arthur, that probably explains why history, in Winston Churchill's phrase, is just one damn thing after another. Increasing returns can take a trivial happenstance—who bumped into whom in the hallway, where the wagon train happened to stop for the night, where trading posts happened to be set up, where Italian shoemakers happened to emigrate— and magnify it into something historically irreversible. Did a certain young actress become a superstar on the basis of pure talent? Hardly: the luck of being in a single hit movie sent her career into hyperdrive on name recognition alone, while her equally talented contemporaries went nowhere. Did British colonists flock to cold, stormy, rocky shores of Massachusetts Bay because New England had the best land for farms? No: They came because Massachusetts Bay was where the Pilgrims got off the boat, and the Pilgrims got off the boat there because the *Mayflower* got lost looking for Virginia. Them that has gets—and once the colony was established, there was no turning back. Nobody was about to pick up Boston and move it someplace else.

Increasing returns, lock-in, unpredictability, tiny events that have immense historical consequences—"These properties of increasing-returns economics shocked me at first," says Arthur. "But when I recognized that each property had a counterpart in the nonlinear physics I was reading, I got very excited. Instead of being shocked, I became fascinated." Economists had actually been talking about such things for generations, he learned. But their efforts had always been isolated and scattered. He felt as though he were recognizing for the first time that all these problems were the *same* problem. "I found myself walking into Aladdin's cave," he says, "picking up one treasure after another."

By the autumn, everything had fallen into place. On November 5, 1979, he poured it all out. At the top of one page of his notebook he wrote the words "Economics Old and New," and under them listed two columns:

Old Economics	New Economics
• Decreasing returns	• Much use of increasing returns
• Based on 19th-century physics (equilibrium, stability, deterministic dynamics)	• Based on biology (structure, pattern, self-organization, life cycle)
• People identical	• Focus on individual life; people separate and different
• If only there were no externalities and all had equal abilities, we'd reach Nirvana	• Externalities and differences become driving force. No Nirvana. System constantly unfolding.
• Elements are quantities and prices	• Elements are patterns and possibilities
• No real dynamics in the sense that everything is at equilibrium	• Economy is constantly on the edge of time. It rushes forward, structures constantly coalescing, decaying, changing.

- Sees subject as structurally simple
- Economics as soft physics

- Sees subject as inherently complex
- Economics as high-complexity science

And so it went, for three pages. It was his manifesto for a whole new kind of economics. After all those years, he says, "I finally had a point of view. A vision. A solution." It was a vision much like that of the Greek philosopher Heraclitus, who observed that you can never step into the same river twice. In Arthur's new economics, the economic world would be part of the human world. It would always be the same, but it would never be the same. It would be fluid, ever-changing, and alive.

What's the Point?

To say that Arthur was bubbling over with enthusiasm for his vision would be an understatement. But it didn't take him too long to realize that his enthusiasm was less than infectious, especially to other economists. "I thought that if you did something different and important—and I did think increasing returns made sense of a lot of phenomena in economics and gave a direction that was badly needed—people would hoist me on their shoulders and carry me in triumph. But that was just incredibly naive."

Before the month of November was out he found himself walking in the park near IIASA's Hapsburg palace, excitedly explaining increasing returns to a visiting Norwegian economist, Victor Norman. And he was suddenly taken aback to realize that Norman, a distinguished international trade theorist, was looking at him in bafflement: What was the point of all this? He heard much the same reaction when he began to give talks and seminars on increasing returns in 1980. About half his audience would typically be very interested, while the other half ranged from puzzled to skeptical to hostile. What was the point? And what does any of this increasing-returns stuff have to do with *real* economics?

Questions like that left Arthur at a loss. How could they not see it? The *point* was that you have to look at the world as it is, not as some elegant theory says it ought to be. The whole thing reminded him of medical practice in the Renaissance, when doctors of medicine were learned in matters of theory and rarely deigned to touch a real patient. Health was simply a matter of equilibrium back then: If you were a sanguine person, or a choleric person, or whatever, you merely needed to have your fluids brought back into balance. "But what we know from 300 years worth of medicine, going from Harvey's discovery of the circulation of the blood on through molecular biology, is that the human organism is profoundly complicated. And that means that we now listen to a doctor who puts a stethoscope to a patient's chest and looks at each individual case." Indeed, it was only when medical researchers started paying attention to the real complications of the body that they were able to devise procedures and drugs that actually had a chance of doing some good.

He saw the increasing-returns approach as a step down that same path for economics. "The important thing is to observe the actual living economy out there," he says. "It's path-dependent, it's complicated, it's evolving, it's open, and it's organic."

Very quickly, however, it became apparent that what was really getting his critics riled up was this concept of the economy locking itself in to an unpredictable outcome. If the world can organize itself into many, many possible patterns, they asked, and if the pattern it finally chooses is a historical accident, then how can you predict anything? And if you can't predict anything, then how can what you're doing be called science?

Arthur had to admit that *was* a good question. Economists had long ago gotten the idea that their field had to be as "scientific" as physics, meaning that everything had to be mathematically predictable. And it was quite some time before even he got it through his head that physics isn't the only kind of science. Was Darwin "unscientific" because he couldn't predict what species will evolve in the next million years? Are geologists unscientific because they can't predict precisely where the next earthquake will come, or where the next mountain range will rise? Are astronomers unscientific because they can't predict precisely where the next star will be born?

Not at all. Predictions are nice, if you can make them. But the essence of science lies in *explanation*, laying bare the fundamental mechanisms of nature. That's what biologists, geologists, and astronomers do in their fields. And that's what he was trying to do for increasing returns.

Not surprisingly, arguments like that didn't convince anyone who didn't want to be convinced. On one occasion at IIASA in February 1982, for example, as Arthur was answering questions from the audience after a lecture on increasing returns, a visiting U.S. economist got up and demanded rather angrily, "Just give me one example of a technology that we are locked in to that isn't superior to its rivals!"

Arthur glanced at the lecture hall clock because he was running out of time, and almost without thinking said, "Oh! The clock."

The clock? Well, he explained, all our clocks today have hands that move "clockwise." But under his theory, you'd expect there might be fossil technologies, buried deep in history, that might have been just as good as the ones that prevailed. It's just that by chance they didn't get going. "For all I know, at some stage in history there may have been clocks with hands that went backward. They might have been as common as the ones we have now."

His questioner was unimpressed. Another distinguished U.S. economist then got up and snapped, "I don't see that it's locked in anyway. I wear a digital watch."

To Arthur, that was missing the point. But time was up for that day. And besides, it was just a conjecture. About three weeks later, however, he received a postcard from his IIASA colleague James Vaupel, who had been vacationing in Florence. The postcard showed the Florence Cathedral clock, which had been designed by Paolo Uccello in 1443—and which ran backward. (It also displayed all 24 hours.) On the flip side, Vaupel had simply written, "Congratulations!"

Arthur loved the Uccello clock so much that he made a transparency of it so he could show it in overhead projectors in all his future lectures on lock-in. It always produced a reaction. Once, in fact, he was showing the clock transparency during a talk at Stanford when an economics graduate student leaped up, flipped the transparency over so that everything was reversed, and said triumphantly, "You see! This is a hoax! The clock actually goes clockwise!" Fortunately, however, Arthur had been doing a little research into clocks in the meantime, and he had another transparency of a backward clock with a Latin inscription. He put this transparency up, and said, "Unless you assume this is mirror writing done by Leonardo da Vinci, you have to accept that these clocks go backwards."

Actually, by that point Arthur was able to give his audiences any number of lock-in examples. There were Beta-versus-VHS and QWERTY, of course. But there was also the strange case of the internal combustion engine. In the 1890s, Arthur discovered, when the automotive industry was still in its infancy, gasoline was considered the least-promising power source. Its chief rival, steam, was well developed, familiar, and safe; gasoline was expensive, noisy, dangerously explosive, hard to obtain in the right grade, and required a new kind of engine containing complicated new parts. Gasoline engines were also inherently less fuel-efficient. If things had been different and if steam engines had benefited from the same ninety years of development lavished on gasoline engines, we might now be living with considerably less air pollution and considerably less dependence on foreign oil.

But gasoline did win out—largely, Arthur found, because of a series of historical accidents. In 1895, for example, a horseless-carriage competition sponsored by the Chicago *Times-Herald* was won by a gasoline-powered Duryea—one of only two cars to finish out of six starters. This may have been the inspiration for Ransom Olds's 1896 patent of a gasoline engine that he subsequently mass-produced in the "Curved-Dash Olds." This allowed gasoline power to overcome its slow start. Then in 1914 there was an outbreak of hoof-and-mouth disease in North America, leading to the withdrawal of horse troughs—which were the only places where steam cars could fill up with water. By the time the Stanley brothers, makers of the Stanley Steamer, were able to develop a condenser and boiler system that did not need to be refilled every thirty or forty miles, it was too late. The steam car never recovered. Gasoline power quickly became locked in.

And then there was the case of nuclear power. When the United States embarked on its civilian nuclear power program in 1956, a number of designs were proposed: reactors cooled by gas, by ordinary "light" water, by a more exotic fluid known as "heavy" water, and even by liquid sodium. Each design had its technical advantages and disadvantages; indeed, with a perspective of thirty years, many engineers believe that a high-temperature, gas-cooled design would have been inherently safer and more efficient than the others, and may have forestalled most of the public anxiety and opposition to nuclear power. But as it happened, the technical arguments were almost irrelevent to the final choice. When the Soviets launched *Sputnik* in October of 1957, the Eisenhower administration was suddenly eager to get some reactor up and running—any reactor. And at the time, the only reactor that was anywhere near being ready was a highly compact, high-powered version of the light-water reactor, which had been developed by the Navy as a power plant for its nuclear submarines. So the Navy's design was hurriedly scaled up to commercial size and placed into operation. That led to further technical development of the light-water design, and by the mid-1960s, it had essentially displaced all the others in the United States.

Arthur recalls using the light-water reactor example in 1984 during a talk at the Kennedy School of Government at Harvard. "I was saying that here's a simple model that shows the economy can lock in to an inferior outcome, as it appears to have done with the light-water reactor. Whereupon a certain very distinguished economist stood up and shouted, 'Well, under perfect capital markets, it couldn't happen!' He gave a lot of technicalities, but basically, if you wheel up a lot of extra assumptions, then perfect capitalism would restore the Adam Smith world."

Well, maybe he was right. But six months later, Arthur gave the same talk in Moscow. Whereupon a member of the Supreme Soviet who happened to be in the audience got up and said, "What you're describing may happen in Western economies. But with perfect socialist planning this can't happen. We would arrive at the correct outcome."

Of course, so long as QWERTY, steam cars, and light-water reactors were just isolated examples, critics could always dismiss lock-in and increasing returns as something rare and pathological. Surely, they said, the *normal* economy isn't that messy and unpredictable. And at first Arthur suspected that they might be right; most of the time the market *is* fairly stable. It was only much later, as he was preparing a lecture on increasing returns for a group of postgraduate students, that he suddenly realized why the critics were also wrong. Increasing returns isn't an isolated phenomenon at all: the principle applies to everything in high technology.

Look at a software product like Microsoft's Windows, he says. The company spent $50 million in research and development to get the first copy out the door. The second copy cost it—what, $10 in materials? It's the same story in electronics, computers, pharmaceuticals, even aerospace. (Cost for the first B2 bomber: $21 billion. Cost per copy: $500 million.) High technology could almost be defined as "congealed knowledge," says Arthur. "The marginal cost is next to zilch, which means that every copy you produce makes the product cheaper and cheaper." More than that, every copy offers a chance for learning: getting the yield up on microprocessor chips, and so on. So there's a tremendous reward for increasing production—in short, the system is governed by increasing returns.

Among high-tech customers, meanwhile, there's an equally large reward for flocking to a standard. "If I'm an airline buying a Boeing jet," says Arthur, "I want to make sure I buy a lot of them so that my pilots don't have to switch." By the same token, if you're an office manager, you try to buy all the same kind of personal computer so that everyone in the office can run the same software. The result is that high technologies very quickly tend to lock in to a relatively few standards: IBM and Macintosh in the personal computer world, or Boeing, McDonnell Douglas, and Lockheed in commercial passenger aircraft.

Now compare that with standard bulk commodities such as grain, fertilizer, or cement, where most of the know-how was acquired generations ago. Today the real costs are for labor, land, and raw materials, areas where diminishing returns can set in easily. (Producing more grain, for example, may require that farmers start to open up less productive land.) So these tend to be stable, mature industries that are described reasonably well by standard neoclassical economics. "In that sense, increasing returns isn't displacing the standard theory at all," says Arthur, "It's helping complete the standard theory. It just applies in a different domain."

What this means in practical terms, he adds, is that U.S. policy-makers ought to be very careful about their economic assumptions regarding, say, trade policy vis-à-vis Japan. "If you're using standard theory you can get it very badly wrong," he says. Several years ago, for example, he was at a conference where the British economist Christopher Freeman got up and declared that Japan's success in consumer electronics and other high-tech markets was inevitable. Just look at the country's low cost of capital, said Freeman, along with its canny investment banks, its powerful cartels, and its compelling need to exploit technology in the absence of oil and mineral resources.

"Well, I was the next speaker," says Arthur. "So I said, 'Let's imagine that Thailand or Indonesia had taken off and Japan was still languishing. Conventional economists would then be pointing to all the same reasons to explain Japan's backwardness. The low cost of capital means a low rate of return on capital—so there's no reason to invest. Cartels are known to be inefficient. Collective decision-making means molasses-slow decision-making. Banks are not set up to take risks. And economies are hobbled if they lack oil and mineral resources. So how could the Japanese economy possibly have developed?' "

Since the Japanese economy quite obviously did develop, says Arthur, he argued for a different explanation: "I said that Japanese companies weren't successful because they had some magical qualities that U.S. and European companies didn't have. They were successful because increasing returns make high-tech markets unstable, lucrative, and possible to corner—and because Japan understood this better and earlier than other countries. The Japanese are very quick at learning from other nations. And they are very good at targeting markets, going in with huge volume, and taking advantage of the dynamics of increasing returns to lock in their advantage."

He still believes that, says Arthur. And by the same token, he suspects that one of the main reasons the United States has had such a big problem with "competitiveness" is that government policy-makers and business executives alike were very slow to recognize the winner-take-all nature of high-tech markets. All through the 1970s and well into the 1980s, Arthur points out, the federal government followed a "hands-off" policy based on a conventional economic wisdom, which did not recognize the importance of nurturing an early advantage before the other side locks in the market. As a result, high-tech industries were treated exactly the same as low-tech, bulk-commodity industries. Any "industrial policy" that might have given a boost to infant industries was ridiculed as an assault on the free market. Free and open trade on everything remained a national goal. And firms were discouraged from cooperating by antitrust regulations drawn up in an era when the world was dominated by bulk commodities. That approach has begun to change a bit in the 1990s, says Arthur. But only a bit. So he, for one, argues that it is high time to rethink the conventional wisdom in light of increasing returns. "If we want to continue manufacturing our wealth from knowledge," he says, "we need to accommodate the new rules."

Meanwhile, even as he was collecting dozens of real-world examples of increasing returns, Arthur was looking for a way to analyze the phenomenon in rigorous mathematical terms. "I'm certainly not against mathematics per se," he says. "I'm a heavy-duty user. I'm just against mathematics when it's misapplied, when it becomes formalism for its own sake." Used correctly, he says, mathematics can give your ideas a tremendous clarity. It's like an engineer who gets an idea for a device and then builds a working model. The equations can tell you which parts of your theory work and which don't. They can tell you which concepts are necessary and which aren't. "When you mathematize something you distill its essence," he says.

Besides, says Arthur, he knew that if he didn't come up with a rigorous mathematical analysis of increasing returns, the wider economics community would never regard his theory as anything *more* than a collection of anecdotes. Look at what had happened in every previous effort to introduce the concept. Back in 1891, the great English economist Alfred Marshall actually devoted quite a bit of space to the increasing returns in his *Principles of Economics*—the book in which he also introduced the concept of diminishing returns. "Marshall thought very deeply on increasing returns," says Arthur. "But he didn't have the mathematical tools to do much with it in an analytical way. In particular, he says, Marshall recognized even then that increasing returns could lead to multiple possible outcomes in the economy, which meant that the fundamental problem for economists was to understand precisely how one solution rather than another came to be selected. And economists ever since have gotten hung up on the same point. "Wherever there is more than one equilibrium point possible, the outcome was deemed to be indeterminate," he says. "End of story. There was no *theory* of how an equilibrium point came to be selected. And without that, economists couldn't bring themselves to incorporate increasing returns."

Something similar happened in the 1920s, when a number of European economists tried to use increasing-returns concepts to explain why cities grew and concentrated the way they did, and why different cities (and different countries) would specialize in, say, shoes or chocolates or fine violins. The basic concepts were correct, says Arthur. But again the mathematical tools just weren't there. "In the face of indeterminacy," he says, "economics came to a halt."

So Arthur sharpened his pencils and went to work. What he wanted was a mathematical framework that incorporated *dynamics*—that showed explicitly, step by step, how the marketplace chose among the multiple possible outcomes. "In the real world, outcomes don't just happen," he says. "They build up gradually as small chance events become magnified by positive feedbacks." What he finally came up with in 1981, after many consultations with friends and colleagues, was a set of abstract equations based on a sophisticated theory of nonlinear, random processes. The equations were actually quite general, he says, and applied to essentially any kind of increasing-returns situation. Conceptually, however, they worked something like this: Suppose you are buying a car. (At the time, lots of people at IIASA were buying Volkswagens and Fiats.) And suppose, for the sake of simplicity, that there are just two models to choose from. Call them A and B. Now, you've read the brochures on both cars, says Arthur, but they're pretty similar, and you still aren't sure which to buy. So what do you do? Like any sensible person, you start asking your friends. And then it so happens, purely by chance, that the first two or three people you talk to say that they've been driving car A. They tell you that it works fine. So you decide to buy one, too.

But notice, says Arthur, there is now one more A-type driver in the world: you. And that means that the next person to come along asking about cars is just a little more likely to encounter an A-type driver. So that person will be just a little bit more likely to choose car A than you were. With enough lucky breaks like this, car A can come to dominate the market.

On the other hand, he says, suppose the breaks had gone the other way. Then you might have chosen to go with model B, and then car B would have gotten the edge and come to dominate.

In fact, says Arthur, under some conditions you can even show mathematically that with a few lucky breaks either way in the beginning, this kind of process can produce any outcome at all. The car sales might eventually come to lock in at a ratio of 40 percent A to 60 percent B, or 89 percent A to 11 percent B, or anything else. And it all works purely by chance. "Showing how chance events work to select one equilibrium point from many possible in random processes was the most challenging thing I've ever done," says Arthur. But by 1981, working in collaboration with his IIASA colleagues Yuri Ermoliev and Yuri Kaniovski of the Skorokhod School in Kiev—"two of the best probability theorists in the world"—he had it. The three of them published the first of their several papers on the subject in the Soviet journal *Kibernetika* in 1983. "Now," says Arthur, "economists could not only follow the entire process by which *one* outcome emerged, they could see mathematically how different sets of historical accidents could cause radically different outcomes to emerge."

And most important, he says, increasing returns was no longer, in the words of the great Austrian economist Joseph Schumpeter, "a chaos that is not under analytical control."

Violating Sacred Ground

In 1982, Arthur suddenly found IIASA to be a far less hospitable place than it had been, courtesy of the rapidly chilling Cold War. The Reagan administration, eager to avoid any further taint by association with the Evil Empire, had abruptly pulled the United States out of the organization. Arthur was sorry to go. He'd greatly enjoyed working with his Soviet colleagues, and how could you beat an office in a Hapsburg palace? But things worked out well enough, as it happens. As a stopgap, Arthur took up a one-year visiting professorship at Stanford, where his reputation for demography seemed to stand him in good stead. And shortly before his year there drew to a close he got a call from the dean: "What would it take to keep you here?"

"Well," Arthur replied, secure in the knowledge that he already had a fistful of job offers from the World Bank, the London School of Economics, and Princeton, "I see there's this endowed chair coming open. . . ."

The dean was shocked. Endowed professorships are very prestigious. They are generally only awarded to the most distinguished researchers. In effect, they are sinecures for life. "We don't negotiate with endowed chairs!" she declared.

"I wasn't negotiating," said Arthur. "You just asked me what it would take to keep me here."

So they gave it to him. In 1983, at age thirty-seven, Arthur became the Dean and Virginia Morrison Professor of Population Studies and Economics. "My first permanent job in academia!" he laughs. He was one of the youngest endowed professors in Stanford's history.

It was a moment to savor—which in retrospect, turned out to be a good thing. He wasn't destined to have many such moments for a long while. However much his fellow economists may have liked his work in demography, many of them seemed to find his ideas on increasing-returns economics outrageous.

To be fair, he says, many of them were also quite receptive, even enthusiastic. But it was true that his most virulent critics had almost always been Americans. And being at Stanford brought him face to face with that fact. "I could talk about these ideas in Caracas, no sweat whatever. I could talk about them in Vienna, no sweat. But whenever I talked about these ideas in the United States, there was hell to pay. People got angry at the very notion that anything like this could happen."

Arthur found the Americans' hostility both mystifying and disturbing. Some of it he put down to their well-known fondness for mathematics. After all, if you spend your career proving theorems about the existence of market equilibrium, and the uniqueness of market equilibrium, and the efficiency of market equilibrium, you aren't likely to be very happy when someone comes along and tells you that there's something fishy about market equilibrium. As the economist John R. Hicks had written in 1939, when he looked aghast at the implications of increasing returns, "The threatened wreckage is that of the greater part of economic theory."

But Arthur also sensed that the hostility went deeper than that. American economists are famous for being far more passionately devoted to free-market principles than almost anyone else in the world. At the time, in fact, the Reagan administration was busily cutting taxes, junking federal regulations, "privatizing" federal services, and generally treating free-market capitalism as a kind of state religion. And the reason for that passion, as Arthur slowly came to realize, was that the free-market ideal had become bound up with American ideals of individual rights and individual liberty:

both are grounded in the notion that society works best when people are left alone to do what they want.

"Every democratic society has to solve a certain problem," says Arthur: "If you let people do their own thing, how do you assure the common good? In Germany, that problem is solved by everybody watching everybody else out the windows. People will come right up to you and say, 'Put a cap on that baby!' "

In England, they have this notion of a body of wise people at the top looking after things. "Oh, yes, we've had this Royal Commission, chaired by Lord So-and-So. We've taken all your interests into account, and there'll be a nuclear reactor in your backyard tomorrow."

But in the United States, the ideal is maximum individual freedom— or, as Arthur puts it, "letting everybody be their own John Wayne and run around with guns." However much that ideal is compromised in practice, it still holds mythic power.

But increasing returns cut to the heart of that myth. If small chance events can lock you in to any of several possible outcomes, then the outcome that's actually selected may *not* be the best. And that means that maximum individual freedom—and the free market—might *not* produce the best of all possible worlds. So by advocating increasing returns, Arthur was innocently treading into a minefield.

Well, he had to admit that he'd had fair warning.

It was in 1980, he recalls. He had been invited to give a series of talks on economic demography at the Academy of Sciences in Budapest. And one evening, at the bar of the Budapest Intercontinental Hotel, he found himself chatting with academician Maria Augusztinovics. Standing there with a scotch in one hand and a cigarette in the other, she was a most formidable lady. Not only had she married, in succession, most of the top economists in Hungary but she was a very perceptive economist herself. Moreover, she was an influential politician, with a post high in the Hungarian government. She was rumored to eat bureaucrats for breakfast. Arthur saw no reason to doubt it.

What are you working on these days? she asked. Arthur enthusiastically launched into a discourse about increasing returns. "It explains so many problems," he concluded, "all these processes and patterns."

Augusztinovics, who knew exactly what the philosophical stakes were for Western economists, simply looked at him with a kind of pity. "They will crucify you," she said.

"She was right," says Arthur. "The years from 1982 through 1987 were dreadful. That's when my hair turned gray."

Arthur has to admit that he brought a lot of that agony on himself. "If I had been the kind of person who forms inside allegiances in the profession, then the whole thing might have gone smoother," he says. "But I'm not an insider by nature. I'm just not a joiner."

With that Irish streak of rebelliousness, he was also not in a mood to dress up his ideas in a lot of jargon and phony analysis just to make them palatable to the mainstream. And that's what led him to make a critical tactical blunder: in the summer of 1983, when he was preparing his first paper on increasing returns for official publication, he wrote the thing in more or less plain English.

"I was convinced that I was onto something crucial in economics," he explains. "So I decided that I should write it at a very intelligible level, where it could be understood even by undergraduates. I thought that fancy mathematics would just get in the way of the argument. I also thought, 'Gee, I've published heavily mathematical papers before. I don't need to prove anything.'"

Wrong. If he hadn't known it before, he says, he learned it soon enough. Theoretical economists use their mathematical prowess the way the great stags of the forest use their antlers: to do battle with one another and to establish dominance. A stag who doesn't use his antlers is nothing. It was fortunate that Arthur circulated his manuscript informally that autumn as an IIASA working paper. The official, published version wasn't to see the light of day for another six years.

The most prestigious U.S. journal, *The American Economic Review*, sent the paper back in early 1984 with a letter from the editor saying, in essence, "No way!" *The Quarterly Journal of Economics* sent the paper back saying that its reviewers could find no technical fault, but that they just didn't think the work was worth anything. *The American Economic Review*, under a new editor this time, tentatively accepted the paper on its second submittal, bounced it around internally for two and a half years while demanding innumerable rewrites, and then rejected it again. And *The Economic Journal* in Britain simply said, "No!" (After some fourteen rewrites, the paper was finally accepted by *The Economic Journal* and published in March 1989 as "Competing Technologies, Increasing Returns, and Lock-In by Historical Events.")

Arthur was left in helpless rage. Martin Luther could nail his ninety-five theses to the church door of Wittenberg to be read by one and all. But in modern academia, there are no church doors; an idea that hasn't been published in an established journal doesn't officially exist. And what he found doubly frustrating, ironically, was the fact that the idea of increasing returns was finally beginning to catch on. It was becoming something of a movement in economics—and so long as his paper was in limbo, he couldn't take part in it.

Take the economic historians, for example—the people who did empirical studies on the history of technologies, the origin of industries, and the development of real economies. Stanford had a first-class group of them, and they had been among Arthur's earliest and most enthusiastic supporters. For years, they had suffered from the fact that neoclassical theory, if really taken seriously, says that history is irrelevant. An economy in perfect equilibrium exists outside of history; the marketplace will converge to the best of all possible worlds no matter what historical accidents intervene. And while very few economists took it quite that seriously, a lot of economics departments around the country were thinking of scrapping their required courses in economic history. So the historians *liked* lock-in. They *liked* the idea that small events could have large consequences. They saw Arthur's ideas about increasing returns as providing them with a rationale for their existence.

No one was a more effective advocate of that point of view than Arthur's Stanford colleague Paul David, who had independently published some thoughts about increasing returns and economic history back in the mid-1970s. But from Arthur's point of view, even David's support backfired. At the national meeting of the American Economics Association in late 1984, David participated in a panel discussion on "What Is the Use of History?" and used the QWERTY keyboard example to explain lock-in and path dependence to 600 economists at once. The talk created a sensation. Even the hard-core mathematical economists were impressed: here was a *theoretical* reason for thinking that history was important. Even the Boston *Globe* wrote about it. And Arthur was soon hearing people ask him, "Oh, you're from Stanford. Have you heard about Paul David's work on lock-in and path dependence?"

"It was simply dreadful," Arthur recalls. "I felt I had something to say, and I couldn't say it—and the ideas were getting credited to other people. It appeared that I was following rather than leading. I felt like I was in some doomed fairy story."

The Berkeley debacle with Fishlow and Rothenberg in March 1987 was arguably his lowest moment—but not by much. He began to have nightmares. "About three times a week I'd have this dream of a plane taking off—and I was not on it. I felt I was definitely getting left behind." He seriously began to think of abandoning economics and devoting himself full time again to his demographic research. His academic career seemed to be turning to ashes.

All that kept him going was stubbornness. "I just pushed, and pushed, and pushed," he says. "I just kept believing that the system had to give somewhere."

Actually, he was right. And as it happens, he didn't have too much longer to wait.

2
The Revolt of the Old Turks

About a month after his ill-starred trip to Berkeley, as Brian Arthur was walking across the Stanford University campus on a sunny California day in April 1987, he was startled to see a bicycle pull up in front of him bearing a distinguished figure in sports coat, tie, and battered white bicycle helmet. "Brian," said Kenneth Arrow, "I was just going to call you."

Arrow. Arthur was instantly on the alert. It wasn't that he was afraid of Arrow, exactly. True, Arrow had pretty much invented the kind of hyper-mathematical economics that he was rebelling against. But he knew Arrow to be an affable, open-minded man who loved nothing better than a good debate, and who could still be your friend after tearing your arguments to shreds. No, it was just that—well, talking to Arrow was like talking to the pope. Arrow was arguably the finest living economist in the world. He had won the Nobel Prize in economics more than a decade before. At age sixty-five he still possessed a lightning-fast intellect and a legendary impatience with sloppy reasoning. He could change the whole tone of a seminar just by walking into the room: The speaker would start walking on eggs. People in the audience would quit joking around and straighten up. Everyone would focus intently on the subject at hand. They would frame their questions and comments very, very carefully. Nobody, but nobody, wanted to look dumb in front of Ken Arrow.

"Oh, hi," said Arthur.

Arrow, clearly in a hurry, quickly explained that he was helping to put together a meeting of economists and physicists at a small institute in New Mexico. It would be held toward the end of the summer, he said. The plan was that he would invite ten economists and that Phil Anderson, the condensed-matter physicist, would invite ten physical scientists. "So could you come and give a paper on mode-locking?" he asked.

"Certainly I could," Arthur heard himself say. Mode-locking? What the devil was mode-locking? Could Arrow be talking about his work on lock-in and increasing returns? Did Arrow even *know* about his work on increasing returns? "Umm—where is this institute?"

"It's in Santa Fe, up in the foothills of the Rockies," Arrow said, climbing back on his bicycle. With a quick good-bye and a promise to send more information later, he pedaled off, his white helmet making him visible down Stanford's palm-shaded walkways for quite a long distance.

Arthur stared after him, trying to figure out what in the world he had just committed himself to. He didn't know which surprised him more: that physicists would want to talk to economists—or that Arrow would want to talk to *him*.

A few weeks later, in May of 1987, Arthur got a telephone call from a soft-spoken man who introduced himself as George Cowan, from the Santa Fe Institute. Cowan thanked him for agreeing to come to the economics meeting that fall. He and his colleagues took this meeting very seriously indeed, he explained. The institute was a small, private organization set up by the physicist Murray Gell-Mann and others to study aspects of complex systems, by which they meant everything from condensed-matter physics to society as a whole—anything with lots of strongly interacting parts. The institute had no faculty or students. But it was interested in building as wide a network of researchers as possible. And economics was very much on its agenda.

But what he was really calling for, added Cowan, was that Ken Arrow had suggested that the institute invite Arthur to be a visiting fellow that fall. This meant that Arthur could come out several weeks before the economics meeting and then stay for several weeks afterward, so that he would have the time to talk and work with other researchers in residence at the institute. Would he be interested?

"Certainly," said Arthur. Six weeks in Santa Fe in the autumn, with all expenses paid—why not? Besides, he had to admit that he was impressed by the academic firepower. After Arrow and Anderson, Gell-Mann was the third Nobel laureate in a row that he'd heard of in connection with this Santa Fe Institute. Gell-Mann was the fellow who'd invented the idea of "quarks," the little thingies that are supposed to run around inside of protons and neutrons. Arthur still had no clear idea of what this guy Cowan meant by "complex systems." But the whole thing was beginning to sound just crazy enough to be interesting.

"Oh, by the way," Arthur said, "I'm afraid no one's mentioned your name to me before. What do you do there?"

There was a pause and a kind of cough at the other end of the line. "I'm the president," said Cowan.

George

Actually, Arthur wasn't the only one baffled by the Santa Fe Institute. The first encounter was always a bit of a shock to everybody. The place violated stereotypes wholesale. Here was an outfit founded by aging academics rich with privilege, fame, and Nobel Prizes—the very people you'd expect to be smugly content with the status quo. And yet they were using it as a platform to foment a self-proclaimed scientific revolution.

Here was an institute populated largely by hard-core physicists and computer jocks from Los Alamos, the original Shangri La of nuclear weaponry. And yet the hallways were full of excited talk about the new sciences of "complexity": a kind of Grand Unified Holism that would run the gamut from evolutionary biology to fuzzy subjects like economics, politics, and history—not to mention helping us all to build a more sustainable and peaceful world.

In short, here was a total paradox. If you tried to imagine the Santa Fe Institute happening in the business world, you'd have to imagine the director of corporate research for IBM going off to start a little New Age Karmic counseling service in his garage—and then talking the chairmen of Xerox, Chase Manhattan, and GM into joining him.

What makes it even more remarkable is that the entrepreneur in this picture—George A. Cowan, the former head of research at Los Alamos—was about as un–New Age as anyone could imagine. At age sixty-seven he was a retiring, soft-spoken man who managed to look a bit like Mother Teresa in a golf shirt and unbuttoned sweater. He was not noted for his charisma; in any given group he was usually the fellow standing off to one side, listening. And he was certainly not known for his soaring rhetoric. Anyone who asked him why he had organized the institute was liable to get a precise, high-minded discussion of the shape of science in the twenty-first century and the need to take hold of scientific opportunities—the sort of recitation that might do well as an earnest guest editorial in *Science* magazine.

Only slowly, in fact, would it begin to dawn on the listener that Cowan, in his own cerebral way, was a fervent and determined man indeed. He didn't see the Santa Fe Institute as a paradox at all. He saw it as embodying a purpose far more important than George A. Cowan, Los Alamos, or any of the other accidents of its creation—and for that matter, far more important than the institute itself. If things didn't work out this time, he often said, then somebody would just have to do it all over again twenty years down the road. To Cowan, the Santa Fe Institute was a mission. To Cowan, it was a chance for science as a whole to achieve a kind of redemption and rebirth.

There was a time, distant as it seems now, when it was perfectly possible for an idealistic young scientist to devote himself to the creation of nuclear weapons for the sake of a better world. And George Cowan has never found cause to regret that devotion. "I've had second thoughts my whole life," he says. "But regrets on moral grounds? No. Without nuclear weapons we might have been on an even more ruinous road to destruction through biological and chemical weapons. I suspect that the history of the past fifty years has been a lot kinder to human beings than if the 1940s hadn't occurred."

Indeed, he says, in those days, work on nuclear weapons was almost a moral imperative. During the war, of course, Cowan and his fellow scientists saw themselves in a desperate race against the Nazis, who still had some of the best physicists in the world, and who were thought—wrongly, as it turned out—to be way ahead on bomb designs. "We knew that if we didn't get cracking, then Hitler would get the bomb," says Cowan, "and that would be the end."

He actually found himself swept up in the bomb effort before the Manhattan Project even existed. In the fall of 1941, when he was a twenty-one-year-old with a fresh undergraduate degree in chemistry from the Worcester Polytechnic Institute in his hometown of Worcester, Massachusetts, he had gone to work on the cyclotron project at Princeton, where physicists were studying the newly discovered process of nuclear fission and its effects on an isotope known as uranium-235. His intention had been to start taking graduate courses in physics on the side. But that intention got put on indefinite hold as of December 7, 1941, when the laboratory suddenly went on a seven-day workweek. Even then it was feared that the Germans were working on an atomic bomb, he says, and the physicists were frantic to find out if such a thing was even possible. "The measurements we were making were absolutely essential to deciding whether you could achieve a chain reaction in uranium," says Cowan. The answer, it turned out, was yes. And the federal government suddenly found itself to be much in need of Mr. Cowan's services. "That particular mix of chemistry and nuclear physics made me an expert on a number of things that were needed in the bomb project."

From 1942 until the end of the war he worked out of the Metallurgy Lab at the University of Chicago, where the Italian physicist Enrico Fermi was leading the effort to build the first atomic "pile"—a stack of uranium and graphite blocks that could demonstrate a controllable chain reaction. As a very junior member of that team, Cowan became something of a jack-of-all-trades, casting uranium metal, machining the graphite blocks that would control the pile's reaction rate, and anything else that needed doing. But by the time Fermi's atomic pile successfully went critical in December 1942, Cowan found that his experience there had made him one of the Manhattan Project's experts on the chemistry of radioactive elements. So the project managers started sending him off to places like Oak Ridge, Tennessee, where he helped the engineers at the hastily constructed nuclear facility there figure out exactly how much plutonium they were producing. "I was single, so they transferred me all around the country," he says. "Whenever there was a bottleneck, I was one of the guys who was likely to be sent off to help fix it." Indeed, Cowan was one of the very select group of people who were allowed to travel back and forth between different components of the project, which was kept tightly compartmentalized for security reasons. "I don't know why they trusted me," he laughs. "I drank as much as anyone else." He still has a souvenir of that period: a letter from the Chicago personnel office to his local draft board in Worcester, stating that Mr. Cowan possessed skills that were uniquely useful to the war effort, that he had been granted a deferment by the president himself, and would they *please* quit trying to reclassify him 1-A?

After the war, the scientists' desperate race against Hitler was transformed into an anxiety-ridden race against the Russians. It was a decidedly nasty time, says Cowan. Stalin's seizure of Eastern Europe, the Berlin blockade, and then Korea—the Cold War seemed all too close to becoming a very hot war indeed. The Soviets were known to be working on their own nuclear capabilities. It seemed that the only way to maintain the precarious balance of power—and not incidentally, to defend the cause of democracy and human freedom—was to continue improving the nuclear weapons on our side. That sense of urgency is what led Cowan to return to Los Alamos in July of 1949, having spent the previous three years getting his doctorate in physical chemistry at Carnegie Tech in Pittsburgh. It wasn't an automatic choice. In fact he made it only after considerable thought and soul searching. But the decision was reinforced almost immediately.

A week or two after he arrived, Cowan recalls, the director of radiochemistry research dropped by and, in a hush-hush, oblique way, asked him if his new laboratory was totally free of radioactive contamination. When Cowan said yes, he and his facility were immediately commandeered for a crash-priority, top-secret analysis job. The air samples arrived that very night. He wasn't told where they came from, but he could guess that they had been obtained from somewhere near the borders of Russia. And once he and his colleagues had detected the telltale signs of radioactive fallout, there was no getting around it: the Soviets had exploded an atomic bomb of their own.

"So they eventually put me on this panel in Washington," says Cowan. "Very covert." Cryptically known as the Bethe Panel—its first chairman was Cornell University physicist Hans Bethe—it was actually a group of atomic scientists convened to track Soviet nuclear weapons development. Cowan was thirty years old. High-level government officials believed at the start that the fallout detected by the chemists couldn't possibly mean what it obviously did mean. The officials *knew* that it would be years before Stalin had an atomic bomb; the Soviets must have had a reactor blow up. "But the nice thing about radiochemistry is that you can tell exactly what happened," says Cowan. The distribution of radioactive isotopes produced in a reactor is very different from the distribution coming out of a bomb explosion. "It took a lot of debate to convince them." But in the end, the older and wiser officials had no choice but to accept the hard evidence. The Soviet bomb was dubbed "Joe-1" in honor of Joseph Stalin, and the nuclear arms race was on.

So no, says Cowan, he has no apologies to make for his work on nuclear weapons. But he does have one very large regret about those years: his sense that the scientific community collectively abdicated responsibility for what it had done.

Oh, not immediately, of course, and not completely. In 1945 a number of the Manhattan Project scientists at Chicago circulated a petition urging that the bomb be demonstrated on an uninhabited island instead of on Japan itself. Then after the bombs were dropped on Hiroshima and Nagasaki and the war was over, many of the scientists on the project started forming political activist groups lobbying for the strictest possible control of nuclear weaponry—*civilian* control, not military. Those years saw the founding of *The Bulletin of the Atomic Scientists*, a magazine for dealing with the social and political consequences of this new form of power, and the formation of activist organizations such as the Federation of Atomic Scientists (now the Federation of American Scientists), of which Cowan was a member. "The people from the Manhattan Project who went to Washington were listened to very carefully," says Cowan. "In the 1940s, after the bomb, physical scientists were looked to as miracle workers. They had a lot to do with drafting the McMahon bill that created the Atomic Energy Commission and put atomic energy under civilian control.

"But that effort wasn't as completely supported by scientists as it might have been," says Cowan. And after the McMahon bill passed in July 1946, the scientists' activism largely died away. It was probably inevitable, he says. The culture of science does not mix well with the culture of politics. "Scientists who go to Washington as scientists generally leave screaming," he says. "It's totally alien to them. They want policy to be made on the basis of logic and scientific facts, and that's probably just a will-o'-the-wisp." But for whatever reason, the researchers went happily back to their labs, leaving war to the generals and politics to the politicians. And in so doing, says Cowan, they blew a chance for access and influence that they may never have again.

Cowan doesn't exempt himself from this indictment, although he actually remained more involved than most. In 1954, for example, he became president of an association of Los Alamos scientists who met with Atomic Energy Commission chairman Lewis Strauss at the height of the McCarthy uproar, when the senator from Wisconsin seemed to be convincing everyone that the country was riddled with communists. Cowan and his colleagues protested the anticommunist witch hunts and made the case for greater freedom of information and less secrecy at the lab. They also tried—without much success—to defend former Manhattan Project director J. Robert Oppenheimer, who was even then being stripped of his security clearance on the grounds that he might have associated with some people who might have attended a Communist party meeting back in the 1930s.

With his ongoing service on the Bethe Panel, meanwhile—a task he continued for some three decades—Cowan had come to realize what a disturbingly simpleminded place Washington could be. In the aftermath of World War II, he says, the United States had emerged from its prewar isolationism with a clear understanding that, yes, military power was extremely important. But having learned that lesson, all too many officials seemed oblivious to anything else. "Their view was, 'You gotta grab 'em by the balls.'" he says. "I felt then that power is a symphony orchestra, and too many people could only play bull fiddle."

In fact, Cowan had the distressing sense that the Soviets understood the intricate harmonies of power much better than Washington did. "They seemed to pay a great deal of attention to power's intellectual appeals, to its emotional and ideological aspects. And at the time, I thought they were paying a great deal of attention to its scientific aspects. It turns out that we thought they were ten feet tall and they weren't. But I was thinking in terms of the contrast between the Russian approach and ours. They were playing it as though it were a big chess game, with lots of moves. We played it as a more one-dimensional sort of game."

Even at the time, says Cowan, he wondered if this was yet another area where scientists were failing in their responsibility. "I felt, although I didn't have it spelled out in my own mind as well as I do now, that scientists were in a position to take a more general view of the nature of the postwar world." But the fact is that they didn't. And more to the point, *he* didn't. There wasn't time. After the Soviets fired Joe-1 in August 1949, Los Alamos had gone full-speed-ahead on designs for a much more powerful thermonuclear weapon: the hydrogen bomb. And then, after the first H-bomb was tested in the fall of 1952, the lab continued full-speed-ahead in the effort to make the things smaller, lighter, more reliable, and easier to handle. Played out against the backdrop of Korea and the continuing confrontation in Europe, says Cowan, "There was a tremendous feeling that nuclear weapons were going to tip the balance of power one way or another. This was a tremendously important mission."

On top of that, Cowan was being drawn more and more into management responsibility at Los Alamos, which didn't leave him much time for science. As a team leader, he was reduced to doing his own experiments on weekends. "So I've had a very undistinguished scientific career," he says, with a trace of sadness.

The issues of power and responsibility continued to haunt him, however. And in 1982, after Cowan had stepped down as head of research at Los Alamos and accepted a seat on the White House Science Council, they returned full force—even as he began to see the possibility of a second chance.

If nothing else, Cowan's meetings with the White House Science Council were a vivid reminder of just why all those researchers had so eagerly gone back to their labs in 1946. There he would sit with his fellow council members: a bunch of august scientists gathered around some conference table in the New Executive Office Building in Washington. Then the president's science adviser—George (Jay) Keyworth II, who had been named to the post the year before while he was a young division leader working under Cowan at Los Alamos—would lay out a series of issues for their comment. And Cowan would have to admit to himself that he didn't have a clue as to what to say about them.

"The AIDS thing was still quiet back then," he says, "but there was a sense of sudden alarm. It was coming up every meeting. And frankly, I was very puzzled how to respond." Was it a public health issue? Was it a moral issue? What? The answer wasn't so obvious at the time.

"Another issue was manned space flight versus unmanned space probes. We were told that Congress wasn't going to vote a dime for the unmanned space program without the manned component. But I had no idea if that was true or not. It was a political issue much more than a scientific issue."

Then there was President Reagan's "Star Wars" Strategic Defense Initiative, the vision of a space-based shield to protect the country against a massive nuclear missile attack. Was it technically feasible? Could it be built without bankrupting the country? And even if it could be, was it wise? Wouldn't it destabilize the balance of power and spin the world into another ruinous arms race?

And what about nuclear power? How did you balance the risk of a reactor meltdown and the difficulty of disposing of nuclear wastes against the virtual certainty of greenhouse warming due to the burning of fossil fuels?

And so it went. Cowan found the experience distressing. "These were very provocative lessons in the interlinked aspects of science, policy, economics, the environment, even religion and morality," says Cowan. Yet he felt incapable of giving relevant advice. Nor did the other academic types on the Science Council seem to be doing much better. How could they? These issues demanded expertise over a broad range. Yet as scientists and as administrators, most of them had spent their entire lives being specialists. The corporate culture of science demanded it.

"The royal road to a Nobel Prize has generally been through the reductionist approach," he says—dissecting the world into the smallest and simplest pieces you can. "You look for the solution of some more or less idealized set of problems, somewhat divorced from the real world, and constrained sufficiently so that you can find a solution," he says. "And that leads to more and more fragmentation of science. Whereas the real world demands—though I hate the word—a more holistic approach." Everything affects everything else, and you have to understand that whole web of connections.

Even more distressing was his sense that things were only getting worse for the younger generation of scientists. Judging from what he'd seen of the ones coming through Los Alamos, they were impressively bright and energetic—but conditioned by a culture that was enforcing more and more intellectual fragmentation all the time. Institutionally (as opposed to politically), universities are incredibly conservative places. Young Ph.D.'s don't dare break the mold. They have to spend the better part of a decade in the desperate pursuit of tenure in an existing department, which means that they had better be doing research that the department's tenure committee will recognize. Otherwise, they're going to hear something like, "Joe, you've been working hard over there with the biologists. But how does that show you're a leader over here in physics?" Older researchers, meanwhile, have to spend all their waking hours in the desperate pursuit of grants to pay for their research, which means that they had better tailor their projects to fit into categories that the funding agencies will recognize. Otherwise, they're going to hear something like, "Joe, this is a great idea—too bad it's not our department." And everybody has to get papers accepted for publication in established scholarly journals—which are almost invariably going to restrict themselves to papers in a recognized specialty.

After a few years of this, says Cowan, the enforced tunnel vision becomes so instinctive that people don't even notice it anymore. In his experience, the closer any of his Los Alamos researchers were to the academic world, the harder it was to get them to participate in team efforts. "I've wrestled with it for thirty years," he sighs.

As he thought about it, however, he began to feel that the most distressing thing of all was what this fragmentation process had done to science as a whole. The traditional disciplines had become so entrenched and so isolated from one another that they seemed to be strangling themselves. There were rich scientific opportunities everywhere you looked, and too many scientists seemed to be ignoring them.

If you wanted an example, Cowan thought, just look at the kind of opportunities opening up in—well, he didn't really have a good name for it. But if what he'd seen around Los Alamos was any indication, something big was brewing. More and more over the past decade, he'd begun to sense that the old reductionist approaches were reaching a dead end, and that even some of the hard-core physical scientists were getting fed up with mathematical abstractions that ignored the real complexities of the world. They seemed to be half-consciously groping for a new approach—and in the process, he thought, they were cutting across the traditional boundaries in a way they hadn't done in years. Maybe centuries.

One of their inspirations, ironically enough, seemed to be molecular biology. That's not the sort of thing that most people would expect a weapons laboratory to be interested in. But in fact, says Cowan, physicists have been deeply involved with molecular biology from the beginning. Many of the pioneers in the field had actually started out as physicists; one of their big motivations to switch was a slim volume entitled *What Is Life?*, a series of provocative speculations about the physical and chemical basis of life published in 1944 by the Austrian physicist Erwin Schrödinger, a coinventor of quantum mechanics. (Having fled from Hitler, Schrödinger spent the war safely ensconced in Dublin.) One of those who was influenced by the book was Francis Crick, who deduced the molecular structure of DNA along with James Watson in 1953—using data obtained from x-ray crystallography, a kind of submicroscopic imaging technique developed by physicists decades earlier. Crick, in fact, had originally trained as an experimental physicist. George Gamow, a Hungarian theoretical physicist who was one of the original proponents of the Big Bang theory of the origin of the universe, became intensely interested in the structure of the genetic code in the early 1950s and helped inspire still more physicists into the field. "The first really perceptive lecture I heard on the subject was by Gamow," says Cowan.

Molecular biology had fascinated him ever since, he says, especially after the discovery of recombinant DNA technology in the early 1970s gave biologists the power to analyze and manipulate life-forms almost molecule by molecule. So when he became head of research at the laboratory in 1978, he had quickly thrown his support behind a major research initiative in the field—officially to study radiation damage in cells, but actually to get Los Alamos involved in molecular biology on a broader front. It was a particularly good time to do so, he recalls. Under director Harold Agnew, Los Alamos had nearly doubled its size in the 1970s, and had opened itself up to much more nonclassified basic and applied research. Cowan's emphasis on molecular biology fit right in. And that program, in turn, had had a tremendous impact on people's thinking at the laboratory. Especially his.

"Almost by definition," he says, "the physical sciences are fields characterized by conceptual elegance and analytical simplicity. So you make a virtue of that and avoid the other stuff." Indeed, physicists are notorious for curling their lips at "soft" sciences like sociology or psychology, which try to grapple with real-world complexity. But then here came molecular biology, which described incredibly complicated living systems that were nonetheless governed by deep principles. "Once you're in a partnership with biology," says Cowan, "you give up that elegance, you give up that simplicity. You're *messy*. And from there it's so much easier to start diffusing into economics and social issues. Once you're partially immersed, you might as well start swimming."

But at the same time, says Cowan, scientists were also beginning to think about more and more complex systems simply because they *could* think about them. When you're stuck with solving mathematical equations by paper and pencil, how many variables can you handle without bogging down? Three? Four? But when you have enough computer power, you can handle as many variables as you like. And by the early 1980s, computers were everywhere. Personal computers were booming. Scientists were loading up their desktops with high-powered graphics workstations. And the big corporate and national labs were sprouting supercomputers like mushrooms. Suddenly, hairy equations with zillions of variables didn't look quite so hairy anymore. Nor did it seem quite so impossible to drink from the firehose of data. Columns of figures and miles of data tapes could be transformed into color-coded maps of crop yields or of oil-bearing strata lying under miles of rock. "Computers," says Cowan with considerable understatement, "are great bookkeeping machines."

But they could also be much more than that. Properly programmed, computers could become entire, self-contained worlds, which scientists could explore in ways that vastly enriched their understanding of the real world. In fact, computer simulation had become so powerful by the 1980s that some people were beginning to talk about it as a "third form of science," standing halfway between theory and experiment. A computer simulation of a thunderstorm, for example, would be like a theory because nothing would exist inside the computer but the equations describing sunlight, wind, and water vapor. But the simulation would also be like an experiment, because those equations are far too complicated to solve by hand. So the scientists watching the simulated thunderstorm on their computer screens would see their equations unfold in patterns they might never have predicted. Even very simple equations can sometimes produce astonishing behavior. The mathematics of a thunderstorm actually describes how each puff of air pushes on its neighbors, how each bit of water vapor condenses and evaporates, and other such small-scale matters; there's nothing that explicitly talks about "a rising column of air with rain freezing into hailstones" or "a cold, rainy downdraft bursting from the bottom of the cloud and spreading along the ground." But when the computer integrates those equations over miles of space and hours of time, that is exactly the behavior they produce. Furthermore, that very fact allows the scientists to experiment with their computer models in ways that they could never do in the real world. What really causes these updrafts and downdrafts? How do they change when I vary the temperature and humidity? Which factors are really important to the dynamics of this storm, and which aren't? And are the same factors equally important in other storms?

By the beginning of the 1980s, says Cowan, such numerical experiments had become almost commonplace. The behavior of a new aircraft design in flight, the turbulent flow of interstellar gas into the maw of a black hole, the formation of galaxies in the aftermath of the Big Bang—at least among physical scientists, he says, the whole idea of computer simulation was becoming more and more accepted. "So you could begin to think about tackling *very* complex systems."

But the fascination with complexity went still deeper than that, says Cowan. In part because of their computer simulations, and in part because of new mathematical insights, physicists had begun to realize by the early 1980s that a lot of messy, complicated systems could be described by a powerful theory known as "nonlinear dynamics." And in the process, they had been forced to face up to a disconcerting fact: the whole really can be greater than the sum of its parts.

Now, for most people that fact sounds pretty obvious. It was disconcerting for the physicists only because they had spent the past 300 years having a love affair with linear systems—in which the whole is precisely *equal* to the sum of its parts. In fairness, they had had plenty of reason to feel this way. If a system is precisely equal to the sum of its parts, then each component is free to do its own thing regardless of what's happening elsewhere. And that tends to make the mathematics relatively easy to analyze. (The name "linear" refers to the fact that if you plot such an equation on graph paper, the plot is a straight line.) Besides, an awful lot of nature does seem to work that way. Sound is a linear system, which is why we can hear an oboe playing over its string accompaniment and recognize them both. The sound waves intermingle and yet retain their separate identities. Light is also a linear system, which is why you can still see the *Walk/Don't Walk* sign across the street even on a sunny day: the light rays bouncing from the sign to your eyes are *not* smashed to the ground by sunlight streaming down from above. The various light rays operate independently, passing right through each other as if nothing were there. In some ways even the economy is a linear system, in the sense that small economic agents can act independently. When someone buys a newspaper at the corner drugstore, for example, it has no effect on your decision to buy a tube of toothpaste at the supermarket.

However, it's also true that a lot of nature is *not* linear—including most of what's really interesting in the world. Our brains certainly aren't linear: even though the sound of an oboe and the sound of a string section may be independent when they enter your ear, the emotional impact of both sounds together may be very much greater than either one alone. (This is what keeps symphony orchestras in business.) Nor is the economy really linear. Millions of individual decisions to buy or not to buy can reinforce each other, creating a boom or a recession. And that economic climate can then feed back to shape the very buying decisions that produced it. Indeed, except for the very simplest physical systems, virtually everything and everybody in the world is caught up in a vast, nonlinear web of incentives and constraints and connections. The slightest change in one place causes tremors everywhere else. We can't help but disturb the universe, as T. S. Eliot almost said. The whole is almost always equal to a good deal more than the sum of its parts. And the mathematical expression of that property—to the extent that such systems can be described by mathematics at all—is a *nonlinear* equation: one whose graph is curvy.

Nonlinear equations are notoriously difficult to solve by hand, which is why scientists tried to avoid them for so long. But that is precisely where computers came in. As soon as scientists started playing with these machines back in the 1950s and 1960s, they realized that a computer couldn't care less about linear versus nonlinear. It would just grind out the solution either way. And as they started to take advantage of that fact, applying that computer power to more and more kinds of nonlinear equations, they began to find strange, wonderful behaviors that their experience with linear systems had never prepared them for.

The passage of a water wave down a shallow canal, for example, turned out to have profound connections to certain subtle dynamics in quantum field theory: both were examples of isolated, self-sustaining pulses of energy called solitons. The Great Red Spot on Jupiter may be another such soliton. A swirling hurricane bigger than Earth, it has sustained itself for at least 400 years.

The self-organizing systems championed so vociferously by the physicist Ilya Prigogine were also governed by nonlinear dynamics; indeed, the self-organized motion in a simmering pot of soup turned out to be governed by dynamics very similar to the nonlinear formation of other kinds of patterns, such as the stripes of a zebra or the spots on a butterfly's wings.

But most startling of all was the nonlinear phenomenon known as chaos.

In the everyday world of human affairs, no one is surprised to learn that a tiny event over *here* can have an enormous effect over *there*. For want of a nail, the shoe was lost, et cetera. But when the physicists started paying serious attention to nonlinear systems in their own domain, they began to realize just how profound a principle this really was. The equations that governed the flow of wind and moisture looked simple enough, for example—until researchers realized that the flap of a butterfly's wings in Texas could change the course of a hurricane in Haiti a week later. Or that a flap of that butterfly's wings a millimeter to the left might have deflected the hurricane in a totally different direction. In example after example, the message was the same: everything is connected, and often with incredible sensitivity. Tiny perturbations won't always remain tiny. Under the right circumstances, the slightest uncertainty can grow until the system's future becomes utterly unpredictable—or, in a word, chaotic.

Conversely, researchers began to realize that even some very simple systems could produce astonishingly rich patterns of behavior. All that was required was a little bit of nonlinearity. The drip-drip-drip of water from a leaky faucet, for example, could be as maddeningly regular as a metronome—so long as the leak was slow enough. But if you ignored the leak for a while and let the flow rate increase ever so slightly, then the drops would soon start to alternate between large and small: DRIP-drip-DRIP-drip. If you ignored it a while longer and let the flow increase still more, the drops would soon start to come in sequences of 4—and then 8, 16, and so forth. Eventually, the sequence would become so complex that the drops would seem to come at random—again, chaos. Moreover, this same pattern of ever-increasing complexity could be seen in the population swings of fruit flies, or in the turbulent flow of fluids, or in any number of domains.

It was no wonder the physicists were disconcerted. They had certainly known that there were some funny things going on with quantum mechanics and black holes and such. But in the 300 years since the time of Newton, they and their predecessors had gotten used to thinking of the everyday world as a fundamentally tidy and predictable place obeying well-understood laws. Now it was as if they had spent the past three centuries living on a tiny desert island and ignoring what was all around them. "The moment you depart from the linear approximation," says Cowan, "you're navigating on a very broad ocean."

Los Alamos, as it happened, was nearly an ideal environment for nonlinear research. Not only had the laboratory been a leader in advanced computing since the 1950s, says Cowan, but the researchers there had been grappling with nonlinear problems from the day the place was founded.

High-energy particle physics, fluid dynamics, fusion energy research, thermonuclear blast waves, you name it. By the early 1970s, in fact, it was clear that a good many of these nonlinear problems were the same problems deep down, in the sense of having the same mathematical structure. So people could obviously save themselves a lot of effort if they would just start working on those problems together. The result, with the enthusiastic support of the Los Alamos theory group, was a vigorous program for non-linear science within the theory division, and eventually a Center for Nonlinear Systems operating entirely on its own.

And yet, as intriguing as molecular biology and computer simulation and nonlinear science were separately, Cowan had a suspicion that they were only the beginning. It was more a gut feeling than anything else. But he sensed that there was an underlying unity here, one that would ultimately encompass not just physics and chemistry, but biology, information processing, economics, political science, and every other aspect of human affairs. What he had in mind was a concept of scholarship that was almost medieval. If this unity were real, he thought, it would be a way of knowing the world that made little distinction between biological sciences and physical sciences—or between either of those sciences and history or philosophy. Once, says Cowan, "The whole intellectual fabric was seamless." And maybe it could be that way again.

To Cowan, it seemed like an incredible opportunity. So why weren't scientists out in the universities jumping on it? Well, to a certain extent they were, here and there. But this really broad view he was looking for seemed to be falling through the cracks. By its very nature, it lay outside the purview of any one academic department. True, universities were full of "interdisciplinary research institutes." But so far as Cowan could tell, these institutes were rarely much more than a bunch of people who occasionally shared a common office space. Professors and graduate students still had to give their loyalty to their home departments, which held the power to grant degrees, tenure, and promotions. Left to themselves, thought Cowan, the universities weren't going to pick up on complexity research for a generation at least.

Unfortunately, Los Alamos didn't seem likely to pick up on it, either. And that was too bad. Ordinarily, a weapons laboratory is a much better environment for this kind of broad, multidisciplinary research than the universities are. That fact is something that visiting academics always find startling. But it goes right back to the laboratory's founding, says Cowan. The Manhattan Project started with a specific research challenge—building the bomb—and brought together scientists from every relevant specialty to tackle that challenge as a team. Granted, it was a pretty remarkable team. Robert Oppenheimer, Enrico Fermi, Niels Bohr, John von Neumann, Hans Bethe, Richard Feynman, Eugene Wigner—one observer at the time called them the greatest gathering of intellects since ancient Athens. But that's been the laboratory's approach to research ever since. The big job for management was to make sure that the right specialists were talking to each other. "I sometimes felt like a marriage broker," says Cowan.

The only problem was that Cowan's grand synthesis just wasn't part of the laboratory's basic mission. Indeed, it was about as far from nuclear weapons development as you could get. And things that weren't part of the laboratory's mission had essentially zero chance of getting funded there. The laboratory would certainly go on doing bits and pieces of complexity research the way it already had, thought Cowan. But it would never do much more than that.

No, he thought, there was really only one way. Cowan began to imagine a new, independent institute. Ideally, that institute would combine the best of both worlds: it would be a place having the broad charter of a university, while retaining Los Alamos' ability to mingle the separate disciplines. It would almost certainly have to be physically separate from Los Alamos, he knew. But if possible, it should be close enough to share some of the laboratory's personnel and computer power. Presumably that meant Santa Fe, which was only thirty-five miles away and which was the nearest city of any size. But wherever you put it, he thought, this institute ought to be a place where you could take very good scientists—people who would really know what they were talking about in their own fields—and offer them a much broader curriculum than they usually get. It ought to be a place where senior researchers could work on speculative ideas without being laughed at by their colleagues, and where the brightest young scientists could come and work alongside world-class figures who would give them credibility back home.

It ought to be a place, in short, that could educate the kind of scientist that had proved all too rare after World War II: "a kind of twenty-first-century Renaissance man," says Cowan, "starting in science but able to deal with the real messy world, which is not elegant, which science doesn't really deal with."

Naive? Of course. But Cowan thought it just might work, if only he could entice people with the vision of an incredible scientific challenge. As he put the question to himself, "What kind of science *had* to be taught to brilliant scientists in the 1980s and 1990s?"

So, who might be willing to listen? And, not incidentally, who might have the clout to make this thing work? As a trial run one day when he was in Washington, Cowan tried explaining the institute idea to the science adviser, Jay Keyworth, and his fellow science board member David Packard, cofounder of Hewlett-Packard. Amazingly enough, they didn't laugh. In fact, they were both quite encouraging. So in the spring of 1983, Cowan decided to take the idea to his weekly lunch companions, the Los Alamos senior fellows.

They loved it.

The Fellows

From the outside, it would be easy enough to dismiss the senior fellows as a collection of old geezers who had been put out to pasture at a ridiculously high salary. From the outside, that's just about what they looked like. The group was composed of about half a dozen longtime Los Alamites who, like Cowan, had done yeoman service at the lab and who had been rewarded with research positions free from any administrative chores or other bureaucratic busywork. Their only duties as a group were to meet for lunch in the cafeteria once a week and to occasionally advise the laboratory director on various policy issues.

But, in fact, the fellows were a remarkably frisky group, the kind of guys whose response to their new status was to say, "Thank God I can finally get some real work done." And since many of them had had heavy administrative responsibility at one time or another, they were not shy about telling the laboratory director exactly what he ought to be doing, whether he wanted to hear from them or not. So when Cowan laid out his institute idea for them, looking for advice and maybe allies, he got both.

Pete Carruthers, for example, immediately resonated with Cowan's sense that something new was in the air—and with his sense that the opportunity was going begging. Under a rumpled and cynical exterior, Carruthers was passionately enthusiastic about "complex" systems—"the next major thrust in science," he declared. He had reason to be. Brought in from Cornell University to head the Los Alamos theory division in 1973—at the recommendation of a search committee chaired by Cowan—he had managed to hire nearly 100 new researchers and start half a dozen new research groups even as the laboratory's budget for such things was going down. Among other things, he had insisted on hiring a handful of young wild men back in 1974 to work on what was then an obscure subfield of nonlinear dynamics. ("What am I supposed to pay them with?" asked his deputy

director, Mike Simmons. "Find the money somewhere," said Carruthers.)
And it was under Carruthers that that subfield had blossomed, making Los
Alamos into a world center for what was soon being called chaos theory.
So if Cowan wanted to build on that foundation, Carruthers was ready to
help.

Another senior fellow, astrophysicist Stirling Colgate, declared his fervent
support for a different reason: "We needed anything that could organize
and reinforce the intellectual capability in the state," he says. Los Alamos,
despite all its efforts to open up to the outside world, was still a scientific
enclave sitting up on its mesa in splendid isolation. In his ten years as
president of the New Mexico Institute of Mining and Technology down in
Socorro, 200 miles to the south, Colgate had learned all too well that the
rest of the state remained beautiful, but backward. All the billions of federal
dollars that had poured into the region since the 1940s had had depressingly
little impact on its schools and industrial base. Its universities were mediocre
at best. And largely because of that, high-tech entrepreneurs looking to
relocate from overcrowded California would routinely fly right over the Rio
Grande valley en route to Austin and points east.

Along with Carruthers, Colgate had recently tried to get New Mexico
to dramatically upgrade its university system. And they had quickly given
it up as hopeless: the state was just too poor. So Cowan's institute looked
like a last, best hope. "Anything that could raise the intellectual extremum
of our environment was not only in our personal interest, but in the laboratory's interest, and most of all in the national interest," declares Colgate.

Senior fellow Nick Metropolis liked the idea because of Cowan's emphasis
on computation. And he had good reason to: Metropolis was pretty much
Mr. Computer at Los Alamos. It was he who had supervised the construction of the laboratory's first computer back in the late 1940s, basing it on
a design pioneered by the legendary Hungarian mathematician John von
Neumann of Princeton's Institute for Advanced Study, who was a consultant
and frequent visitor at Los Alamos. (The machine was dubbed the Mathematical Analyzer, Numerator, Integrator, And Computer: MANIAC.) It
was Metropolis, along with the Polish mathematician Stanislaus Ulam,
who had pioneered the art of computer simulation. And in no small measure
it was Metropolis who was responsible for Los Alamos' now having some
of the biggest and fastest supercomputers on the planet.

And yet, Metropolis felt that the laboratory was not being sufficiently
innovative even in this arena. Along with Gian-Carlo Rota, a mathematician from MIT who was a Los Alamos visiting fellow and who often came

for an extended stay, Metropolis pointed out to the assembled fellows that computational science was undergoing just as much ferment as biology and nonlinear sciences. There were revolutionary changes going on in hardware design alone, he said. The existing one-step-at-a-time computers had gotten about as fast as they were ever going to get, and designers were beginning to investigate new kinds of computers that could do hundreds, or thousands, or even millions of computational steps in parallel. It was a good thing, too: anyone who seriously wanted to tackle the kind of complex problems Cowan was talking about was probably going to need such a machine.

But computational science went much further than that. Rota, in particular, thought of it as extending all the way to the study of the mind—based on the idea that thinking and information processing were fundamentally the same thing. Also known as cognitive science, this was a hot area and getting hotter. When done properly, it combined the talents of neuroscientists studying the detailed wiring of the brain, cognitive psychologists studying the second-by-second process of high-level thinking and reasoning, artificial intelligence researchers trying to model those thinking processes in a computer—even linguists studying the structure of human languages and anthropologists studying human culture.

Now that, Rota and Metropolis told Cowan, was an interdisciplinary topic worthy of his institute.

Another visitor was David Pines, who had started sitting in on the discussions at Metropolis's invitation in the midsummer of 1983. A theoretical physicist from the University of Illinois, Pines was editor of the journal *Reviews of Modern Physics* and chairman of the advisory board for the Los Alamos Theory Division. He also turned out to be someone who resonated strongly with Cowan's idea of a grand synthesis in science. After all, much of his own research, starting with his Ph.D. dissertation in 1950, had been focused on innovative ways of understanding "collective" behavior in systems of many particles; examples ranged from the vibration modes of certain massive atomic nuclei to the quantum flow of liquid helium. And Pines had been known to speculate aloud that a similar analysis might lead to a better understanding of collective human behavior in organizations and societies. "So I had an intellectual predisposition to the idea," he says. Pines was likewise an enthusiast for Cowan's vision of a new institute. He'd had quite a bit of experience along those lines himself, having been founding director of Illinois' Center for Advanced Study, and one of the founders of the Aspen Center for Physics in Colorado. Go for it, he told Cowan; he could hardly wait to get going on this one. "I always find it great fun to bring together very able scientists to talk about something quite new," says Pines. "It can be as much fun to start an institution as to write a good scientific paper."

And so it went. The fellows had a great time with the institute idea, to the point of occasionally getting a bit giddy. There was the day, for example, when they all got very excited about the thought that they might be founding "the New Athens"—a center for intellectual inquiry on a par with the city-state that gave us Socrates, Plato, and Aristotle. On a more practical level, they debated innumerable questions. How big should the place be? How many students should it have—or should it have any? How closely should it be tied to Los Alamos? Should it have a permanent faculty, or should people rotate through and then go back to their own institutions? And gradually, before they fully realized it, this hypothetical institute began to become more and more real in their minds.

The only problem, unfortunately, was that everybody had something different in mind. "Every week," sighs Cowan, "We'd go back to first base, and go round and round again."

The most serious bone of contention was also the most fundamental: What should the institute be about?

On one side were Metropolis and Rota, who felt the place should focus exclusively on computational science. A grand "synthesis" was nice, they argued. But if nobody here could quite define it, how could they ever hope to get somebody out there to drop $400 million on it? That's about what you would need to endow a facility on the scale of, say, the Rockefeller Institute in New York. Of course, it wasn't going to be easy to raise that kind of money in any case. But at least if you focused on information processing and cognitive science, you would cover a lot of what George was talking about, and you might conceivably get an endowment from one of these new teen-age computer zillionaires.

On the other side were Carruthers, Pines, and most of the others. Computers were nice, they felt. And Metropolis and Rota certainly had a point about the money. But damn it, *another* computer research center? Was that really going to set anybody on fire? The institute ought to be something more than that—even if they couldn't figure out precisely what. And that was just the problem. As senior fellow Darragh Nagle points out, "We didn't articulate the alternative very well." Everyone felt that Cowan was right, that something new was brewing out there. But no one could do much better than vague talk about "new ways of thinking."

Cowan himself kept a low profile on this issue. He knew where *he* was coming from: he privately thought of the place as an "institute on the art of survival." And to him that meant a program as broad as it possibly could be, and as free of strings as it possibly could be. At the same time, however, he was convinced that getting a consensus on the direction of the institute was far more important than money or any of the rest of the details. If this institute were just a one-man show, he felt, then it wasn't going anywhere. After thirty years as an administrator, he was convinced that the only way to make something like this happen was to get a lot of people excited about it. "You have to persuade *very* good people that this is an important thing to do," he says. "And by the way, I'm not talking about a democracy. I'm talking about the top one-half of one percent. An elite. But once you do that, then the money is—well, not easy, but a smaller part of the problem."

It was something of a slow-motion debate, since everyone was working more than full time on their various research projects. (Cowan, in particular, was immersed in an experiment to detect solar neutrinos, which are near-invisible particles emitted from the core of the sun.) But that couldn't last forever. On August 17, 1983, Cowan called the fellows together in one of the laboratory administration building's fourth-floor conference rooms and suggested it was time to get serious. Some friends of his were talking about offering 50 or 100 acres of land as a campus for the institute, he told them. But at a minimum, they would want to know what the institute was going to be about.

No go. The fellows were amicably, but firmly, divided into two camps. They ended that meeting no closer to a resolution than before—which was probably just as well, since the couple who had promised Cowan the land got divorced a few months later and had to rescind the offer. But Cowan had to wonder if this was ever going to go anywhere.

Murray

It was Murray who really broke the logjam. Professor Murray Gell-Mann of Caltech, the fifty-five-year-old *enfant terrible* of particle physics.

Gell-Mann had called up Cowan about a week before the August 17 meeting, saying that Pines had told him about the institute idea. Gell-Mann thought it was fantastic. He'd been wanting to do something like this all his life, he said. He wanted to tackle problems like the rise and fall of ancient civilizations and the long-term sustainability of our *own* civilization—problems that would transcend the disciplinary boundaries in a big way. He'd had no success whatsoever getting anything started at Caltech. So could he join the institute discussions the next time he was in Los

Alamos? (Gell-Mann had been a consultant to the laboratory since the 1950s, and came quite often.)

Cowan couldn't believe his luck: "By all means, come on by!" If ever there were someone who belonged to that top one-half of one percent, it was Murray Gell-Mann. Born and raised in New York City, his dark-rimmed glasses and white crew-cut hair giving him the look of a cherubic Henry Kissinger, Gell-Mann was brash, brilliant, charming, and incessantly verbal—not to mention being self-confident to the point of arrogance. In fact, more than one person found him insufferable. He had spent a lifetime being the smartest kid in class. At Caltech, where the late, irrepressible physicist Richard Feynman had entitled his best-selling memoirs *Surely You're Joking, Mr. Feynman!*, it was said that Gell-Mann would have to call his own memoirs *Well, You're Right Again, Murray!* On those rare occasions when he didn't get his own way, Gell-Mann could also be remarkably childish: colleagues had observed his lower lip extending outward in what looked suspiciously like a pout.

But for all of that, Murray Gell-Mann was clearly one of the major figures of twentieth-century science. When he arrived on the scene as a young Ph.D. in the early 1950s, the subatomic world seemed a senseless mess—a hodgepodge of pi particles, sigma particles, rho particles, and on and on through an endless list of Greek alphabetical names assigned at random. But two decades later, largely because of concepts that Gell-Mann had pioneered, physicists were drawing up Grand Unified Theories of all the interparticle forces and were confidently classifying that hodgepodge of particles as various combinations of "quarks"—simple subatomic building blocks that Gell-Mann had named after a made-up word in James Joyce's *Finnegans Wake*. "For a generation," says a theoretical physicist who has known him for twenty years, "Murray defined the centroid of the research effort in particle physics. What Murray was thinking about was what everyone else *should* be thinking about. He knew where the truth lay, and he led people to it."

On the face of it, this thirty-year preoccupation with the inner reaches of protons and neutrons made Gell-Mann an odd recruit to Cowan's vision of scientific holism; it's hard to imagine anything more reductionist. But, in fact, Gell-Mann's interests were legion. He was driven by an omnivorous curiosity. He had been known to turn to strangers sitting next to him on an airplane and grill them about their life stories for hours. He had first come to science through a love of natural history, which he started learning at age five when his older brother took him on nature walks through the Manhattan parks. "We thought of New York as a hemlock forest that had been overlogged," he says. Ever since, he had been an ardent bird-watcher and conservationist. As chairman of the committee on World Environment and Resources at the John D. and Catherine T. MacArthur Foundation, he had helped found a Washington environmental think tank known as the World Resources Institute, and he was deeply involved in efforts to preserve tropical forests.

Gell-Mann likewise had a lifelong fascination with psychology, archeology, and linguistics. (He originally enrolled as a physics major at Yale only to satisfy his father, who feared he would starve if he majored in archeology.) When mentioning a foreign scientist he pronounces the name with a lovingly precise accent—in any of several dozen languages. One colleague remembers mentioning that he would soon be visiting his sister in Ireland.

"What's her name?" asked Gell-Mann.

"Gillespie."

"What does it mean?"

"Well, in Gaelic I think it means 'servant of a bishop.' "

Gell-Mann thought for a moment. "No—in medieval Scots-Gaelic it means more like 'religious follower of a bishop.' "

And if anyone at Los Alamos didn't already know it, Gell-Mann could use that verbal ability with immensely persuasive effect. "Murray can improvise, on the spot, an inspirational speech that may not be Churchillian," says Carruthers, "but the clarity and brilliance of it are overwhelming." As soon as he joined the institute discussions, his arguments for a broad-based institute gave the majority of the fellows something to rally around, and the Metropolis-Rota concept of a computer-focused institute quickly lost altitude.

Gell-Mann got his real chance to shine just after Christmas 1983. Taking advantage of the fact that Gell-Mann, Rota, and Pines loved to spend the holidays in New Mexico—in fact, Gell-Mann had just finished building a house in Santa Fe—Cowan called yet another meeting of the fellows to try to get this institute moving.

Gell-Mann pulled out all the stops. These narrow conceptions weren't grand enough, he told the fellows. "We had to set ourselves a really big task. And that was to tackle the great, emerging syntheses in science—ones that involve many, many disciplines." Darwin's theory of biological evolution had been just such a grand synthesis in the nineteenth century, he said. It combined evidence from biology, which revealed that different species of plants and animals were clearly related; from the emerging science of geology, which showed that the earth was incredibly ancient and that the past afforded immense vistas of time; and from paleontology, which proved that the plants and animals who dwelled in that immense past had been very different from those alive today. More recently, he said, there had been the grand synthesis known as the Big Bang theory, which detailed how all the matter in all the stars and galaxies had come into being in an unimaginably vast cosmic explosion some fifteen billion years ago.

"I said I felt that what we should look for were great syntheses that were emerging today, that were highly interdisciplinary," says Gell-Mann. Some were already well on their way: Molecular biology. Nonlinear science. Cognitive science. But surely there were other emerging syntheses out there, he said, and this new institute should seek them out.

By all means, he added, choose topics that could be helped along by these huge, big, rapid computers that people were talking about—not only because we can use the machines for modeling, but also because these machines themselves were examples of complex systems. Nick and Gian-Carlo were perfectly correct: computers might very well turn out to be part of such a synthesis. But don't put blinders on before you start. If you're going to do this at all, he concluded, do it right.

To his listeners it was spellbinding stuff. "I had said it before," says Gell-Mann, "but perhaps not so convincingly."

Gell-Mann's rhetoric pretty much carried the day. Here, in compelling terms, was the vision that Cowan and the majority of the fellows had been trying to articulate for nearly a year. After that it became more or less unanimous: The fellows would try to build an institute with the broadest possible charter. And if Gell-Mann was willing to go out and knock the potential donors dead—as apparently he was—then maybe it was time to move.

With that settled, however, the group then had to deal with a less exalted question: Exactly who was going to do the work? Who was going to make this institute happen?

Everyone looked in the obvious direction.

Actually, this was about the last job that Cowan himself wanted. Yes, the institute had been his idea. He believed in it. He thought it ought to be done. He thought it *had* to be done. But, damn it, he'd been an administrator practically all his adult life. He was tired of it: tired of always scrambling for funds, tired of telling his friends that he'd have to cut their budgets, tired of trying to sneak in his own scientific work on the weekends. He was sixty-three years old and he had notebooks crammed with ideas he'd never had time to work on. Searching for solar neutrinos, investigating an extremely rare and intriguing form of radioactivity known as double beta decay—this was the kind of science he had always wanted to do. And now this was what he was *going* to do.

So, of course, when Pines nominated him to spearhead the effort, he said, "Yes." Cowan had already given it some thought, since Pines had talked to him about the nomination beforehand. And what had finally persuaded him was the same thing that had always lured him into management positions at Los Alamos: "Management was stuff that other people could do—but I always felt that maybe they were doing it wrong." Besides, nobody else was exactly frothing at the mouth to step forward.

Okay, he told the group. He was willing to be the Little Red Hen and get everything done, at least until they could convince someone else to step in. But just one thing: in the meantime he wanted Murray out front doing the talking.

"When you're looking for funds," says Cowan, "people want to hear how you're going to solve the energy crisis tomorrow. But we were starting much more modestly. I thought it was going to be years before we produced anything terribly useful, other than a new way of looking at the world. So what you say is, 'Here's professor so and so, who's giving up his preoccupation with quarks in order to work on something a little more related to your daily concerns.' They aren't quite certain of what you're talking about. But they listen."

The fellows agreed. Cowan would be the institute's president and man on the spot. Gell-Mann would be the chairman of the board.

George

Reticence aside, Cowan was actually well suited to be the man on the spot. He had contacts everywhere. Of course, he could hardly have avoided it. New Mexico has such a comparatively tiny population that any Los Alamos administrator quickly gets to know all the powers that be. But it helps if that Los Alamos administrator also happens to have made himself a millionaire several times over.

Cowan usually won't bring up that subject himself, and seems almost embarrassed when asked. "Anybody who tells me there's anything difficult about that—well, I just don't agree."

It happened back in the early 1960s, he explains. "Los Alamos was the ideal example of a kind of socialist economy: There was no private property. People were assigned housing according to their rank and importance.

Junior people were assigned to what were essentially shacks. They looked like army barracks."

"Well, I was trying to hire people—in those days it was usually a man—but it wasn't easy. There would be immediate friction with his wife about having to live in those shacks. So we persuaded the government to make real estate available. But the banks wouldn't lend to a government installation. So we said to ourselves, 'We'll start our own savings and loan.' I remember telling my wife that we were probably going to lose our investment. She said, 'Okay.' But we didn't! The savings and loan turned out to be very profitable, so we decided to start a bank, the Los Alamos National Bank. It was an immediate success."

"All it took," he says, "was a good lawyer and a couple of friendly senators."

Cowan had already foreseen the need for seed money for the institute back in the summer of 1983, and he'd gone for help to an old friend of his: Art Spiegel of the Spiegel Catalog fortune. He and Spiegel had been members of the group that founded the Santa Fe Opera, and he knew that Spiegel and his wife were principal fund-raisers for the New Mexico Symphony Orchestra. Spiegel, for his part, had no clear idea what Cowan was talking about with this institute thing. But it sounded to him like a great idea, if only as a much-needed response to the growing Japanese leadership in high technology. So he began helping Cowan to canvas the assorted rich people in Santa Fe, of whom there are many.

By the spring of 1984, Spiegel had been able to raise a bit of cash from Mountain Bell and one of the more prosperous local savings and loans (which has since gone broke). It wasn't a lot. But, then, Cowan didn't consider fund-raising his top priority yet, either. He felt it was more important to lay some groundwork. Around Easter 1984, for example, Cowan laid out $300 of his own money for a lunch for community leaders in Santa Fe. "We felt it was politically desirable to let them know what we had in mind and to invite their interest and support. We didn't pursue it very strongly. We just didn't want them to read about it in the papers that a bunch of eggheads from Los Alamos were suddenly appearing in Santa Fe to do something they didn't know about."

This lunch didn't bring in any money, either. But it was good practice. Gell-Mann came and gave a speech. The crowd loved it: a Nobel Prize winner!

Meanwhile, there was the matter of incorporation: if you're going to start asking people for money, you really ought to have something besides your own personal checking account to put it into. So Cowan and Nick Metropolis went to Jack Campbell, an old friend who had once been governor of the state and who was now head of a very prosperous law firm in Santa Fe. Campbell was enthusiastic. He'd wanted to do something like this all the time he was governor, he said; the universities in New Mexico were too damn isolated from real-world problems. Campbell agreed to provide his firm's services pro bono to draw up the incorporation papers and bylaws. He also advised Cowan on how to persuade the IRS that the fledgling institute really did deserve a not-for-profit status. (The IRS is notoriously skeptical about such things; Cowan had to fly to Dallas and make the argument in person.)

In May 1984, the Santa Fe Institute was incorporated. It didn't have a location or a staff. It had essentially zero money. In fact, it wasn't much more than a post office box and a telephone number that rang in Art Spiegel's office in Albuquerque. And it didn't even have the right name: "Santa Fe Institute" was already copyrighted by a therapy service, so Cowan and the fellows had had to settle for "The Rio Grande Institute." (The Rio Grande flows a few miles west of town.) But it existed.

However, there was still that nagging question of content. Gell-Mann's visionary rhetoric was all very well. Gell-Mann was a very smart guy. But nobody was going to plunk down several hundred million dollars until they heard precisely what the institute was going to do—or, for that matter, until they had some evidence that it was going to work. "Herb, how do we get this thing started?" Cowan asked Los Alamos fellow Herb Anderson that spring. Well, said Anderson, his favorite formula was to bring a bunch of very good people together in a workshop, and have each one talk about whatever was nearest and dearest to his heart. You could get the coverage of all the different disciplines by the kind of people you invited, he said. And if there were really a convergence between the disciplines, you would see it start to emerge from the debate.

"So I said, 'Fine, you start developing that,' " says Cowan, and that's what he did." Shortly thereafter, Pines volunteered to put the workshops together—he had been thinking along much the same lines—and Anderson happily turned it over to him.

Phil

At Princeton, Philip Anderson got the note from Pines on June 29, 1984: Would he like to attend a workshop that fall on "Emerging Syntheses" in science?

Hmm. Maybe. Anderson was skeptical, to say the least. He'd heard rumors about this outfit. Gell-Mann had been talking up the institute everywhere he went, and so far as Anderson could tell, the place was shaping up as a cushy retirement home for aging Nobel laureates from Caltech—complete with megabuck endowments and lots of scientific glitz and glamor.

Well, Anderson could match credentials with Murray Gell-Mann any day of the week, thank you. He'd won his own Nobel Prize in 1977 for his work in condensed-matter physics, and for thirty years he'd been as much a centroid in that field as Gell-Mann had in his. But personally, Anderson despised glitz and glamor. He didn't even like working on fashionable problems. Whenever he felt other theorists crowding in on a subject he was working on, his instinct was to move to something else.

He particularly found it insufferable the way so many young hotshots went around wearing their specialty like a badge of academic rank—whether they'd accomplished anything or not: "Look at me, I'm a particle physicist! Look at me, I'm a cosmologist!" And he was outraged at the way Congress lavished money on shiny new telescopes and fantastically expensive new accelerators while smaller-scale projects—and, in Anderson's opinion, more scientifically productive projects—were starving. He had already spent more than his share of time in front of congressional committees denouncing the particle physicists' recently announced plans for a multibillion-dollar Superconducting Supercollider.

Besides, he thought, this Santa Fe bunch sounded like a pack of amateurs. What did Murray Gell-Mann know about putting together an interdisciplinary institute? He'd never worked on an interdisciplinary project in his life. Pines had at least spent some time working with astrophysicists and trying to apply solid-state physics to the structure of neutron stars. Indeed, he and Anderson were working on that little problem together. But what about the rest of them? Anderson had spent most of his own research career at Bell Labs, an interdisciplinary environment if ever there was one. And he knew how tricky such undertakings could be. The academic landscape is littered with the corpses of fancy new institutes that failed miserably; if they didn't get taken over by crackpots, they generally just sank into high-minded stagnation. In fact, Anderson had a close-up view of a sad example right there in Princeton: the august Institute for Advanced Study, the home of Oppenheimer, Einstein, and von Neumann. It did do some things very well, like math. But as an interdisciplinary institute he considered it an abject failure, a collection of very bright people who each did their own thing and barely talked to one another. Anderson had seen a lot of good scientists go in there and never live up to their promise.

And yet, Anderson was intrigued with this Santa Fe Institute in spite of himself. Reversing the tide of reductionism—now that was his kind of language. He had personally been fighting a guerilla war against reductionism for decades.

What first incited him to action, he recalls, was reading a lecture back in 1965 by the particle physicist Victor Weisskopf. In it, Weisskopf seemed to imply that "fundamental" science—that is, particle physics and some parts of cosmology—was somehow different from and better than more applied disciplines such as condensed-matter physics. Deeply annoyed, and scathing as only an insulted condensed-matter physicist can be, Anderson had immediately prepared a lecture of his own in rebuttal. In 1972 he had published it as an article in *Science* magazine entitled "More Is Different." And he had been pushing the argument at every opportunity since then.

To begin with, he says, he is the first to admit that there is a "philosophically correct" form of reductionism: namely, a belief that the universe is governed by natural law. The vast majority of working scientists accept that assertion wholeheartedly, says Anderson. Indeed, it's hard to imagine how science could exist if they didn't. To believe in natural law is to believe that the universe is ultimately comprehensible—that the same forces that determine the destiny of a galaxy can also determine the fall of an apple here on Earth; that the same atoms that refract the light passing through a diamond can also form the stuff of a living cell; that the same electrons, neutrons, and protons that emerged from the Big Bang can now give rise to the human brain, mind, and soul. To believe in natural law is to believe in the unity of nature at the deepest possible level.

However, says Anderson, this belief does *not* imply that the fundamental laws and the fundamental particles are the only things worth studying—and that everything else could be predicted if you only had a big enough computer. A lot of scientists certainly do seem to think that way, he says. Back in 1932, the physicist who discovered the positron—the antimatter version of the electron—declared, "The rest is chemistry!" More recently, Murray Gell-Mann himself had been known to dismiss condensed-matter theory as "dirt physics." But that was precisely the kind of arrogance that Anderson found so infuriating. As he wrote in his 1972 article, "The ability to reduce everything to simple fundamental laws does not imply the ability to start from those laws and reconstruct the universe. In fact, the more the elementary particle physicists tell us about the nature of the fundamental laws, the less relevance they seem to have to the very real problems of the rest of science, much less society."

This everything-else-is-chemistry nonsense breaks apart on the twin shoals of scale and complexity, he explains. Take water, for example. There's nothing very complicated about a water molecule: it's just one big oxygen atom with two little hydrogen atoms stuck to it like Mickey Mouse ears. Its behavior is governed by well-understood equations of atomic physics. But now put a few zillion of those molecules together in the same pot. Suddenly you've got a substance that shimmers and gurgles and sloshes. Those zillions of molecules have collectively acquired a property, liquidity, that none of them possesses alone. In fact, unless you know precisely where and how to look for it, there's nothing in those well-understood equations of atomic physics that even hints at such a property. The liquidity is "emergent."

In much the same way, says Anderson, emergent properties often produce emergent behaviors. Cool those liquid water molecules down a bit, for example, and at 32°F they will suddenly quit tumbling over one another at random. Instead they will undergo a "phase transition," locking themselves into the orderly crystalline array known as ice. Or if you were to go the other direction and heat the liquid, those same tumbling water molecules will suddenly fly apart and undergo a phase transition into water vapor. Neither phase transition would have any meaning for one molecule alone.

And so it goes, says Anderson. Weather is an emergent property: take your water vapor out over the Gulf of Mexico and let it interact with sunlight and wind, and it can organize itself into an emergent structure known as a hurricane. Life is an emergent property, the product of DNA molecules and protein molecules and myriad other kinds of molecules, all obeying the laws of chemistry. The mind is an emergent property, the product of several billion neurons obeying the biological laws of the living cell. In fact, as Anderson pointed out in the 1972 paper, you can think of the universe as forming a kind of hierarchy: "At each level of complexity, entirely new properties appear. [And] at each stage, entirely new laws, concepts, and generalizations are necessary, requiring inspiration and creativity to just as great a degree as in the previous one. Psychology is not applied biology, nor is biology applied chemistry."

No one reading that 1972 article or talking to its author could have any doubt where his sympathies lay. To Anderson, emergence in all its infinite variety was the most compelling mystery in science. Next to that, quarks just seemed so—boring. That's why he had gone into condensed-matter physics in the first place: It was a wonderland of emergent phenomena. (The Nobel Prize he received in 1977 honored his theoretical explanation

of a subtle phase transition in which certain metals went from being con-
ductors of electricity to being insulators.) And that's also why condensed-
matter physics was never quite enough to contain him. By the time Pines's
invitation reached him in June 1984, Anderson was busily applying techniques he'd developed in physics to understanding the three-dimensional
structure of protein molecules, and to analyzing the behavior of neural
networks—arrays of simple processors that try to do computation in much
the same way that networks of neurons do in the brain. He had even grappled
with one of the ultimate mysteries, suggesting a model of how the first life-
forms on Earth might have arisen from simple chemical compounds
through collective self-organization.

So if this Santa Fe outfit was for real, thought Anderson, he was ready
to listen. *If* it was for real.

A few weeks after getting Pines's invitation, he had his chance to find
out. That summer, as it happens, he was serving as chairman of the board
for the Aspen Center for Physics, a summer retreat for theoretical physicists
located across a broad meadow from the Aspen Institute. Anderson had
already planned to meet with Pines there to discuss some calculations about
the innards of neutron stars. So at their first encounter in Pines's office he
got right to the point: "Okay, Dave, is this thing flaky or is it for real?" He
knew exactly what Pines was going to say—"It's for real"—but he wanted
to hear how the answer *sounded*.

Pines did his best to make it sound good. He badly wanted Anderson in
on this. For all of Anderson's skepticism, he had a breadth of interest and
insight that was at least the equal of Gell-Mann's. He would serve as a
much-needed counterbalance and, not incidentally, his Nobel Prize would
give the institute an additional quantum leap in credibility.

So Pines assured Anderson that, yes, the institute really was going to
look at the intersections between disciplines, not just look at a few fashionable topics. And no, it was not going to be a front for Murray Gell-
Mann. Nor, for that matter, was it going to be just an appendage of Los
Alamos—which Anderson, Pines knew, would have nothing to do with.
Cowan was playing a lead role. Pines was playing a lead role. And if
Anderson would come aboard, he, Pines, would see to it that *he* played a
lead role. In fact—did Anderson have any speakers to suggest for these
workshops?

That did it. As soon as Anderson heard himself mulling over names and
topics, he knew he was hooked. The opportunity to make his presence felt
was just too tempting. "It was the sense that I could have some influence

on the institute," he says. "If it was really going to happen, I was eager to be there trying to contribute to the way things went, to avoid the mistakes of the past, to have it happen more or less right."

The discussions about the workshops and the institute continued throughout the summer, since Gell-Mann and Carruthers were also in Aspen. And as soon as Anderson got back to Princeton at the end of the summer, he jotted down three or four pages of suggestions for how to organize the institute so as to avoid the pitfalls. (The main point: Don't have separate departments!)

And he made reservations to travel to Santa Fe in the autumn.

"What am *I* doing here?"

Putting together the workshops proved to be a tricky business. Actually, it hadn't been too difficult to find funding. Gell-Mann had used his contacts to wangle $25,000 from the Carnegie Foundation. IBM kicked in $10,000. And Cowan had gotten another $25,000 from the MacArthur Foundation. (Gell-Mann, who was on the MacArthur board, had felt it was improper to ask himself.)

Much tougher, however, was the issue of whom to invite. "The question was," says Cowan, "could you get people to talk to one another and mutually stimulate one another about what was happening at the boundariès between disciplines? And could we develop a community that would actually nurture this kind of thing?" It was all too easy to imagine such a meeting dissolving into mutual incomprehension, with everyone talking right past each other—if they didn't walk out first in utter boredom. The only way to guard against that was to invite people with the right quality of mind.

"We didn't want the reclusive types, the ones who shut themselves off to write their book in some office," says Cowan. "We needed communication, we needed excitement, we needed mutual intellectual stimulation."

In particular, he says, they needed people who had demonstrated real expertise and creativity in an established discipline, but who were also open to new ideas. That turned out to be a depressingly rare combination, even among (or especially among) the most prestigious scientists. Gell-Mann suggested a number of people who might do. "He has great taste about intellectual strengths," says Cowan. "And he knows everybody." Herb Anderson suggested some others, as did Pines and Phil Anderson. "Phil has a hell of a lot of common sense," says Cowan. "He comes down hard on people he feels are fakers." Finding a mix that covered a broad enough range took a summer's worth of cross-country telephone calls and brain-storming, says Cowan. But in the end, he feels that what they came up with was "an astonishing list of good people," ranging from physicists to archeologists to clinical psychologists.

Of course, neither Cowan nor anyone else had the slightest idea what would happen when all these people got together.

Actually, there turned out to be no way to get all of them together in any case. Scheduling conflicts forced Pines to split the workshops over two separate weekends, October 6–7 and November 10–11, 1984. But Cowan remembers that for a while, at least, even this truncated group had trouble getting started. Gell-Mann had led off the October 6 session with a forty-five-minute talk, "The Concept of the Institute"—essentially an enlarged version of his "emerging syntheses" exhortation to the fellows the previous Christmas. And then there had followed an extended discussion of how to turn that concept into a real scientific agenda and a real institute. "There was a little sparring around," says Cowan. It wasn't entirely obvious at first how to find common ground.

For example, University of Chicago neuroscientist Jack Cowan (no relation) argued that it was high time for molecular biologists and neuro-scientists to start paying more attention to theoretical considerations, as a way of making sense out of the immense amounts of data they were collecting about individual cells and individual molecules. There were immediate objections that cells and biomolecules are too much the product of random evolution for theory to do much good. But Jack Cowan had heard that argument before, and he stood his ground. As an example, he pointed to the visual hallucinations caused by peyote or LSD. These come in a variety of patterns, including lattices, spirals, and funnels, he said. And every one of them could be explained as linear waves of electrical activity marching across the visual cortex of the brain. Might it be possible, he suggested, that these waves could be modeled with the kind of mathematical field theories used by physicists?

Douglas Schwartz of the School for American Research, the Santa Fe–based archeology center that was hosting the workshop, argued that archeology was a subject that was especially ripe for interactions with other disciplines. Researchers in the field were confronted with three fundamental mysteries, he said. First, when did nonhuman primates first begin to acquire the essence of humanity, including complex language and culture? Did it happen nearly a million years ago, with the rise of *Homo erectus*? Or only a few tens of thousands of years ago, as the Neanderthals gave way to fully modern humans, *Homo sapiens sapiens*? And either way, what caused the change? Millions of species have gotten along just fine without brains as large as ours. Why was our species different?

Second, said Schwartz, why did agriculture and fixed settlements replace nomadic hunting and gathering? And third, what forces triggered the development of cultural complexity, including specialization of crafts, the rise of elites, and the emergence of power based on factors such as economics and religion?

None of these mysteries had any real answer yet, said Schwartz, although the archeological record left by the rise and fall of the Anasazi civilizations in the American Southwest offered a wonderful field laboratory for investigating the last two. The only hope of finding some answers, he felt, lay in achieving much more cooperation between archeologists and other specialists than any of them have been used to. Field researchers needed increased input from physicists, chemists, geologists, and paleontologists to help to reconstruct the ups and downs of climates and ecosystems in those ancient times. And more than that, he said, they needed input from historians, economists, sociologists, and anthropologists to help them understand what motivations may have driven these peoples.

That kind of talk certainly resonated with Robert McCormack Adams, a University of Chicago archeologist who had been sworn in as the secretary of the Smithsonian Institution only weeks before. For at least the past decade, he said, he had been getting more and more impatient with anthropologists' gradualist approach to the evolution of civilization. When he went out digging in Mesopotamia, he said, he saw those ancient cultures undergoing chaotic oscillations and upheaval. Increasingly, he said, he was beginning to think of the rise and fall of civilizations as a kind of self-organizing phenomenon, in which human beings chose different clusters of cultural alternatives at different moments in response to different perceptions of environment.

This self-organization theme was also taken up in a quite different form by Stephen Wolfram of the Institute for Advanced Study, a twenty-five-year-old *wunderkind* from England who was trying to investigate the phenomenon of complexity at the most fundamental level. Indeed, he was already negotiating with the University of Illinois to found a Center for Complex Systems Research there. Whenever you look at very complicated systems in physics or biology, he said, you generally find that the basic components and the basic laws are quite simple; the complexity arises because you have a great many of these simple components interacting simultaneously. The complexity is actually in the organization—the myriad possible ways that the components of the system can interact.

Recently, Wolfram said, he and many other theorists had begun to study complexity using cellular automata, which are essentially programs for generating patterns on a computer screen according to rules specified by the programmer. Cellular automata have the virtue of being precisely defined, so that they can be analyzed in detail. And yet they are still rich enough for very simple rules to generate patterns of startling dynamism and complexity. The challenge for theorists, he said, is to formulate universal laws that describe when and how such complexities emerge in nature. And while the answer wasn't in yet, he remained optimistic.

In the meantime, he added, whatever else you do with this institute, make sure that every researcher in the place is equipped with a state-of-the-art computer. Computers are the essential tool in complexity research.

And so it went. How should you organize the institute? Robert Wilson, the founding director of the Fermi National Accelerator Laboratory outside of Chicago, said it was crucial for the institute to keep close tabs with experimenters; too much theory and you could end up gazing into your navel. Louis Branscomb, chief scientist of IBM, strongly endorsed the idea of an institute without departmental walls, where people could talk and interact creatively. "It's important to have people who steal ideas!" he said.

By lunchtime on the first day, says Cowan, the participants were beginning to warm to their task. As luck would have it, Santa Fe was showing off with one of its characteristically marvelous autumn days; people went through the buffet line and carried their plates outside to continue talking and arguing out on the American School grounds. (The school is located on an estate once owned by an eccentric heiress who had buried 220 dogs there.) "They began to realize that something was going on, and they opened up," says Cowan. By the second day, Sunday, he adds, "it became a very exciting thing." And by the time the participants headed home on Monday morning, it was clear to everyone that there really could be a core of science here.

Carruthers, for one, spent the weekend in heaven. "Here was a collection of many of the most creative people in the whole world, in many fields," he says. "And they turned out to have a lot to say to each other. They basically had the same world view, in the sense that they all seemed to feel that 'emerging syntheses' really meant a restructuring of science—that the overlapping themes of different parts of science would be put together in a new way. I can remember discussions with Jack Cowan [Stanford population biologist], Marc Feldman, various mathematicians, all of us coming from a different research culture, and discovering that our problems had enormous overlap, in both technique and structure. Now, some of this may be that the human mind only works in certain ways. [But] those workshops turned all of us into true believers. I won't quite call it a religious experience, but it was close enough."

For Ed Knapp, a Los Alamite who had gone off to Washington to serve a term as director of the National Science Foundation, and who had sat in on some of the early institute discussions, it was overwhelming to find himself among so many accomplished people at once. At one point he came up to Carruthers and said, "Hey, what am *I* doing here?"

And the Smithsonian's Bob Adams had much the same reaction. "It was a wonderful array of papers," he says. "When things are in the air, and you are beginning to make little linkages anyway, and then you go out to something like that symposium in Santa Fe, and suddenly there are gropings in neurobiology and cosmology and ecosystem theory and whatever—Jesus, you want to be in on it."

The second workshop, held a month later with a whole new crew of participants, turned out to be just as effective as the first. Even Anderson was impressed. "You couldn't help but be enthusiastic," he says. The event removed any last vestiges of doubt in his mind: this outfit really was going to be different from all the other advanced research institutes he knew about. "It was going to be much more interdisciplinary," he says. "They really *were* going to focus on the spaces between the fields." Moreover, there really was something there. "It wasn't clear that all these things would be on the agenda, but it was clear that many of them could be."

More than that, however, the workshops gave some much needed clarity to what Cowan's vision of a unified science might actually be about. As Gell-Mann recalls, "We had *fantastic* amounts of similarities. There were a huge number of common features in the things that were presented among various fields. You had to look carefully, but once you got past the jargon of all these things, it was there."

In particular, the founding workshops made it clear that every topic of interest had at its heart a system composed of many, many "agents." These agents might be molecules or neurons or species or consumers or even corporations. But whatever their nature, the agents were constantly organizing and reorganizing themselves into larger structures through the clash of mutual accommodation and mutual rivalry. Thus, molecules would form cells, neurons would form brains, species would form ecosystems, consumers and corporations would form economies, and so on. At each level, new emergent structures would form and engage in new emergent behaviors. Complexity, in other words, was really a science of emergence.

And the challenge that Cowan had been trying to articulate was to find the fundamental laws of emergence.

By no coincidence, it was also about this time that the new, unified science acquired a name: the sciences of complexity. "It seemed a much better canopy for everything we were doing than any other phrase we were using, including 'emerging syntheses,' " says Cowan. "It embraced everything I was interested in, and probably everything that anyone else at the institute was interested in."

So after the two founding workshops, Cowan and company were on their way. All they needed was for that fabled donor to step forward and give them the money.

John

Fifteen months later, they were still waiting. Looking back on that period, Cowan maintains that he was still confident that the money would follow the excitement. "It was an incubation period," he says. "I had the feeling things were moving rather rapidly." But others in the group were biting their fingernails to the elbow. "We had a growing sense of urgency," says Pines. "If we didn't keep a certain momentum, then we were going to lose support."

Granted, the time hadn't been totally unproductive. In many ways, in fact, those fifteen months had gone rather well. Cowan and his colleagues had come up with enough money to run a few workshops. They had hammered out an infinite number of organizational details. They had persuaded Mike Simmons, Pete Carruthers' former right-hand man at the Los Alamos theory division, to come in part-time as a vice-president and thereby take a lot of the administrative problems off Cowan's shoulders. And they had even gotten back the name they wanted. After more than a year of existence as "The Rio Grande Institute," which the fellows had accepted only out of necessity, they had been approached by a local firm wanting *that* name. So they said, "Sure—if you can get us the name we want." The firm had accordingly bought out the name "Santa Fe Institute" from the moribund therapy business that owned it, and the trade was made.

Perhaps most important, however, Cowan and his group had finessed a potentially explosive situation with Gell-Mann. Gell-Mann continued to be a superbly inspirational speaker. Moreover, he had drawn on his contacts to recruit a number of new members for the institute board. "I always expect that they'll say, 'No. I'm busy,' " says Gell-Mann. "But they almost always say, 'Oh, my God, yes! When can I come? I love this idea. I've been waiting for this all my life!' "

And yet as chairman of the board—the fund-raiser in chief—Gell-Mann was simply not getting anything done. The politest way to say it was that he was not a natural-born administrator. Cowan was exasperated: "Murray was always somewhere else." Gell-Mann had his fingers in a dozen pies, not all of them located in Santa Fe. Paperwork was piling up on his desk, he was not returning phone calls, and people were going crazy. The situation was only resolved to everyone's satisfaction with an executive meeting at Pines' house in Aspen in July 1985; Gell-Mann agreed to step down as head of the board of trustees and instead be head of a new Science Board, where he could happily plot the intellectual agenda of the institute. The new chairman of the board of trustees would be Ed Knapp, who had just finished his term at the National Science Foundation.

But for all of that, the hoped-for $100-million angel had not yet materialized, despite any number of feelers put out by Cowan and others. The major foundations weren't exactly eager to pour money into a flaky-sounding idea like this when the established research programs desperately needed their help to survive the Reagan budget cuts. "We were going to solve all of the outstanding problems of the modern world," says Carruthers. "A lot of people just laughed."

Meanwhile, the federal funding agencies were a giant question mark. Although they certainly weren't going to put up $100 million or anything like it, Eric Bloch, Knapp's successor at the National Science Foundation, seemed to be sympathetic to the idea of giving the institute some desperately needed seed money at, say, the $1 million level. So was Cowan's old friend Alvin Trivelpiece, who was now head of research at the Department of Energy. Bloch had even suggested the possibility of joint funding from both agencies. The problem was that nothing was going to happen until the institute could put together a formal proposal and get it approved—a process that could easily take a couple of years, considering that everyone was still working part time. And until then, Cowan barely had operating funds. The Santa Fe Institute seemed to be floundering.

So it was that the board of trustees meeting on March 9, 1986, was largely devoted to brainstorming for names of people who might give them money. Lots of ideas were batted around. In fact, it was only toward the end that Bob Adams, sitting at the far end of the conference table near the back of the room, rather diffidently put up his hand.

By the way, he said, he'd recently been up in New York at a meeting of the board of the Russell Sage Foundation, which gives away a lot of money for social science-type research. And while he was there he'd talked to a friend of his, John Reed, the new chief executive officer of Citicorp. Now, Reed was a pretty interesting guy, said Adams. He had just turned forty-seven, which made him one of the youngest CEOs in the country. He'd grown up in Argentina and Brazil, where his father had worked as an executive for Armour and Company. He had a bachelor's degree in liberal arts from Washington and Jefferson University, another bachelor's degree in metallurgy from MIT, and a master's degree in business from the Sloan School at MIT. He was very knowledgeable about science, and he genuinely seemed to enjoy kicking around ideas with the academic types at the Russell Sage board.

Anyway, said Adams, during one of the coffee breaks he'd told Reed about the institute, as best as he could explain it, and Reed had been very interested. He certainly didn't have $100 million to give away. But he was wondering if the institute might help him understand the world economy. When it came to world financial markets, Reed had decided that professional economists were off with the fairies. Under Reed's predecessor, Walter Wriston, Citicorp had just taken a bath in the Third World debt crisis. The bank had lost $1 billion in profits in one year, and was still sitting on $13 billion of loans that might never be paid back. And not only had the in-house economists not predicted it, their advice had made matters worse.

So Reed thought that a whole new approach to economics might be necessary, said Adams, and he had asked him to find out if the Santa Fe Institute might be interested in taking a crack at the problem. Reed had said he'd even be willing to come out to Santa Fe himself and talk about it. What do you think?

When Adams was finished, says Pines, "I thought about the proposition for about six microseconds and said, 'That's a great idea!' " Cowan wasn't far behind. "Get him out here," he said, "and I'll find some money to pay for it." Gell-Mann and the others chimed in with their approval. So far as any of them could tell, it was about twenty years too early to be tackling anything nearly as complex as economics—"It almost set the boundary condition for difficulty," says Cowan. "It involved human behavior." But what the hell. At the rate they were going, they were in no position to say no to anybody. It was worth a shot.

. . .

Yes, Dave, Phil Anderson told Pines over the phone. Yes, he was interested in economics. It was a bit of a hobby, in fact. And yes, this meeting with Reed sounded interesting. But no, Dave, I can't come. I'm too busy.

But, Phil, said Pines, who knew that Anderson hated to travel, if you work it right you can ride out on Reed's private plane. You can bring your wife, and you both can have the fun of taking a private jet. It's incredible. Those jets go right to your destination. It cuts six hours off the door-to-door time. It'll give you a chance to get to know John, and discuss the program with him. You can . . .

All right, said Anderson. All right. I'll come.

And so late in the afternoon of Wednesday, August 6, 1986, Anderson and his wife, Joyce, climbed aboard the Citicorp Gulfstream jet and rocketed off toward Santa Fe. Well, Anderson had to admit it was fast. It was also freezing. The Citicorp jet flew at about 50,000 feet, well above commercial airspace, and its heaters didn't seem quite able to handle the chill. Joyce Anderson huddled in back under a blanket while Anderson himself sat up front talking economics with Reed and three of his assistants: Byron Knief, Eugenia Singer, and Victor Menezes. Also along was Carl Kaysen of MIT, an economist who had once been head of the Institute for Advanced Study, and who now served on the boards of both the Russell Sage Foundation and the Santa Fe Institute.

Anderson found Reed to be pretty much as Adams had described him: smart, direct, and articulate. Around New York he had a fierce reputation for firing people en masse. But in person he struck Anderson as easygoing and unpretentious—the kind of CEO who likes to chat with one leg draped over an arm of his seat. He clearly wasn't intimidated by Nobel laureates. In fact, he said he'd been looking forward to this meeting, for exactly the same reason he enjoyed the meetings of the Russell Sage board and all the other academic boards he was on. "That sort of thing is fun for me," he says. "It gives me an opportunity to talk to an academic-intellectual group of folks who tend to look at the world quite differently from my day-to-day job. I think I benefit from seeing it both ways." In this particular case, Reed recalls having had a great time thinking about how to explain his admittedly biased view of the world economy to a set of scholars. "It was obviously different from the way one would explain it to a bunch of bankers."

For Anderson, the trip to Santa Fe proved to be a marvelous bull session on physics, economics, and the vagaries of global capital flows. He also found that one of Reed's assistants, in particular, was not about to be left out of the conversation. Shivering under several layers of sweaters, Eugenia Singer was coming along to talk about a survey she'd done for Reed on econometric models: the big computer simulations of the world economy used by the Federal Reserve Bank, the Bank of Japan, and others. Anderson liked her immediately.

Singer, as it happens, was not shivering just because of the cabin temperature. "I was terrified of what John had gotten me into!" she laughs. Here she was with nothing more than a master's degree in mathematical statistics, and essentially no recent experience working in that area. "And based on that, John had me running out there and talking to all these Nobel Prize-winning physicists! I didn't feel up to that technical level, to put it mildly.

"It was the only time I'd ever tried to say no to an assignment from John," she says. "But he'd said, in a very casual, offhand way, 'Ah, Eugenia, you'll do fine. You know more about it than they do.' " So she'd come. And Reed was right.

The encounter, jointly chaired by Adams and Cowan, started at 8 A.M. the next morning at Rancho Encantado, a kind of dude ranch about ten miles north of Santa Fe. Only a dozen people were present—among them Cowan's old friend Jerry Geist, chairman of the Public Service Company of New Mexico and the man who had put up the money for this meeting. The event was not really intended to be a scientific interchange. It was a show-and-tell, with each side trying to persuade the other side to do what it very much wanted to do anyhow.

Reed, armed with a fistful of overhead transparencies, went first. Basically, he said, his problem was that he was up to his eyeballs in a world economic system that defied economic analysis. The existing neoclassical theory and the computer models based on it simply did not give him the kind of information he needed to make real-time decisions in the face of risk and uncertainty. Some of these computer models were incredibly elaborate. One, which Singer would be talking about in more detail later, covered the whole world in 4500 equations and 6000 variables. And yet none of the models really dealt with social and political factors, which were often the most important variables of all. Most of them assumed that the modelers would put in interest rates, currency exchange rates, and other such variables by hand—even though these are precisely the quantities that a banker wants to *predict*. And virtually all of them tended to assume that the world was never very far from static economic equilibrium, when in fact the world was constantly being shaken by economic shocks and upheavals. In short, the big econometric models often left Reed and his colleagues with little more to go on than gut instinct—with results that might be imagined.

A case in point was the most recent world economic upheaval, which was symbolized by President Carter's 1979 appointment of Paul Volker to head the Federal Reserve Board. The story of that upheaval actually began in the 1940s, explained Reed, at a time when governments around the world found themselves struggling to cope with the economic consequences of two World Wars and a Great Depression in between. Their efforts, which culminated in the Bretton Woods agreements of 1944, led to a widespread recognition that the world economy had become far more interconnected than ever before. Under the new regime, nations shifted away from isolationism and protectionism as instruments of national policy; instead, they agreed to operate through international institutions such as the World Bank, the International Monetary Fund, and the General Agreements on Tariffs and Trade. And it worked, said Reed. In financial terms, at least, the world remained remarkably stable for a quarter of a century.

But then came the 1970s. The oil shocks of 1973 and 1979, the Nixon administration's decision to let the price of the dollar float on the world currency market, rising unemployment, rampant "stagflation"—the system cobbled together at Bretton Woods began to unravel, said Reed. Money began flowing around the world at an ever-increasing rate. And Third World countries that had once been starving for investment capital now began borrowing heavily to build their own economies—helped along by U.S. and European companies that were moving their production offshore to minimize costs.

Following the advice of their in-house economists, said Reed, Citicorp and many other international banks had happily lent billions of dollars to these developing countries. No one had really believed it when Paul Volker came to the Fed vowing to reign in inflation no matter what it took, even if it meant raising interest rates through the roof and causing a recession. In fact, the banks and their economists had failed to appreciate the similar words being voiced in ministerial offices all over the world. No democracy could tolerate that kind of pain. Could it? And so, said Reed, Citicorp and the other banks had continued loaning money to the developing nations throughout the early 1980s—right up until 1982, when first Mexico, and then Argentina, Brazil, Venezuela, the Philippines, and many others revealed that the worldwide recession triggered by the anti-inflation fight would make it impossible to meet their loan payments.

Since becoming CEO in 1984, said Reed, he'd spent the bulk of his time cleaning up this mess. It had already cost Citibank several billion dollars—so far—and had caused worldwide banking losses of roughly $300 billion.

So what kind of alternative was he looking for? Well, Reed didn't expect that any new economic theory would be able to predict the appointment of a specific person such as Paul Volker. But a theory that was better attuned to social and political realities might have predicted the appointment of someone *like* Volker—who, after all, was just doing the politically necessary job of inflation control superbly well.

More important, he said, a better theory might have helped the banks appreciate the significance of Volker's actions as they were happening. "Anything we could do that would enhance our understanding and tease out a better appreciation for the dynamics of the economy in which we live would be well worth having," he said. And from what he'd heard about modern physics and chaos theory, the physicists had some ideas that might apply. Could the Santa Fe Institute help?

The Santa Fe contingent was fascinated; to most of them this was brand new stuff. They were equally intrigued with Eugenia Singer's detailed review of global computer models. These included Project Link (the one with 6000 variables), the Federal Reserve Multi-Country Model, the World Bank Global Development Model, the Whalley Trade Model, and the Global Optimization Model. None of them could quite fill the bill, she concluded, especially when it came to dealing with change and upheaval.

So, again, could Santa Fe help?

Well, maybe. A lot of the afternoon was given over to the institute's side of the show and tell. Anderson talked about mathematical models of emergent, collective behavior. Others talked about the use of advanced computer graphics to convert mountains of data into vivid and comprehensible patterns; the use of artificial intelligence techniques to model agents that could adapt, evolve, and learn with experience; and the possible use of chaos theory to analyze and predict the gyrations of stock-market prices, weather records, and other such random-seeming phenomena. And in the end, not surprisingly, the consensus on both sides of the table was that, yes, an economics program was worth a try. As Anderson recalls, "We all said there's a possible intellectual agenda here. What was missing in modern equilibrium economics that permits the kind of upheavals that John was talking about?"

However, the Santa Fe contingent also played it very cagey. As dearly as Cowan and company wanted to see some of Citicorp's money, they also wanted to make it very clear to Reed that they couldn't promise him a miracle. Yes, they had some ideas that might be helpful. But this was a high-risk enterprise that might not lead anywhere. The last thing the fledgling institute needed was a lot of inflated expectations and hype; it would be suicide if they seemed to be promising something they couldn't deliver.

Reed said he understood completely. "My view was that I didn't think we were going to get something hard and concrete," he recalls. He just wanted some new ideas. So he promised not to put a time limit on the product, or even to define a deliverable product. If Santa Fe just got started on the job and made visible progress from year to year, that would be enough.

"That fueled my enthusiasm for doing this thing," says Anderson. The next thing to do, they agreed, was to hold another meeting—an extended workshop where you would have a significant number of economists and physical scientists sitting down together to thrash out the issues and set a real agenda. If Reed was prepared to kick in a few thousand dollars to support an effort along those lines, the Santa Fe Institute would be prepared to undertake it.

So the deal was done. The next morning. Reed had the East Coast crew routed out of bed at 5 A.M. and piled into limousines for the Santa Fe airport. He wanted to be back in New York as soon as possible to put in a full day's work.

Ken

No, Dave, said Anderson. I don't have time to organize this new economics workshop.

But, Phil, said Pines over the phone, you said a lot of interesting things when we met with Reed. And this new workshop is going to be an incredible opportunity. You'll invite the physical scientists, and we'll get a top-notch economist to invite the economics half.

No.

Look, said Pines, I know this is one more thing you're taking on, but I think you'll find it really interesting. Think about it. Talk to Joyce about it. And if you say yes, I'll help. You won't be out there all alone.

All right, Anderson sighed. All right, Dave. I'll do it.

Having said yes, Anderson was at a loss how to proceed. He'd never organized anything quite like this. Who had? Well, the first thing to do, obviously, was to find someone to head up the economics half of the meeting. He did know at least one economist: James Tobin of Yale, who had been a few years ahead of him at University High School in Champaign-Urbana, Illinois—and who, as it happens, had won a Nobel Prize in economics. Jim, he said over the phone, would you be interested in such a thing?

No, said Tobin after he'd heard Anderson explain what he wanted. He wasn't the right person. But Ken Arrow out at Stanford might be. In fact, he would be happy to call Arrow for him if he'd like.

Tobin had apparently been glowing in his description. When Anderson called, Arrow proved to be very interested indeed. "Ken and I chatted over the phone quite a bit," says Anderson. "It turned out that we had very similar ideas." For all that Arrow was one of the founders of establishment economics, he had also, like Anderson, remained a bit of an iconoclast himself. He knew full well what the drawbacks of the standard theory were. In fact, he could articulate them better than most of the critics could. Occasionally he even published what he called his "dissident" papers, calling for new approaches. He'd urged economists to pay more attention to real human psychology, for example, and most recently he had gotten intrigued with the possibility of using the mathematics of nonlinear science and chaos theory in economics. So if Anderson and the Santa Fe crowd thought they could strike out in new directions—"Well," he says, "it sounded like something that couldn't be uninteresting."

So Anderson and Arrow each started drawing up a list of names, using pretty much the same criteria as had been used for the founding workshops: they wanted people with superb technical backgrounds coupled with open minds.

Arrow, in particular, felt that he needed people who had a very strong command of the orthodox view of economics. He didn't mind people criticizing the standard model, but they'd better damn well understand what it was they were criticizing. He thought a bit and wrote down a few names.

And then he wanted to mix in a few people with an empirical bent. It wouldn't be healthy to have a solid phalanx of neoclassical theorists, he thought; you needed somebody to remind you of things the standard theory had trouble with. Let's see—maybe the young fellow he'd heard give that seminar last year, the one who'd done all that work in demography and who was always going on about increasing returns. Nice stuff.

Arthur, he wrote on the list. Brian Arthur.

In the autumn of 1986, even as Phil Anderson and Ken Arrow were drawing up names for the economics meeting, George Cowan was making a deal with the Archdiocese of Santa Fe for a three-year lease on the Christo Rey Convent: a one-story adobe structure located just past the pricey stretch of art galleries lining the crooked little lane known as Canyon Road.

It was about time. By that point Cowan and his colleagues had begun to hire a small cadre of staffers for the institute, thanks to operating funds that were beginning to trickle in from sources such as the MacArthur Foundation. And those staffers desperately needed a space to call their own. Furthermore, what with the economics meeting coming up and several other workshops being planned, the institute desperately needed a little office space where it could keep its academic visitors happy with desks and telephones. Cowan decided that the convent was small, but workable— and came at a price that was too good to pass up. So in February 1987, the institute staff moved in. And within days they had filled the tiny space to overflowing.

Chaos

The crowding never got any better. When Brian Arthur walked through the front entrance for the first time on Monday, August 24, 1987, he practically fell over the receptionist's desk; it was crammed into a sort of entry alcove and left the door with only about an inch of clearance to swing open. The corridors were lined with boxes full of books and papers. The copier machine was tucked into a closet. One staffer's "office" was in a hallway. The place was chaos. And Arthur fell in love with it immediately.

"I couldn't have designed a place that was better suited to my interests and temperament," he says. The chaotic convent somehow managed to convey a sense of intellectual ferment in the midst of peace, shelter, and serenity. When the institute's director of programs, Ginger Richardson, came out to welcome him and show him around, she and Arthur walked over creased linoleum floors and looked at lovingly crafted doors, polished mantles, and intricately decorated ceilings. She showed him the way to the coffeepot in the Eisenhower-era kitchen; to get there you had to walk through the mother superior's office, where Cowan was now ensconced as president. She showed him the former chapel, which now served as the large conference room; on the far wall, where the altar had once been, a blackboard full of equations and diagrams was washed by the ever-shifting light from stained-glass windows. She showed him the row of cramped little offices for the visiting scholars; formerly the nuns' bedroom cubicles, the rooms were now filled with cheap metal desks and typist chairs, and had windows looking out onto a sun-drenched patio and the Sangre de Cristo Mountains.

As a first-time visitor to New Mexico, Arthur was already in a mood to be enchanted. The mountains, the clear desert sunlight, the crystalline desert vistas had all had the same effect on him as they had had on generations of painters and photographers. But he felt at once that the convent had a special magic. "The whole atmosphere was unbelievable," says Arthur. "As I looked at the sort of books that were out on display, the kind of articles that were lying around, the freedom of the atmosphere, the informality—I couldn't believe that such a place existed." He was beginning to think that this economics workshop might be very exciting indeed.

Accommodations being what they were, academic visitors to the institute were crammed into the offices two or three at a time, with their names handwritten on pieces of paper taped up by each door. One office featured a name that Arthur was very interested to see: Stuart Kauffman of the University of Pennsylvania. Arthur had briefly met Kauffman two years earlier at a conference in Brussels, where he had been immensely impressed by Kauffman's talk on cells in a developing embryo. The idea was that the cells send out chemical messengers to trigger the development of other cells in the embryo in a self-consistent network, thus producing a coherent organism instead of just a lump of protoplasm. It was a concept that re-

sonated strongly with Arthur's ideas on the self-consistent, mutually supportive webs of interactions in human societies, and he remembers coming
back from that conference and telling his wife, Susan, "I've just listened
to the best talk I've ever heard!"

So as soon as he had gotten settled in his own office, Arthur wandered
down the hall to Kauffman's cubicle: Hello, he said, do you remember
that we met two years ago . . . ?

Well, no, actually, Kauffman didn't. But come on in! Tanned, curly-
haired, and California casual, the forty-eight-year-old Kauffman was nothing if not affable. But then, so was Arthur; he was in a mood to love
everybody that morning. The two men found themselves hitting it off
immediately. "Stu is an immensely warm person," says Arthur, "someone
you feel you have to hug—and I don't go around hugging people. He's
just such a lovable character."

They quickly fell to discussing economics, of course. With the meeting
coming up, the subject was very much on their mind—and, no, neither
of them had the slightest idea of what to expect. Arthur started to tell
Kauffman a bit about his work on increasing returns. "And that," he laughs,
"was a good excuse for Stuart to corner me and tell me about his latest
ideas."

It always was. Kauffman, as Arthur quickly learned, was an immensely
creative man, like a composer whose mind was endlessly aboil with melody.
He emitted ideas nonstop. He also displayed a very high ratio of talking to
listening. Indeed, that seemed to be the way he thought things through: by
talking about his ideas out loud. And talking about them. And talking about
them.

It was a trait that was already well known around the Santa Fe Institute.
In the course of the previous year, Kauffman had become the ubiquitous
man there. As the son and heir of a Romanian immigrant who had accumulated a minor fortune in real estate and insurance, he was one of the
few scientists who could afford to set himself up in a second home in Santa
Fe and live there half the year. At any given institute planning session,
Kauffman was to be found spinning out a steady stream of suggestions in
a mellifluous, confident baritone. At any given seminar he could be heard
thinking out loud during the question and answer period about how to
conceptualize the subject at hand: "Imagine a network of light bulbs hooked
together at random, okay, and . . ." And at any given moment in between,
he could be heard trying out his latest ideas on anyone who would listen;
rumor had it that he was once overheard explaining some of the finer points

of theoretical biology to the copier repair man. Or if visitors weren't around, he would soon be explaining the ideas to the nearest available colleague about a hundred times in a row. At length. In detail.

It was enough to drive his best friends away screaming. Worse, it had given Kauffman a widespread reputation for possessing an oversized ego combined with a nagging insecurity, even among colleagues who say in the next breath that they care about him deeply. It came across as a craving to be told, "Yes, Stuart, that's a great idea. You're very smart." But whatever truth there may have been to that perception, it was also true that Kauffman couldn't help it. For nearly a quarter of a century he had been a man in the grip of a vision—a vision that he found so powerful, so compelling, so overwhelmingly beautiful that he simply could not hold it in.

The closest English word for it is "order." But that word doesn't begin to capture what he meant by it. To hear Kauffman talk about order was to hear the language of mathematics, logic, and science being used to express a kind of primal mysticism. For Kauffman, order was an answer to the mystery of human existence, an explanation for how we could possibly come to exist as living, thinking creatures in a universe that seems to be governed by accident, chaos, and blind natural law. For Kauffman, order told us how we could indeed be an accident of nature—and yet be very much more than *just* an accident.

Yes, Kauffman always hastened to add, Charles Darwin was absolutely right: human beings and all other living things are undoubtedly the heirs of four billion years of random mutation, random catastrophes, and random struggles for survival; we are not here as the result of divine intervention, or even space aliens. But, he would emphasize, neither was Darwinian natural selection the whole story. Darwin didn't know about self-organization—matter's incessant attempts to organize itself into ever more complex structures, even in the face of the incessant forces of dissolution described by the second law of thermodynamics. Nor did Darwin know that the forces of order and self-organization apply to the creation of living systems just as surely as they do to the formation of snowflakes or the appearance of convection cells in a simmering pot of soup. So the story of life is, indeed, the story of accident and happenstance, declared Kauffman. But it is also the story of order: a kind of deep, inner creativity that is woven into the very fabric of nature.

"I love it," he says. "I do truly love it. My whole life has been an unfolding of this story."

Order

Walk down the corridors of almost any scientific institution in the world, and you won't get far before a glance through an open office door reveals a poster of Albert Einstein: Einstein bundled in an overcoat, absentmindedly walking through the snows of Princeton. Einstein gazing soulfully at the camera, a fountain pen clipped to the neck of his ratty sweater. Einstein with a maniacal grin, sticking his tongue out at the world. The creator of relativity is very nearly the universal scientific hero, the very emblem of profound thought and free creative spirit.

Back in the early 1950s, in Sacramento, California, Einstein was certainly a hero to a teenaged boy named Stuart Kauffman. "I admired Einstein enormously," he says. "No—admired is the wrong word. Loved. I loved his image of theory as the free invention of the human mind. And I loved his idea that science was a quest for the secrets of the Old One"—Einstein's metaphor for the creator of the universe. Kauffman especially remembers his first contact with Einstein's ideas in 1954, when he was fifteen and read a popular book on the origins of relativity by Einstein and his collaborator Leopold Infeld. "I was so thrilled that I could understand it, or thought I could understand it. Somehow, by being powerfully inventive, and free, Einstein had been able to create a world in his head. I remember thinking that it was absolutely beautiful that anyone could do that. And I remember crying when he died [in 1955]. It was as if I'd lost an old friend."

Until reading that book, Kauffman had been a good, if not spectacular, student earning As and Bs. Afterward, he was inflamed by a passion for—well, not science exactly. He didn't feel that he had to follow in Albert's footsteps that closely. But he certainly felt that same fierce desire to peer into the depths. "When you look at a cubist painting and see the structure hidden within it—that's what I wanted." The most immediate manifestation, in fact, was not scientific at all. The teenaged Kauffman developed an all-consuming desire to be a playwright, to fathom the forces of light and dark within the human soul. His first effort, a musical written in collaboration with his high-school English teacher, Fred Todd, was "absolutely atrocious." And yet the thrill of being taken seriously by a real adult—Todd was twenty-four at the time—was for Kauffman a crucial step in his intellectual awakening. "If Fred and I could write a musical when I was sixteen, even if it wasn't a very good musical, then why not the world?"

So the Stuart Kauffman who entered Dartmouth as a freshman in 1957 was every inch a playwright. He even smoked a pipe, because a friend of his had told him that if you wanted to be a playwright, you *had* to smoke a pipe. And, of course, he continued to write plays: three more of them that year with his freshman roommate and high-school buddy, Mac Magary.

But Kauffman soon began to notice something about his plays: the characters pontificated a great deal. "They blabbered about the meaning of life and what it means to be a good person—talking about it instead of doing it." He began to realize that he was less interested in the plays per se than in the ideas his characters were grappling with. "I wanted to find my way to something hidden and powerful and wonderful—without being able to articulate what it was. And when I found out that my friend Dick Green at Harvard was going to major in philosophy, I was terribly upset. I wished I could be a philosopher. But, of course, I had to be a playwright. To give it up meant giving up an identity I had begun to assume for myself."

It took about a week of struggle, he recalls, before he reached a profound revelation: "I didn't *have* to be a playwright—I could be a philosopher! So I spent the next six years studying philosophy with enormous passion." He started out in ethics, of course. He had wanted to understand the problem of good and evil as a playwright, and so what else could he possibly do as a philosopher? Yet he quickly found himself attracted elsewhere, toward the philosophy of science and the philosophy of mind. "It seemed to me they harbored the depths," he says. What is it about science that allows it to discover the nature of the world? And what is it about the mind that allows it to *know* the world?

With that passion to carry him, Kauffman graduated third in his class at Dartmouth in 1961, and then went on to a Marshall scholarship at Oxford from 1961 to 1963. As it happens, he didn't travel by a very direct route. "I had eight months before I had to be at Oxford, so I did the only rational thing possible: I bought a Volkswagen bus and lived in the Alps, skiing. I had the most prestigious possible address in St. Anton in Austria. I parked in the parking lot of the Post Hotel, and I used their restroom all winter long."

But once he arrived at Oxford he felt in his element. He can recall having been in three incredibly exciting intellectual environments in his life, and Oxford was the first. "It was the first time in my life that I was surrounded by people who were smarter than I was. The Americans there were just spectacular. The Rhodes scholars, the Marshall scholars. Some of them are now fairly well known. David Souter, who was in our group at Magdalene, is now on the Supreme Court. And George F. Will and I used to go have Indian meals all the time to escape college cooking."

Kauffman's passion for understanding science and the mind led him to study a curriculum that Oxford called Philosophy, Psychology, and Physiology. It included not just traditional philosophy, but a much more modern emphasis on the neural anatomy of the visual system and on more general models of neural wiring in the brain. In short, it dealt with what science could tell us about how the mind *really* works. His tutor in psychology was named Stuart Sutherland, who turned out to be another influential figure. Sutherland had a penchant for sitting behind his desk and subjecting his students to a nonstop volley of brain teasers: "Kauffman! How could a visual system possibly make a discrimination between two points of light that project onto adjacent cones in the retina?" Kauffman discovered that he loved this kind of challenge. He found that he had a facility for thinking up models on the spot to provide at least a plausible answer. ("Well, the eye isn't still, but it jiggles. So maybe you spread out the sensation over several rods and cones, and . . .") Indeed, he admits that this impromptu model-making got to be a habit; in one way or another he's been doing it ever since.

And yet, he also has to acknowledge a certain irony in it all. It was this very ability to make up models that led him to give up philosophy in favor of something much more down to earth: medical school.

"I decided in a way that proved I was never a great philosopher," he laughs. "The argument was: 'I'll never be as smart as Kant. There's no point in being a philosopher unless you are as smart as Kant. Therefore, I should go to medical school.' You'll notice it's not a syllogism."

Seriously, he says, the fact was that he was getting impatient with philosophy. "It wasn't that I didn't love philosophy. It's that I distrusted a certain facileness in it. Contemporary philosophers, or at least those of the 1950s and 1960s, took themselves to be examining concepts and the implications of concepts—not the facts of the world. So you could find out if your arguments were cogent, felicitous, coherent, and so on. But you couldn't find out if you were *right*. And in the end I felt dissatisfied with that." He wanted to delve into reality, to know the secrets of the Old One. "If I had to choose, I'd rather be Einstein than Wittgenstein."

Furthermore, he distrusted a certain facileness in himself. "I've always had a conceptual facility," he says. "At its best, it is the deepest part of me, God's greatest gift to me. But at its worst, it's glib. Shallow. Because of that concern, I said to myself, 'I'll go to med school and those sons of bitches won't let me be glib and swivel-hipped intellectually—because I'll have to take care of people. They'll force me to learn a lot of facts.' "

· · ·

They did. But somehow, they couldn't deflect Kauffman from his fascination with the play of ideas. In fact, they never really got a chance to. Since he had taken no premed courses whatsoever, he arranged to attend Berkeley starting in the fall of 1963 for a year of preparatory work before going on to medical school across the bay at the University of California, San Francisco. So it was in Berkeley that he took his first course in developmental biology.

He was thunderstruck. "Here was this absolutely *stunning* phenomenology," he says. "Here you start with a fertilized egg, and the damn thing unfolds, and it gives rise to an ordered newborn and adult." Somehow, that single egg cell manages to divide and differentiate into nerve cells and muscle cells and liver cells—hundreds of different kinds. And it does so with the most astonishing precision. The strange thing isn't that birth defects happen, as tragic as they are; the strange thing is that most babies are born perfect and whole. "This still stands as one of the most beautiful mysteries in biology," says Kauffman. "Well, I became absolutely enthralled with the problem of cellular differentiation, and set straight away to thinking hard about it."

It was a good time to be doing so: Jacob and Monod were publishing their first papers on genetic circuits in 1961 through 1963. It was the work for which they later won the Nobel Prize (and which Brian Arthur was to discover sixteen years later on the beach at Hauula). So Kauffman soon came across their work showing that any cell contains a number of "regulatory" genes that act as switches and can turn one another on and off. "That work was a revelation for all biologists. If genes can turn one another on and off, then you can have genetic circuits. Somehow, the genome has to be some kind of biochemical computer. It is the computing behavior— the orderly behavior—of this entire system that somehow governs how one cell can become different from another."

The question was how?

Actually, says Kauffman, most researchers at the time (or for that matter, now) weren't terribly bothered by this question. They talked about the "developmental program" of the cell as if the DNA computer were really carrying out its genetic instructions in the same way that an IBM mainframe executes a program written in FORTRAN: step by step by step. Moreover, they seemed to believe that these genetic instructions were organized with exquisite precision, having been as thoroughly debugged by natural selection as any piece of computer code ever devised by humans. How could it be otherwise? The slightest error in the genetic program could turn a developing cell cancerous or kill it entirely. That's why hundreds of molecular geneticists were already hard at work in their laboratories deciphering the precise biochemical mechanisms by which gene A switched on gene B, and how that switching process was affected by activities of genes C, D, and E. Detail, they felt, was everything.

And yet, the more Kauffman thought about this picture, the more he found this question of How? looming large. The genome was a computer, all right. But it wasn't anything at all like the machines that IBM was turning out. In a real cell, he realized, a great many regulatory genes could be active at the same time. So instead of executing its instructions step by step by step, the way human-built computers do, the genomic computer must be executing most or all of its genetic instructions simultaneously, in parallel. And if that was the case, he reasoned, then what mattered was not whether *this* regulatory gene activated *that* regulatory gene in some precisely defined sequence. What mattered was whether the genome as a whole could settle down into a stable, self-consistent *pattern* of active genes. At most, the regulatory genes might be going through a cycle of two or three or four configurations—a small number, anyhow; otherwise, the cell would just thrash around chaotically, with genes switching each other on and off at random. Of course, the pattern of active genes in a liver cell would be very different from the pattern in a muscle cell or a brain cell. But maybe that was just the point, Kauffman thought. The fact that a single genome can have many stable patterns of activation might be what allows it to give rise to many different cell types during development.

Kauffman was also troubled by people's tacit assumption that detail was everything. The biomolecular details were obviously important, he knew. But if the genome really had to be organized and fine-tuned to exquisite perfection before it could work at all, then how could it have arisen through the random trial and error of evolution? That would be like shuffling an honest deck of cards and then dealing yourself a bridge hand of thirteen spades: possible, but not very likely. "It just didn't feel right," he says. "You don't want to ask that much of either God or selection. If we had to explain the order in biology by lots of detailed, incredibly improbable bits of selection, and ad hocery, if everything we see was a hard struggle in the beginning, we wouldn't be here. There simply was not world enough and time for chance to have brought it forth."

There had to be more to it than that, he thought. "Somehow, I wanted it to be true that the order emerged in the first place, without having to be built in, without having to be evolved. I intentionally wanted it to be true that the order in a genetic regulatory system was *natural*, that it was quasi-inevitable. Somehow, the order would just be there for free. It would be spontaneous." If that was the case, he reasoned, then this spontaneous, self-organizing property of life would be the flip side of natural selection. The precise genetic details of any given organism would be a product of random mutations and natural selection working just as Darwin had described them. But the organization of life itself, the order, would be deeper and more fundamental. It would arise purely from the structure of the network, not the details. Order, in fact, would rank as one of the secrets of the Old One.

"Where that impulse comes from, I don't know," he says. "Why should Stu Kauffman have happened to come along and wonder that? It's an absolutely wonderful puzzle. I think it's a bizarre and wonderful thing that a mind can come afresh to a problem and can ask a question like that. But I've felt that way all my life. And all the science that I've done, that I truly love, is an effort to understand that vision."

Indeed, for the twenty-four-year-old premed student, the question of order was like an itch that wouldn't go away. What would it really mean for genetic order to be there "for free," he wondered? Well, look at the genetic circuits you find in real cells. They've obviously been refined by millions of years of evolution. But other than that, is there anything really special about them? Out of all the zillions of possible genetic circuits, are they the only ones that can produce orderly, stable configurations? If so, then they would be the analog of the bridge hand with thirteen spades, and it would truly be a miracle that evolution was ever lucky enough to produce them. Or are stable networks actually as common as the usual mix of spades, hearts, diamonds, and clubs? Because if that was the case, then it would be easy for evolution to stumble on a useful one; the networks in real cells would just be the ones that happened to survive natural selection.

The only way to find the answer, Kauffman decided, was to shuffle the deck, so to speak, deal out a bunch of "utterly typical" genetic circuits, and see if they did indeed produce stable configurations. "So I immediately started thinking about what would happen if you just took thousands of genes and hooked them together at random—what would they do?"

Now here was a problem he knew how to think about: he had studied neural circuitry ad nauseam at Oxford. Real genes were pretty complicated, of course. But Jacob and Monod had shown that the regulatory genes, at least, were essentially just switches. And the essence of a switch is that it flips back and forth between two states: active or inactive. Kauffman liked to think of them as light bulbs (on-off) or as a statement in logic (true-false). But whatever the image, he felt that this on-off behavior captured the essence of the regulatory gene. What remained was the network of interactions between genes. So, as the Berkeley free speech movement was unfolding down on campus, he spent his spare time sitting on the rooftop of his apartment in Oakland, obsessively drawing little diagrams of his regulatory genes hooked up in wiring diagrams, and trying to understand how they turned each other on and off.

The obsession didn't let up, even after he finished his premed courses at Berkeley and started going to medical school in San Francisco full time. It wasn't that he was bored in medical school. Quite the opposite: he found medical school very, very difficult. When his teachers weren't demanding mountainous quantities of memorization, they would go through an infinitely painstaking systems analysis of things like the physiology of the kidney. At that point, moreover, he still had every intention of practicing medicine. It appealed to the Boy Scout in him: it was a combination of doing good and knowing exactly what to do in any given situation, like pitching a tent in a storm.

No, Kauffman kept playing with the networks because he almost couldn't help himself. "I passionately wanted to be doing this bizarre science about these random nets." He got a C in pharmacology: "My notes from that class are all full of diagrams of genetic circuitry," he says.

He found that circuitry terribly confusing at first. He knew a lot about abstract logic, but almost nothing about mathematics. And the computer textbooks he found in the library told him almost nothing that was helpful. "Automata theory was well established by then, and that was all about logical switching nets. But those books told me how to synthesize a system that would do something, or what the general limits were of the capacity of complex automata. What I was interested in were the natural laws of complex systems. Whence cometh the order? And nobody was thinking about that at all. Certainly nobody that I knew." So he kept on drawing reams of diagrams, trying to get an intuitive feel for how these networks might behave. And whatever mathematics he needed, he invented for himself as best he could.

He quickly convinced himself that if the network became as densely tangled as a plate of spaghetti, so that every gene was controlled by lots of other genes, then the system would just thrash around chaotically. Using the light bulb analogy, it would be like a giant Las Vegas–style billboard gone haywire, so that all the lights twinkled at random. No order here.

Kauffman likewise convinced himself that if each gene were controlled by at most *one* other gene, so that the network was very sparsely connected, then its behavior would be too simple. It would be like a billboard where most of the bulbs just pulsed on and off like mindless strobe lights. That wasn't the kind of order Kauffman had in mind; he wanted his genetic light bulbs to organize themselves into interesting patterns analogous to waving palm trees or dancing flamingos. Besides, he knew that very sparsely connected networks were unrealistic: Jacob and Monod had already demonstrated that real genes tended to be controlled by several other genes. (Today, the number is known to be typically two to ten.)

So Kauffman started to concentrate on networks in between, where the connections were sparse, but not too sparse. To keep things simple, in fact, he looked at networks with precisely two inputs per gene. And here he began to find hints of something special. He already knew that densely connected networks were hypersensitive in the extreme: if you went in and flipped the state of any one gene from say, *on* to *off*, then you would trigger a whole avalanche of changes that would cascade back and forth through the network indefinitely. That's why densely connected networks tended to be chaotic. They could never settle down. But in his two-input networks, Kauffman discovered that flipping one gene would typically *not* produce an ever-expanding wave of change. Most often, the flipped gene would simply unflip, going back to what it was before. In fact, so long as the two different patterns of gene activation were not too different, they would tend to converge. "Things were simplifying," says Kauffman. "I could see that light bulbs tended to get into states where they got stuck on or off." In other words, the two-input networks were like a billboard where you could start the lights blinking at random, and yet they would always organize themselves into a flamingo or a champagne glass.

Order! Stealing whatever time he could from his medical courses, Kauffman filled his notebooks with more and more of his random two-input networks, analyzing the behavior of each one in detail. It was both tantalizing and frustrating work. The good news was that the two-input networks almost always seemed to stabilize very quickly. At most they would cycle over and over through a handful of different states. That's exactly what you wanted for a stable cell. The bad news was that he couldn't tell if his models had anything at all to do with real genetic regulatory networks. Real networks in real cells involved tens of thousands of genes. And yet Kauffman's pencil-and-paper networks were already getting out of hand when they contained only five or six genes. Keeping track of all the possible states and state transitions of a seven-gene network meant filling in a matrix containing 128 rows and 14 columns. Doing it for an eight-gene network would have required a matrix twice as big as that, and so on. "The chances for making an error by hand were just awfully large," says Kauffman. "I kept looking longingly at my seven-element networks. I just couldn't stand the idea of having to do eight."

"Anyway," he says, "somewhere in my sophomore year in medical school I couldn't take it anymore. I'd been playing long enough. So I went across the street to the computer center, and I asked if someone would help me program it. They said, 'Sure, but you have to pay for it.' So I whipped out my wallet. I was ready to pay for it."

Having decided to take the plunge into computers, Kauffman vowed to go all out: he would simulate a network with 100 genes. Looking back on it now, he laughs, it was a good thing he didn't quite know what he was doing. Think of it this way. One gene by itself can have only two states: *on* and *off*. But a network of two genes can have 2×2, or four states: *on-on, on-off, off-on, off-off*. A network of three genes can have $2 \times 2 \times 2$, or eight states, and so on. So the number of states in a network of 100 genes is 2 multiplied by itself 100 times, which turns out to be almost exactly equal to one million trillion trillion: 1 followed by 30 zeros. That's an immense space of possibilities, says Kauffman. In principle, moreover, there was no reason why his simulated network shouldn't have just wandered around in that space at random; after all, he was deliberately wiring it up at random. And that would have meant that his idea of cell cycles was hopeless: the computer would have had to go through roughly one million trillion trillion transitions before it ever started retracing its steps. It would be a cell cycle of sorts, but vast beyond imagining. "If it takes a microsecond for the computer to go from one state to another," says Kauffman, "and if it had to keep running for something like a million trillion trillion microseconds, you'd have billions of times the history of the universe. I'd have never made it through medical school!" Indeed, the computer charges alone would have bankrupted him long before graduation.

Fortunately, however, Kauffman hadn't done that calculation at the time. So, with the aid of a very helpful programmer at the computer center, he coded up a simulated two-input network with 100 genes in it, and then blithely turned in his deck of punch cards at the front desk. The answer came back ten minutes later, printed out on wide sheets of fan-fold paper. And exactly as he had expected, it showed his network quickly settling into orderly states, with most of the genes frozen on or off and the rest cycling through a handful of configurations. These patterns certainly didn't look like flamingos or anything recognizable; if his 100-gene network had been a Las Vegas billboard with a hundred light bulbs, then the orderly states would have looked like oscillating blobs. But they were there, and they were stable.

"I was just *unbelievably* thrilled!" says Kauffman. "And I felt then and feel now that it was rather profound. I had found something that no one would have intuited then." Instead of wandering through a space of one million trillion trillion states, his two-input network had quickly moved to an infinitesimal corner of that space and stayed there. "It settled down and oscillated through a cycle of five or six or seven or, more typically it turned out, about ten states. That's an amazing amount of order! I was just stunned."

That first simulation was only the beginning. Kauffman still had no idea of *why* sparsely connected networks were so magical. But they were, and he felt as though they had given him a whole new way of thinking about genes and embryonic development. Using that original program as a template and modifying it as needed, he ran simulations in endless variety. When and why did this orderly behavior occur, he wanted to know. And, not incidentally, how could he test his theory with real data?

Well, he thought, one obvious prediction of his model was that real genetic networks would have to be sparsely connected; densely connected networks seemed incapable of settling down into stable cycles. He didn't expect them to have precisely two inputs per gene, like his model networks. Nature is never quite that regular. But from his computer simulations and his reams of calculations, he realized that the connections only had to be sparse in a certain statistical sense. And when you looked at the data, by golly, real networks seemed to be sparse in exactly that way.

So far so good. Another test of the theory was to look at a given organism with a given set of regulatory genes, and ask how many cell types it was capable of producing. Kauffman knew he couldn't say anything specifically, of course, since he was deliberately trying to study the *typical* behavior of networks. But he could certainly look for a statistical relationship. His presumption all along had been that a cell type corresponded to one of his stable-state cycles. So he began to run bigger and bigger simulations, keeping track of how many state cycles occurred as the size of the model network increased. By the time he got up to networks of 400 to 500 genes, he had determined that the number of cycles scaled roughly as the square root of the number of genes in the network. Meanwhile, he had also been spending every spare hour in the medical school library, pouring through obscure references looking for comparable data on real organisms. And when he finally plotted it all up, there it was: the number of cell types in an organism did indeed scale roughly as the square root of the number of genes it had.

And so it went. "Goddamn it, it worked!" says Kauffman. It was the most beautiful thing he had ever experienced. By the end of his sophomore year at medical school he had run up hundreds upon hundreds of dollars in computer bills. He paid it all without a quiver.

In 1966, at the beginning of his third year of medical school, Kauffman wrote a letter to the neurophysiologist Warren McCulloch of MIT, explaining what he had done with his genetic network models and asking if McCulloch was interested.

It took a certain chutzpah to write that letter, Kauffman admits. Originally trained as an MD himself, McCulloch was one of the grand old men of neurophysiology—not to mention computer science, artificial intelligence, and the philosophy of mind. For the past two decades, he and a band of loyal followers had been working out the implications of an idea first put forward in 1943, when he and an eighteen-year-old mathematician named Walter Pitts had published a paper entitled "A Logical Calculus of the Ideas Immanent in Nervous Activity." In that paper, McCulloch and Pitts had claimed that the brain could be modeled as a network of logical operations such as *and*, *or*, *not*, and so forth. It had been a revolutionary idea at the time, to put it mildly, and had proved to be immensely influential. Not only was the McCulloch-Pitts model the first example of what would now be called a neural network, it was the first attempt to understand mental activity as a form of information processing—an insight that provided the inspiration for artificial intelligence and cognitive psychology alike. Their model was also the first indication that a network of very simple logic gates could perform exceedingly complex computations—an insight that was soon incorporated into the general theory of computing machines.

Grand old man or not, however, McCulloch seemed to be the only scientist Kauffman could share his work with. "McCulloch was the only person I knew who had done a lot of stuff with neural networks," he says. "And it was clear that genetic networks and neural networks were fundamentally the same thing."

Besides, Kauffman badly needed a little outside support by that point. Medical school was shaping up to be a decidedly mixed blessing. He was certainly getting the "facts" that he'd wanted so badly as a philosophy student at Oxford. But they weren't going down very well. "I think I chafed inside at having to take other people's word for what you're supposed to do," he says. "What one had to do in medical school was to master the facts, master the diagnosis, absorb the pearls of diagnostic wisdom, and then execute the appropriate procedures. And while there is a joy in executing those procedures, it didn't have the beauty that I wanted. It wasn't like searching for the secrets of the Old One."

His professors, meanwhile, didn't take kindly to the fact that Kauffman was finding that beauty in his genetic networks. "One of the most profound things you go through in medical school is almost a hazing," says Kauffman. The round-the-clock duty shifts, the endless demands—"The purpose is to make it clear to you that the patient comes first. You *will* get up at four thirty in the morning to do what's necessary. That part I didn't mind at all. But there were some of the faculty in medical school who thought they stood as guardians to the house of medicine. If you didn't have the proper attitude to be a doctor, then you could never be a real doctor."

Kauffman remembers his junior-year surgery professor in particular: "He thought my mind was elsewhere—and he was right," he says. "I remember him telling me that he didn't care if I got an A on the final, he would give me a D for the course. I think I got a B on the final and he still gave me a D.

"So you have to picture being a medical student and being grotty and not very happy, getting a D in surgery—it was an emotional thing for me. I had been a Marshall scholar and a big success academically, and here I was trying to survive in medical school, with my professor of surgery telling me what a loser I was."

About the only bright spot in his life, in fact, was that he had just gotten married to an Italian-American New Yorker named Elizabeth Ann Bianchi, a graduate student in art whom he had met during his Oxford days while she was an undergraduate traveling through Europe. "I was holding the door open for her and thought, 'Gee, that's a pretty girl.' I've been holding the door for her ever since." And yet even she had to wonder about this network stuff. "Liz is much more concrete than I am," say Kauffman. "She was profoundly interested in medicine. She came with me to anatomy class and a bunch of other things. But her response to the networks was 'That's nice—but is it real?' To her it seemed awfully phantasmagorical."

It was in the midst of all that that the reply arrived from McCulloch: "All Cambridge excited about your work," he wrote. Kauffman laughs at the memory. "It took me about a year to figure out that when Warren said that, it meant he had read what I sent him and thought it was kind of interesting."

But at the time, he was both thrilled and astonished at McCulloch's reply. He hadn't expected anything like it. He was emboldened to write back, explaining that UC-San Francisco encouraged its junior-year medical students to go somewhere else for three months to get outside experience. So could he come out to MIT and spend that time working with McCulloch?

Certainly, McCulloch wrote back. And furthermore, Kauffman and Liz should stay with him while they were there.

They accepted instantly. Kauffman will never forget his first meeting with McCulloch: it was about nine o'clock of a winter's evening, as he and Liz were driving around and around in the dark in a strange neighborhood in Cambridge, Massachusetts, hopelessly lost after having driven all the way across country. "And then there was Warren, looming out of the fog with his patriarchal beard, to welcome us into his home." While his wife, Rook, set out cheese and tea for the exhausted travelers, McCulloch called up Marvin Minsky, guru of MIT's artificial intelligence group: "Kauffman's in town."

McCulloch, a devout Quaker, proved to be a considerate and fascinating host. Enigmatic, lyrical, possessed of a mind that wandered freely over a vast intellectual landscape, he was endlessly enthusiastic about the quest for the inner workings of thought. He wrote in a bygone style, filling his scientific articles with allusions to everyone from Shakespeare to Saint Bonaventura—and then giving them titles like "Where Is Fancy Bred?," "Why the Mind Is in the Head," and "Through the Den of the Metaphysician." He loved riddles and wordplay. And he turned out to be one of the few people in the world who could outtalk Kauffman himself.

"Warren tended to corner you into conversations that dragged on," says Kauffman. Former students who had lived with McCulloch told stories of leaving the house through the upper bedroom window to avoid being trapped. McCulloch would habitually follow Kauffman into the bathroom while he was taking a shower, flip down the toilet seat, and sit there happily discussing networks and logical functions of various kinds while Kauffman was trying to get the soap out of his ear.

Most important, however, McCulloch became a mentor, guide, and friend to Kauffman—as he did with virtually all of his students. Knowing that Kauffman's goal at MIT was to run really big computer simulations, so that he could begin to get detailed statistical information about his networks' behavior, McCulloch introduced him to Minsky and his colleague Seymour Papert, who in turn arranged for Kauffman to do his simulations on the powerhouse computers of what was then known as Project MAC: Machine-Aided Cognition. McCulloch likewise arranged for Kauffman to get programming help from an undergraduate who knew a lot more about computer code than he did; they ended up running simulations with thousands of genes.

Meanwhile, McCulloch was introducing Kauffman to the small but intense world of theoretical biology. It was in the living room of McCulloch's house that he met the neurophysiologist Jack Cowan, who had been McCulloch's research assistant back in the late 1950s and early 1960s, and who had just been given the mandate to rejuvenate the theoretical biology group at the University of Chicago. It was in McCulloch's office that Kauffman met Brian Goodwin of the University of Sussex in England, who has been one of his closest friends ever since.

"Warren was like Fred Todd," says Kauffman. "Warren was the first person to take me seriously as a young scientist in my own right, not as a student." Sadly, McCulloch died only a few years later, in 1969. But Kauffman still considers himself, in some small way, his heir. "Warren literally catapulted me into the world that I've lived in ever since."

Indeed he did. Kauffman had decided before he ever came to MIT that he would devote himself to science once he graduated, not medical practice. It was the group of people he met through McCulloch who really brought him into the fold.

"It was through Jack Cowan, Brian Goodwin, and others that I was invited to my first scientific conference in 1967," he says. The event was the third in a series of conferences on theoretical biology run by the late British embryologist Conrad Waddington. "Those conferences were an attempt, in the mid-late 1960s, to be what the institute is like today," says Kauffman. "It was just wonderful. From drawing blood and checking stool samples at four in the morning—talk about getting your hands on reality!—I was flown to northern Italy to the Villa Serbelloni on Lake Como, a site picked out by Pliny the Younger. Absolutely gorgeous. And here were all these amazing people. John Maynard Smith was there. Rene Thom was just inventing catastrophe theory. Dick Lewontin from Chicago was there. Dick Levins from Chicago. Lewis Wolpert from London. These are people who are still friends.

"So I gave my talk about order in these genetic nets, and the numbers of cell types, and so on," he says. "And afterward we went out for coffee on the terrace overlooking three arms of the lake. Jack Cowan walked out and asked me if I wanted to be on the faculty at Chicago. So I thought about it for exactly a nanosecond and said, 'Of course!' I didn't ask Jack for a year and a half what my salary would be."

Death and Life

Around lunchtime on that first day of Arthur's residence at the Santa Fe Institute, he and Kauffman wandered down through the adobe art galleries along Canyon Road to what was then called Babe's, one of Kauffman's favorite watering holes. And almost every day thereafter for the next two weeks they met for lunch again or just to talk.

They did much of their talking on their feet. Kauffman loved the open air even more than Arthur did. As a teenager in the Boy Scouts he had gone on innumerable hiking and camping expeditions in the Sierras. In college he had been an avid skier and mountain climber. And nowadays he still went out hiking whenever possible. So Kauffman and Arthur would talk as they strolled along Canyon Road, or as they walked up behind the convent onto a broad hill, where they could sit on top and gaze out at an immense view of Santa Fe and the mountains.

There was an ineffable sadness about Kauffman, Arthur began to realize. In the midst of his jokes, his wordplay, his omnivorous curiosity, and his incessant talking about his ideas, there would sometimes come a pause. A flash of grief. And one evening shortly after he arrived in Santa Fe, as Arthur and his wife, Susan, were having dinner out with the Kauffmans, Stuart Kauffman told them the story: How he and Liz had come home one Saturday night the previous October to learn that their thirteen-year-old daughter, Merit, had been struck by a hit-and-run driver and was in grave condition at a local hospital. How they had raced to the hospital with their son, Ethan. And how they had learned when they got there that Merit had died fifteen minutes before.

Today, more than half a decade after the fact, Kauffman can tell that story without breaking down. But he couldn't that night. Merit Kauffman had been very much her father's daughter. "It was pulverizing," says Kauffman. "Just agonizing. There's no way of describing it. We went upstairs to see her. There was my daughter's broken body on a table, cooling off. Just unbearable. The three of us huddled in bed together that night, crying. She had a kind of feistiness, but an awareness of people we just marveled at. We all felt that she had been the best of the four of us."

"They say that time heals," he adds. "But that's not quite true. It's simply that the grief erupts less often."

As they walked along the roads and hillsides around the convent, Arthur couldn't help but be intrigued by Kauffman's concept of order and self-organization. The irony of it was that when Kauffman used the word "order," he was obviously referring to the same thing that Arthur meant by the word "messiness"—namely emergence, the incessant urge of complex systems to organize themselves into patterns. But then, maybe it wasn't so surprising that Kauffman was using exactly the opposite word; he was coming at the concept from exactly the opposite direction. Arthur talked about "messiness" because he had started from the icy, abstract world of economic equilibrium, in which the laws of the market are supposed to determine everything as precisely as the laws of physics. Kauffman talked about "order" because he had started from the messy, contingent world of Darwin, in which there are no laws—just accident and natural selection. But by starting from totally different places, they had arrived at essentially the same place.

Kauffman, meanwhile, was both intrigued and perplexed by Arthur's increasing-returns ideas. "I had a hard time understanding why this was new," he says. "Biologists have been dealing with positive feedback for years." It took him a long time to comprehend just how static and changeless the neoclassical world view really is.

He was even more intrigued, however, when Arthur started asking Kauffman about another economic problem that had been bugging him: technological change. This had already become a political hot-button issue, to put it mildly. You could feel the undercurrent of anxiety in almost every magazine or newspaper you picked up: Can America Compete? Have we lost our fabled American ingenuity, the old Yankee know-how? Are the Japanese going to wipe us out, industry by industry?

Good questions. The problem, as Arthur explained to Kauffman, was that economists didn't have any answers—at least, not at the level of fundamental theory. The whole dynamic of technological development was like a black box. "Until about fifteen or twenty years ago," he says, "the notion was that technologies came at random out of the blue, fell from heaven in celestial books of blueprints for making process steel, or silicon chips, or anything like that. And those things were made possible by inventors—smart people like Thomas Edison who sort of got these ideas in their bathtubs and added a page to their book of blueprints." Strictly speaking, in fact, technology wasn't part of economics at all. It was "exogenous"—delivered magically by noneconomic processes. More recently, there had been a number of efforts to model technology as being "endogenous," meaning that it's produced within the economic system itself. But usually that meant regarding technology as the outcome of investment in research and development, almost like a commodity. And while Arthur thought there might be some truth to that, he still didn't think it got to the heart of the matter.

When you look at economic history, as opposed to economic theory, he told Kauffman, technology isn't really like a commodity at all. It is much more like an evolving ecosystem. "In particular, innovations rarely happen in a vacuum. They are usually made possible by other innovations being already in place. For example, a laser printer is basically a Xerox machine with a laser and a little computer circuitry to tell the laser where to etch on the Xerox drum for printing. So a laser printer is possible when you have computer technology, laser technology, and a Xerox reproducing technology. But it is also only possible because people need fancy, high-speed printing."

In short, technologies form a highly interconnected web—or in Kauffman's language, a network. Furthermore, these technological webs are highly dynamic and unstable. They can grow in a fashion that is almost organic, as when laser printers give rise to desktop publishing software, and desktop publishing opens up a new niche for graphics programs. "Technology A, B, and C might make possible technology D, and so on," says Arthur. "So there'd be a network of possible technologies, all interconnected and growing as more things became possible. And therefore the economy could become more complex."

Moreover, these technological webs can undergo bursts of evolutionary creativity and massive extinction events, just like biological ecosystems. Say a new technology like the automobile comes in and replaces an older technology, the horse. Along with the horse go the smithy, the pony express, the watering troughs, the stables, the people who curried horses, and so on. The whole subnetwork of technologies that depended upon the horse suddenly collapses in what the economist Joseph Schumpeter once called "a gale of destruction." But along with the car come paved roads, gas stations, fast-food restaurants, motels, traffic courts and traffic cops, and traffic lights. A whole new network of goods and services begins to grow, each one filling a niche opened up by the goods and services that came before it.

Indeed, said Arthur, this process is an excellent example of what he meant by increasing returns: once a new technology starts opening up new niches for other goods and services, the people who fill those niches have every incentive to help that technology grow and prosper. Moreover, this process is a major driving force behind the phenomenon of lock-in: the more niches that spring up dependent on a given technology, the harder it is to change that technology—until something very much better comes along.

So this idea about technological webs was very much in keeping with his vision of a new economics, Arthur explained. The problem was that the mathematics that he had developed was only good for looking at one technology at a time. What he really needed was a networky kind of model like the ones Kauffman had developed. "So," he asked Kauffman, "could you do a model where a technology is 'switched on' when it is created, maybe, and . . . ?"

Kauffman listened to all this thunderstruck. *Could* he? What Arthur was describing was, in a very different language, a problem that Kauffman had been working on for a decade and a half.

Within minutes Kauffman was off, explaining to Arthur why the process of technological change is exactly like the origin of life.

Kauffman first got the idea back in 1969, around the time he arrived at the theoretical biology group in Chicago.

After medical school, he says, being in Chicago felt like being in heaven. Looking back on it, in fact, Chicago was the second of his three most exciting intellectual environments. "It was an extraordinary place full of extraordinarily able people," he says. "The department in Chicago was the focus in the United States for the same set of friends I'd met in Italy." Jack Cowan was doing his groundbreaking work on cortical tissue, writing down simple equations that described waves of excitation and inhibition moving across two-dimensional sheets of neurons in the brain. John Maynard Smith was doing his equally groundbreaking work on evolutionary dynamics, using a mathematical technique known as game theory to clarify the nature of competition and cooperation among species. Maynard Smith, who was there on sabbatical from the University of Sussex, also gave Kauffman some much-needed help on the mathematical analysis of networks. "John taught me to 'Do sums,' as he put it," says Kauffman, "and I cured his pneumonia one day."

Now that he was surrounded by colleagues and soul mates, Kauffman quickly discovered that he had not been alone in thinking about the statistical properties of networks. In 1952, for example, the English neurophysiologist Ross Ashby had speculated along much the same lines in his book *Design for a Brain*. "He was asking quite similar questions about what is the generic behavior of complex networks," says Kauffman. "But that was completely unknown to me. I got in touch with him as soon as I found out about it."

At the same time, Kauffman discovered that in developing his genetic networks, he had reinvented some of the most avant-garde work in physics and applied mathematics—albeit in a totally new context. The dynamics of his genetic regulatory networks turned out to be a special case of what the physicists were calling "nonlinear dynamics." From the nonlinear point of view, in fact, it was easy to see why his sparsely connected networks could organize themselves into stable cycles so easily: mathematically, their behavior was equivalent to the way all the rain falling on the hillsides around a valley will flow into a lake at the bottom of the valley. In the space of all possible network behaviors, the stable cycles were like basins— or as the physicists put it, "attractors."

After six years of agonizing over these networks, Kauffman was gratified to think that he was finally beginning to understand them so well. And yet—he couldn't help but feel that something was still missing. Talking about self-organization in genetic regulatory networks was all very nice. But at a molecular level, genetic activity depends on the incredibly complex and sophisticated molecules known as RNA and DNA. Where did *they* come from?

How did life get started at all?

Well, according to the standard theory in the biology textbooks, the origin of life was rather straightforward. DNA, RNA, proteins, polysac-charides, and all the other molecules of life must have arisen billions of years ago in some warm little pond, where simple molecular building blocks like amino acids and such had accumulated from the primordial atmos-phere. Back in 1953, in fact, the Nobel laureate chemist Harold Urey and his graduate student Stanley Miller showed experimentally that an early atmosphere of methane, ammonia, and the like could have produced those building blocks quite spontaneously; all it would have taken was the oc-casional lightning bolt to provide energy for chemical reactions. So over time, went the argument, those simple compounds would have collected in ponds and lakes, undergoing further chemical reactions and growing more and more complex. Eventually, there would have arisen a collection of molecules that included the DNA double helix and/or its single-strand cousin, RNA—both of which have the power to reproduce themslves. And once there was self-reproduction, all the rest would then follow from natural selection. Or so went the standard theory.

But Kauffman didn't buy it. For one thing, most biological molecules are enormous objects. To make a single protein molecule, for example, you might have to chain together several hundred amino-acid building blocks in a precise order. That's hard enough to do in a modern laboratory, where you have access to all the latest tools of biotechnology. So how could such a thing form all by itself in a pond? Lots of people had tried to calculate the odds of that happening, and their answers always came out pretty much the same: if the formation were truly random, you would have to wait far longer than the lifetime of the universe to produce even *one* useful protein molecule, much less all the myriads of proteins and sugars and lipids and nucleic acids that you need to make a fully functioning cell. Even if you assumed that all the trillions of stars in all the millions of galaxies in the observable universe had planets like Earth, with warm oceans and an atmosphere, the probability that any of them would bring forth life would still be—infinitesimal. If the origin of life had really been a random event, then it had really been a miracle.

More specifically, however, Kauffman didn't buy the standard theory because it equated the origin of life with the appearance of DNA. It just didn't seem reasonable to Kauffman that the origin of life should depend on something so complicated. The DNA double helix can reproduce itself, all right. But its ability to do so depends critically on its ability to uncoil, unzip its two strands, and make copies of itself. In modern cells, moreover, that process also depends on a host of specialized protein molecules, which serve various helper roles. How could all *that* have happened in a pond? "It was the same impulse that lay behind my question of whether you could find order in genetic regulatory networks," says Kauffman. "There was something too marvelous about DNA. I simply didn't want it to be true that the origin of life depended on something quite as special as that. The way I phrased it to myself was, 'What if God had hung another valence bond on nitrogen? [Nitrogen atoms are abundant in DNA molecules.] Would life be impossible?' And it seemed to me to be an appalling conclusion that life should be that delicately poised."

But then, thought Kauffman, who says that the critical thing about life is DNA? For that matter, who says that the origin of life was a random event? Maybe there was another way to get a self-replicating system started, a way that would have allowed living systems to bootstrap their way into existence from simple reactions.

Okay, then. Think about what that primordial soup must have been like, with all those little amino acids and sugars and such banging around. Obviously, you couldn't expect them to just fall together into a cell. But you could expect them to undergo at least some random reactions with one another. In fact, it's hard to see what could have stopped them from doing so. And while random reactions wouldn't have produced anything very fancy, you could do the calculations and show that, on the average, they would have produced a fair number of smallish molecules having short chains and branches.

Now, that fact in itself wouldn't have made the origin of life any more probable. But suppose, thought Kauffman, just suppose that some of these smallish molecules floating around in the primordial soup were able to act as "catalysts"—submicroscopic matchmakers. Chemists see this sort of thing all the time: one molecule, the catalyst, grabs two other molecules as they go tumbling by and brings them together, so that they can interact and fuse very quickly. Then the catalyst releases the newly wedded pair, grabs another pair, and so on. Chemists also know of a lot of catalyst molecules that act as chemical axe murderers, sidling up to one molecule after another and slicing them apart. Either way, catalysts are the backbone of the modern chemical industry. Gasoline, plastics, dyes, pharmaceuticals—almost none of it would be possible without catalysts.

All right, thought Kauffman, imagine that you had a primordial soup containing some molecule A that was busily catalyzing the formation of another molecule B. The first molecule probably wasn't a very effective catalyst, since it essentially formed at random. But then, it didn't need to be very effective. Even a feeble catalyst would have made B-type molecules form faster than they would have otherwise.

Now, thought Kauffman, suppose that molecule B itself had a weak catalytic effect, so that it boosted the production of some molecule C. And suppose that C also acted as a catalyst, and so on. If the pot of primordial soup was big enough, he reasoned, and if there were enough different kinds of molecules in there to start with, then somewhere down the line you might very well have found a molecule Z that closed the loop and catalyzed the creation of A. But now you would have had more A around, which means that there would have been more catalyst available to enhance the formation of B, which then would have enhanced the formation of C, and on and on.

In other words, Kauffman realized, if the conditions in your primordial soup were right, then you wouldn't have to wait for random reactions at all. The compounds in the soup could have formed a coherent, self-rein- forcing web of reactions. Furthermore, each molecule in the web would have catalyzed the formation of other molecules in the web—so that all the molecules in the web would have steadily grown more and more abun- dant relative to molecules that were *not* part of the web. Taken as a whole, in short, the web would have catalyzed its own formation. It would have been an "autocatalytic set."

Kauffman was in awe when he realized all this. Here it was again: order. Order for *free*. Order arising naturally from the laws of physics and chemistry. Order emerging spontaneously from molecular chaos and manifesting itself as a system that grows. The idea was indescribably beautiful.

But was it life? Well no, Kauffman had to admit, not if you meant life as we know it today. An autocatalytic set would have had no DNA, no genetic code, no cell membrane. In fact, it would have had no real independent existence except as a haze of molecules floating around in some ancient pond. If an extraterrestrial Darwin had happened by at the time, he (or it) would have been hard put to notice anything unusual. Any given molecule participating in the autocatalytic set would have looked pretty much like any other molecule. The essence was not to be found in any individual piece of the set, but in the overall dynamics of the set: its collective behavior.

And yet in some deeper sense, thought Kauffman, maybe an autocatalytic set *would* have been alive. Certainly it would have exhibited some remarkably lifelike properties. It could grow, for example: there was no reason in principle why an autocatalytic set shouldn't be open-ended, producing more and more molecules as time went on—and molecules that were more and more complex. Moreover, the set would have possessed a kind of metabolism: The web molecules would take in a steady supply of "food" molecules, in the form of amino acids and other simple compounds floating all around them in the soup, and catalytically glue them together to form the more complex compounds of the set itself.

An autocatalytic set would even exhibit a primitive kind of reproduction: if a set from one little pond happened to slosh over into a neighboring pond—in a flood, say—then the displaced set could immediately start growing in its new environment. Of course, if another, different set were already in place, then the two would engage in a competition for resources. And that, Kauffman realized, would immediately open the door for natural selection to winnow and refine the sets. It was easy enough to imagine such a process selecting those sets that were more robust to environmental changes or that contained more efficient catalysts and more elaborate reactions or that contained more complex and sophisticated molecules. Ultimately, in fact, you could imagine the winnowing process giving rise to DNA and all the rest. The real key was to get an entity that could survive and reproduce; after that, evolution could do its work in comparatively short order.

Okay, this was admittedly piling up a lot of ifs on top of ifs. But to Kauffman, this autocatalytic set story was far and away the most plausible explanation for the origin of life that he had ever heard. If it were true, it meant the origin of life didn't have to wait for some ridiculously improbable event to produce a set of enormously complicated molecules; it meant that life could indeed have bootstrapped its way into existence from very simple molecules. And it meant that life had not been just a random accident, but was part of nature's incessant compulsion for self-organization.

Kauffman was enthralled. He immediately plunged into calculations and computer simulations and random network models, doing exactly what he had done at Berkeley: he wanted to understand the natural laws of autocatalytic sets. Okay, he thought, you don't know what compounds and what reactions were really involved way back when. But at least you can think about probabilities. Was the formation of an autocatalytic set a wildly unlikely event? Or was it almost inevitable? Look at the numbers. Suppose you have a few different kinds of "food" molecules—amino acids, et cetera—and suppose that in the primordial soup, they start linking up into polymer chains. How many different kinds of polymers can you make this way? How many reactions are there among the polymers, so that you can make a big web of reactions? And what is the likelihood that if you had this big web of reactions it would close on itself and form an autocatalytic set?

"As I thought it through," he says, "it became perfectly obvious to me that the number of reactions went up faster than the number of polymers. So that if there was any fixed probability that a polymer catalyzed a reaction, there'd be some complexity at which this thing would have to become mutually autocatalytic." In other words, it was just like his genetic networks: if the primordial soup passed a certain threshold of complexity, then it would undergo that funny phase transition. The autocatalytic set would indeed be almost inevitable. In a rich enough primordial soup, it would *have* to form—and life would "crystallize" out of the soup spontaneously.

The whole story was just too beautiful, Kauffman felt. It *had* to be true. "I believe in this scenario as strongly now as I did when I first came up with it," he says. "I believe very strongly that this is how life began."

Arthur was ready to believe it too. He thought that this was great stuff, and not only because it was a marvelous idea about the origin of life. The analogies between autocatalysis and economics were just too delicious to pass up. He and Kauffman kicked it around for days, walking through the hills or hunched over the lunch tables at Babe's.

Most obviously, they agreed, an autocatalytic set was a web of transformations among molecules in precisely the same way that an economy is a web of transformations among goods and services. In a very real sense, in fact, an autocatalytic set was an economy—a submicroscopic economy that extracted raw materials (the primordial "food" molecules) and converted them into useful products (more molecules in the set).

Moreover, an autocatalytic set can bootstrap its own evolution in precisely the same way that an economy can, by growing more and more complex over time. This was a point that fascinated Kauffman. If innovations result from new combinations of old technologies, then the number of possible innovations would go up very rapidly as more and more technologies became available. In fact, he argued, once you get beyond a certain threshold of complexity you can expect a kind of phase transition analogous to the ones he had found in his autocatalytic sets. Below that level of complexity you would find countries dependent upon just a few major industries, and their economies would tend to be fragile and stagnant. In that case, it wouldn't matter how much investment got poured into the country. "If all you do is produce bananas, nothing will happen except that you produce more bananas." But if a country ever managed to diversify and increase its complexity above the critical point, then you would expect it to undergo an explosive increase in growth and innovation—what some economists have called an "economic takeoff."

The existence of that phase transition would also help explain why trade is so important to prosperity, Kauffman told Arthur. Suppose you have two different countries, each one of which is subcritical by itself. Their economies are going nowhere. But now suppose they start trading, so that their economies become interlinked into one large economy with a higher complexity. "I expect that trade between such systems will allow the joint system to become supercritical and explode outward."

Finally, an autocatalytic set can undergo exactly the same kinds of evolutionary booms and crashes that an economy does. Injecting one new kind of molecule into the soup could often transform the set utterly, in much the same way that the economy was transformed when the horse was replaced by the automobile. This was the part of autocatalysis that really captivated Arthur. It had the same qualities that had so fascinated him when he first read about molecular biology: upheaval and change and enormous consequences flowing from trivial-seeming events—and yet with deep law hidden beneath.

So Kauffman and Arthur had a grand time spinning out ideas and finding connections. It was like one of the all-time great freshman bull sessions. Kauffman was especially excited. He felt that they were onto something really new. Obviously, a network analysis wouldn't help anybody predict precisely what new technologies are going to emerge next week. But it might help economists get statistical and structural measures of the process. When you introduce a new product, for example, how big an avalanche does it typically cause? How many other goods and services does it bring with it, and how many old ones go out? And how do you recognize when a good has become central to an economy, as opposed to being just another hula-hoop?

Furthermore, Kauffman felt that it might ultimately be possible to apply these ideas far beyond the economy. "I think these kinds of models are the place for contingency and law at the same time," he says. "The point is that the phase transitions may be lawful, but the specific details are not. So maybe we have the starts of models of historical, unfolding processes for such things as the Industrial Revolution, for example, or the Renaissance as a cultural transformation, and why it is that an isolated society, or ethos, can't stay isolated when you start plugging some new ideas into it." You can ask the same thing about the Cambrian explosion: the period some 570 million years ago when a world full of algae and pond scum suddenly burst forth with complex, multicellular creatures in immense profusion. "Why all of a sudden do you get all this diversity?" Kauffman asks. "Maybe you had to get to a critical diversity to then explode. Maybe it's because you've gone from algal mats to something that's a little more trophic and complex, so that there's an explosion of processes acting on processes to make new processes. It's the same thing as in an economy."

Of course, even Kauffman had to admit that none of this was any more than a hope—yet. On the other hand, he told Arthur, it all might be very possible. He'd been laying the groundwork for it since 1982, when he returned to the subject of autocatalysis after a hiatus of more than a decade.

The hiatus began one day in 1971, Kauffman remembers, when Chicago chemist Stuart Rice dropped by the theoretical biology group for a visit. Rice had a major reputation in theoretical chemistry, and Kauffman very much wanted to impress him. "He came in and asked me what I was doing. So I told him. And he said, 'Why in the world are you doing *that*?' I don't know why he said that. I presume he thought there was no point to it. But I thought, 'My goodness, Stuart certainly knows what he's talking about.

I shouldn't be doing this.' So I wrote it all up and published in the *Cybernetics Society Journal* in 1971, and then put it away. And I forgot all about it."

Kauffman's reaction wasn't entirely a matter of insecurity. The fact was that his autocatalysis models were at an impasse. No matter how many calculations and computer simulations he carried out on the origin of life, they were still just calculations and computer simulations. To make a really compelling case he would have had to take the experiments of Miller and Urey one step farther, by demonstrating that their primordial soup could actually give rise to an autocatalytic set in the laboratory. But Kauffman had no idea how to do that. Even if he had had the patience and know-how to do laboratory chemistry, he would have had to look at millions of possible compounds in all conceivable combinations under a wide range of temperatures and pressures. He could have spent a lifetime on the problem and gotten nowhere.

No one else seemed to have any good ideas either. Kauffman wasn't alone in thinking about autocatalytic models. Several years earlier, the Berkeley Nobelist Melvin Calvin had explored several different autocatalytic scenarios for the origin of life in his 1969 book, *Chemical Evolution*. In Germany, meanwhile, Otto Roessler was pursuing autocatalytic ideas quite independently, as was Manfred Eigen in Göttingen. Eigen was even able to demonstrate a form of autocatalytic cycle in the laboratory using RNA molecules. But no one had yet been able to demonstrate autocatalytic sets emerging from the simple molecules of a Miller-Urey primordial soup. Pending a new idea, there seemed nowhere to go.

However, even if Kauffman's reaction to Rice's comment wasn't *all* because of insecurity, a lot of it undoubtedly was. He was feeling a strong need to prove himself in his newfound profession. Theory, he was discovering, was in very low repute among biologists.

"The people doing math in biology were the lowest of the low," he says. It was exactly the opposite of the situation in physics or economics, where the theorists are kings. Especially in molecular and developmental biology, the experimental tools were still so new, and there was such a vast amount of data to be collected about the details of living systems, that the honor and the glory went to the laboratory. "There is a remarkable certainty among molecular biologists that all the answers will be found by understanding specific molecules," says Kauffman. "There is a great reluctance to study how a system works. For example, the concept of an attractor strikes them as gobbledygook."

The atmosphere was not quite so hostile to theory in neuroscience and evolutionary biology. But Kauffman's network ideas were considered a little weird even there. He was talking about order and the statistical behavior of large networks without being able to say anything specific about this molecule or that molecule. It was hard for most researchers to understand what he meant. "People certainly responded to the genetic network stuff early on," he says. "Waddington liked it. All sorts of folks liked it. That's why I got my first job. And I felt very pleased and proud of myself. But then it sort of quieted down, and it was a low trickle from the early 1970s. The world didn't particularly care."

Instead, Kauffman threw his considerable energies into learning how to do experimental biology. The impulse was much the same as the one that had led him from philosophy to medical school: He distrusted his own glibness and theorizing. "The emotional roots were in part that I needed to be grounded," he says. "But it's also in part that I really want to know how the world works."

In particular, Kauffman focused his attention on the tiny fruit fly, *Drosophila melanogaster*, which had been intensively studied by geneticists during the early part of the twentieth century, and which had now become a favorite experimental subject for biologists doing research into the development process. Among its many other fine qualities, *Drosophila* had the ability to bring forth weird monsters known as homeotic mutants, in which a newly hatched fly turns out to have legs where its antennae are supposed to be, or genitalia where its head is supposed to be, and so on. These mutations offered Kauffman ample play for model-making and for thinking about how the developing embryo organizes itself.

In 1973, his work on *Drosophila* carried him to the National Institutes of Health, just outside of Washington, D.C., where he had managed to swing a two-year appointment that allowed him to fulfill his military requirement without going off to Vietnam. (While he was at Chicago he had arranged a four-year draft deferment under what was known as the "Berry Plan," which allowed physicians to postpone active service while doing medical research.) And in 1975, his work on the fruit fly carried him onward to a tenured appointment at the University of Pennsylvania. "I chose Penn," he jokes, "because they have great Indian restaurants near there." More seriously, he felt that he couldn't bring his family back to live in the Hyde Park neighborhood around the University of Chicago, even though he had been offered tenure there. "The crime rate and racial tension there were so ugly," he says, "and you felt so helpless to do anything good about it."

Kauffman certainly doesn't regret the time he spent on *Drosophila*. His discourses on fruit fly development are as passionate as anything he has to say about network models. And yet, he also remembers a vivid moment in 1982. "I was up in the mountains of the Sierra, and I realized I had not thought a new thought about *Drosophila* for a couple of years. I'd been pushing to do the experiments on nuclear transplantation, and cloning, and other stuff. But I hadn't actually had a new thought. And I felt utterly stuck."

Somehow, he knew at that moment that it was time to return to his original ideas about networks and autocatalysis. If nothing else, damn it, he felt that he had paid his dues: "I've earned my own right in my own guts to think what I want to think. Having gone through medical school, having become a doctor, having delivered sixty babies, and doing spinal taps on infants, and taking care of cardiac arrests, and all the things you do as a young doctor, having run a lab and learned how to do 2-D gels and 1-D gels, and run scintillation counters, having learned how to do *Drosophila* genetics, this, that, and the other thing—even if the biological community still looks askance at theory, I've earned my own right to do what the hell I want. I've answered the Oxford craving—to not be afraid to think glibly. I now just trust myself as a theoretician, profoundly. It doesn't mean that I'm right. But I trust myself."

In particular, he decided that it was time to go back to the autocatalytic set idea and do it right. Back in 1971, he says, all he'd really had was a very simple computer simulation. "I clearly understood that as the number of proteins in the solution goes up, the number of reactions goes up even faster. So if you get a complicated enough system, you get autocatalysis. But I didn't have much in the way of analytic work."

So he plunged back into his calculations—and, as usual, ended up inventing the mathematics that he needed as he went along. "I spent the entire fall of 1983, from October until just after Christmas, proving all sorts of theorems," he says. Numbers of polymers, numbers of reactions, probabilities of polymers catalyzing reactions, phase transitions in this giant graph of reactions—under precisely what conditions would autocatalysis happen. And how could he *prove* that it would happen? He remembers deriving a whole flurry of results in November, during a twenty-four-hour flight home from a meeting in India. "I got back to Philadelphia and collapsed," he says. He scribbled theorems on Christmas Day. And by New Year's 1984 he had it: a proof of this funny phase transition that he could only conjecture back in 1971. If the chemistry was too simple and the complexity of the interactions was too low, then nothing would happen; the system would be "subcritical." But if the complexity of the interactions was rich enough—and Kauffman's mathematics now allowed him to define precisely what that meant—then the system would be "supercritical." Autocatalysis would be inevitable. And the order really would be for free.

Wonderful. The next step, obviously, was to test out these theoretical ideas with a much more sophisticated computer simulation. "I had the idea of subcritical and supercritical systems," he says, "and I was anxious to see if the simulation would behave that way." But it was also important to incorporate something resembling real chemistry and real thermodynamics into the model; if nothing else, a more realistic model might give experimenters some guidance about how to create an autocatalytic set in the laboratory.

Kauffman knew just the two people to help him. He had met one of them, Los Alamos physicist Doyne Farmer, during a conference in Bavaria in 1982. Then only twenty-nine, Farmer had proved to have an imagination just as fertile and as energetic as Kauffman's; he was also just as fascinated with the concept of self-organization. They had spent one marvelous day hiking through the Alps, talking about networks and self-organization all the while. They hit it off so well, in fact, that Farmer arranged for Kauffman to start making periodic visits to Los Alamos as a consultant and lecturer. Shortly thereafter, Farmer had introduced Kauffman to Norman Packard, a young computer scientist at the University of Illinois.

Farmer and Packard had been collaborators since their graduate student days in the physics department at the University of California, Santa Cruz, in the late 1970s. While they were there they had both been members of the self-styled "Dynamical Systems Collective," a small crew of graduate students who devoted themselves to what was then the avant-garde subject of nonlinear dynamics and chaos theory. They had made a number of seminal contributions to the field—a fact that would earn the Dynamical Systems Collective its own chapter in James Gleick's book *Chaos*, which appeared in the fall of 1987 at about the same time that Arthur, Kauffman, and the others were converging on Santa Fe for the economics meeting.

And by the time Kauffman first met them in the early 1980s, Farmer and Packard were both getting downright bored with chaos theory.

As Farmer says, "So what? The basic theory of chaos was already fleshed out." He wanted the excitement of being on the frontier, where things were not well understood. Packard, for his part, wanted to get his hands dirty with some *real* complexity. Chaotic dynamics were complex, all right: think of the way a leaf seems to flutter at random in a steady breeze. But the complexity was pretty simpleminded. There is one set of forces—from the wind, in the case of the leaf. Those forces can be described by one set of mathematical equations. And the system just blindly follows those equations forever. Nothing changes, and nothing adapts. "I wanted to go beyond that, to richer forms of complexity like biology and the mind," says Packard. He and Farmer had been casting around for the right problem to work on. So when Kauffman suggested that they all collaborate on simulating an autocatalytic system, they decided to give it a try.

They got down to work in 1985, after Kauffman had returned from a sabbatical that took him to Paris and Jerusalem. "It became an intense collaboration," says Kauffman. It was one thing to talk about a random network of reactions; such a network can be described in clean mathematical terms. It was quite something else to model those reactions with reasonably realistic chemistry; things got complicated fast.

What Kauffman, Farmer, and Packard finally came up with was a simplified version of polymer chemistry. The basic chemical building blocks—analogous to the amino acids and other simple compounds that you might expect to form in the primordial soup by the Miller-Urey process—were represented in the model by symbols such as a, b, and c. These building blocks could then link up into chains to form larger molecules, such as *accddbacd*. These larger molecules, in turn, could undergo two kinds of chemical reaction. They could split apart:

accddbacd → *accd* + *dbacd*

Or they could do the reverse and join at the ends:

bbcad + *cccba* → *bbcadcccba*

Each of the reactions would have a number associated with it—what a chemist would call a rate constant—and that number would determine how fast the reaction would occur if there were no catalysts around.

But, of course, the whole point of the exercise was to watch what happened when catalysts *did* appear. So Kauffman, Farmer, and Packard had to find a way to specify which molecule could catalyze which reaction. They tried several ways of doing that. One way, which was suggested by Kauffman and which worked as well as any other, was simply to pick a series of molecules, such as *abccd*, and arbitrarily assign each one to a reaction, such as *baba* + *ccda* → *babaccda*.

To run the model, once all the reaction rates and catalytic strengths were specified, Kauffman, Farmer, and Packard would simply tell the computer to start enriching their simulated pond with a steady stream of "food" molecules such as *a*, *b*, and *aa*. And then they would sit back to see what their simulated chemistry would produce.

For quite a while, it didn't produce much of anything. This was frustrating, but not surprising. Reaction rates, catalytic strengths, rate of food supply—any number of parameters could be off. The thing to do was to vary the parameters and see what worked and what didn't. And as they did that, they began to find that occasionally, as they moved the parameters into certain favorable ranges, the simulated autocatalytic sets *did* form. Moreover, they seemed to form under just about the same conditions that Kauffman had predicted from his theorems about abstract networks.

Kauffman and his collaborators published their results in 1986. Farmer and Packard had already moved on to other interests by that point—although Farmer had taken on a graduate student, Richard Bagley, to amplify the model and speed it up substantially. And Kauffman himself had gone on to think about other ways in which self-organization could have occurred in evolution. But after that computer model, he felt more deeply than ever that he had truly come face to face with a secret of the Old One.

He remembers taking a solo hike back up into the Sierras near Lake Tahoe to one of his favorite spots, Horsetail Falls. It was a lovely summer day, he recalls. He sat on a rock by the falls, thinking about his autocatalysis work and what it meant. "And suddenly," he says, "I knew that God had revealed to me a part of how his universe works." Not a personal God, certainly; Kauffman had never been able to believe in such a being. "But I had a holy sense of a knowing universe, a universe unfolding, a universe of which we are privileged to be a part. In fact, it was quite the opposite of a vainglorious feeling. I felt that God would reveal how the world works to anyone who cared to listen.

"It was a lovely moment," he says, "the closest I've ever come to a religious experience."

Santa Fe

As the opening day of the economics workshop drew nearer, and as Arthur began to spend more and more time polishing his talk—he was scheduled to be the first speaker on the first day—Kauffman went out by himself for long walks along the dirt roads near his house. "I remember pacing back and forth, trying to put together the central conceptual structure of my own talk," he says. By agreement, Arthur would focus his presentation on increasing returns, while Kauffman, who had already outlined what he wanted to say about his various network models, would add a discussion of their ideas on technology and autocatalysis. The notion of viewing the economy as an autocatalytic set was just too beautiful to pass up, and Kauffman couldn't wait to share it.

Kauffman's Santa Fe home was a good spot for this kind of thinking and contemplation—in its own way, almost as good as Horsetail Falls. A big rambling structure with floor-to-ceiling picture windows, it was located on a long dirt road out in the desert northwest of town, where it had a spectacular view of the Jemez Mountains across the Rio Grande Valley. The setting was somehow timeless, almost spiritual. He had bought the property less than a year before, for the express purpose of spending more time at the institute.

The Santa Fe Institute, of course, was the third of his great intellectual environments. "As exciting as Oxford and Chicago were," he says, "they were small potatoes compared with what the institute is like. It's just an amazing place." He had heard about it from Doyne Farmer in 1985, while they were both working on the computer model of autocatalysis. But he had only really experienced it in August of 1986, when he attended a Santa Fe Institute workshop, "Complex Adaptive Systems," organized by Jack Cowan and Stanford evolutionary biologist Marc Feldman. Like Arthur, he had just fallen in love with the place, and it took him no time at all to decide that this is where he wanted to be. "The sense of ferment and intellectual excitement and chaos and seriousness and joy is here all the time—along with a sense of 'Thank God, I'm not alone.' "

Liz Kauffman and their two children, Ethan and Merit, had been more than happy with the idea of spending time in Santa Fe. When Kauffman brought them out to see the area, they had immediately fallen in love with it, too. He still remembers the day they all went out hunting for mushrooms in the Sangre de Cristo Mountains. Moreover, Liz was a painter, and there was no place in the world that had light like New Mexico. So the Kauffmans closed on their Santa Fe house October 12, 1986, thinking that they might be spending a month or so in New Mexico every year.

It was less than two weeks later, on October 25, 1986, that Merit Kauffman was killed. And in the aftermath of her death, the Santa Fe house suddenly became a great deal more for the Kauffmans than a vacation home. From then on it was a refuge. Liz and Ethan moved there essentially full time, while Stuart Kauffman himself became a kind of exile moving in between, tethered to Penn only by his students, a salary, and a tenure slot. His department chairman, recognizing that it was a matter of emotional salvation, arranged for Kauffman to spend up to half his time in Santa Fe. "It was extraordinarily kind," he says. "Not many places would allow you to do that."

Kauffman says he remembers very little about the following year. In May 1987 he learned that he had been awarded one of the MacArthur Foundation's no-strings "genius" grants. He was exhilarated—and yet he hardly felt it. "Talk about one of the worst things that can happen and one of the best things that can happen." He basically retreated into his work. "Being a scientist," he says, "is the place I can go to that feels normal." And often he would walk along these dirt roads in the desert, gazing at the mountains and looking for the secret.

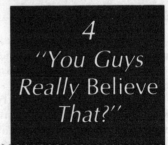

Under ordinary circumstances, Brian Arthur didn't get nervous when he gave a talk. But, then, the Santa Fe Institute's economics conference was no ordinary circumstance.

He'd had a sense that something big was afoot before he even got there. "When Ken stopped me, when I started to hear that names like John Reed and Phil Anderson and Murray Gell-Mann were behind it, when the president of the institute was on the phone—it was clear that this meeting was being regarded by people at Santa Fe as quite a landmark event." As organizers, Arrow and Anderson had scheduled the meeting for ten full days, which was very long by academic standards. And George Cowan had called a press conference for the final day—when John Reed himself was supposed to be there. (Indeed, it was a testament that Anderson was planning to come at all. Seven months earlier, in February 1987, every condensed-matter physicist in the world had been galvanized by the announcement of a new class of grungy-looking ceramic materials that could conduct electricity without resistance at the relatively balmy boiling point of liquid nitrogen, − 321°F. Anderson, like a great many other theorists, was working overtime to figure out how these "high-temperature" superconductors accomplished that feat.)

But for Arthur, the real moment of revelation came when he arrived at the institute in late August, two weeks before the meeting, and finally saw the roster. He already knew about Arrow and Anderson, of course. And he'd known about his Stanford colleague Tom Sargent, who was often mentioned as a Nobel Prize contender because of his analysis of how

"rational" economic decisions in the private sector are intimately intertwined with the economic environment created by the government. But
here were names like Harvard emeritus professor Hollis Chenery, a former
director of research for the World Bank (and owner of the racehorse Sec-
retariat); Harvard whiz-kid Larry Summers, who was currently serving as
economics adviser to the Dukakis campaign; the University of Chicago's
José Scheinkman, a pioneer in applying chaos theory to economics; the
Belgian physicist David Ruelle of the Institut des Hautes Études Scienti-
fiques outside Paris, one of the founders of chaos theory. Arthur was deeply
impressed. There were nearly two dozen names on this list, and they were
all of this caliber.

He could feel his adrenaline level beginning to climb. "I realized that
it might well be a crucial moment for me, a chance to present my ideas
on increasing returns to a group of people I very much wanted to convince.
I instinctively felt that the physicists would be very much at home with my
ideas. But I didn't really know what they would have to say. Or Arrow, for
that matter. And although the economists were high-level, they were mainly
known for conventional theory. So I had no idea how my ideas would be
received. There was no context at all. I didn't know what the tone would
be, whether it would be the kind of bitter attack you run into sometimes,
or whether it would be a very friendly format."

So, as the day approached for the meeting's opening—Tuesday, September 8—Arthur spent less and less time walking and talking with Stuart
Kauffman, and more and more time polishing his presentation. He also
remembers doing lots of tai chi. "Tai chi teaches you to absorb attacks and
immediately come back with a counter hit," he says. "I thought I might
need that. For keeping yourself grounded under fire, there's nothing better
than practicing slow-motion martial arts. Because every time you punch
you can imagine delivering something to an audience."

The meeting convened at 9 A.M. in the convent's chapel-turned-conference room. The participants were seated around a long double row of
collapsible tables. And as always, the light from the western sky came
streaming in through the stained-glass windows.

After a short introductory talk by Anderson, laying out some of the major
themes that he was hoping to see addressed in the workshop, Arthur stood
up to begin the first formal presentation: "Self-Reinforcing Mechanisms in
Economics." And as he began it Arthur somehow had the impression that
Arrow was keyed up, too, as if he was worried that this guy Arthur might

give a very peculiar picture of economics to the physicists. Arrow himself doesn't remember feeling that way. "I know Brian's communicating ability is very high," he says. And besides, so far as he was concerned the meeting was only an experiment. "There was a great deal on the line from the institute's point of view," he says, "but there was nothing on the line intellectually. There was no position to erode. If the experiment failed, it failed."

But whether he was consciously concerned or not, Arrow was certainly being his rigorously analytical self that morning. Only more so. Arthur started out by addressing himself to the physicists. When he used the words "self-reinforcing mechanisms," he explained, he was basically talking about nonlinearity in economics . . .

"Stop!" said Arrow. "In precisely what sense do you mean nonlinear? Aren't all economic phenomena nonlinear?"

Well, umm, yes. To be mathematically precise, said Arthur, the ordinary assumption of decreasing returns corresponds to economic equations with a "second-order" nonlinearity, which drives the economy toward equilibrium and stability. What he was looking at were "third-order" nonlinearities—factors that would drive some sector of the economy *away* from equilibrium. This is what an engineer would call positive feedback.

That seemed to satisfy Arrow. And at various points around the table, Arthur could see Anderson, Pines, and the other physicists beginning to nod. Increasing returns, positive feedback, nonlinear equations—to them it was familiar stuff.

Then, about halfway through the morning, Phil Anderson stuck up his hand and asked, "Isn't the economy a lot like a spin glass?" And Arrow understandably interjected, "What's a spin glass?"

As it happens, Arthur had been reading a lot of condensed-matter physics over the past few years and knew exactly what a spin glass was. The name actually referred to a group of obscure magnetic materials whose practical utility was nil, but whose theoretical properties were fascinating. Anderson had personally been studying them since they were discovered in the 1960s, and had coauthored several of the seminal papers in the field. Just as in more familiar magnetic materials such as iron, the key components of a spin glass are metal atoms whose electrons possess a net whirling motion, or "spin." And just as in iron, these spins cause each atom to produce a tiny magnetic field, which in turn causes it to exert a magnetic force on the spins of its neighbors. Unlike in iron, however, the interatomic forces in a spin glass do not cause all the spins to fall into line with one another and produce a large-scale magnetic field, the kind we see in compass needles and refrigerator magnets.

Instead, the forces in a spin glass are completely random—a state that physicists refer to as "glassy." (The atomic bonds in a pane of window glass are random in the same way. Technically, in fact, it's a toss-up whether ordinary glass should be called a solid or an exceptionally viscous liquid.) Among other things, this atomic-scale disorder means that a spin glass is a complex mixture of positive and negative feedbacks, as each atom tries to align its spin in parallel with certain of its neighbors and opposite to all the rest. In general, there is no way to do this consistently; each atom will always have to endure a certain amount of frustration at having to align with neighbors that it doesn't want to be aligned with. But by the same token, there are a vast number of ways to arrange the spins so that the frustration is reasonably tolerable for everyone—a situation that a physicist would describe as "local equilibrium."

So yes, agreed Arthur. In this sense a spin glass was quite a good metaphor for the economy. "It naturally has a mixture of positive and negative feedbacks, which gives it an extremely high number of natural ground states, or equilibria." That's exactly the point he'd been trying to make all along with his increasing-returns economics.

Arthur could see the physicists nodding some more. Hey, this economics stuff was all right. "I really resonated with Brian," says Anderson. "We found ourselves very impressed."

And so it went, for two solid hours: Lock-in. Path dependence. QWERTY and possible inefficiencies. The origin of Silicon Valleys. "All the time in my talk, the physicists were nodding and beaming," says Arthur. "But every ten minutes or so, Arrow would say, 'Stop,' asking me to expand something, or explaining why he didn't agree. He wanted to know exactly where each step of the reasoning was coming from. And when I started to state precise theorems on the board, he, like several of the economists there, wanted to see the precise proofs. This slowed me down. But it also sewed up the arguments."

When Arthur finally sat down at the end he was drained—but he had the sense that his career was made. "My ideas got legitimized that morning," he says. "Not by me convincing Arrow and the others, but by physicists convincing the economists that what I was doing was bread and butter to them. In effect, they said, 'Oh yes, this guy knows what he's talking about— you economists don't have to worry.' "

Perhaps it was just projection. But to Arthur, Arrow seemed visibly more relaxed.

• • •

If Arthur's talk had given the physicists the impression that they were on the same wavelength as the economists, however, they were soon disabused.

For the first two or three days of the meeting, since few of the physicists had had much exposure to the subject beyond undergraduate Economics 101, Arrow and Anderson had asked several of the economists to give survey talks on the standard neoclassical theory. "We were fascinated by this structure," says Anderson, for whom economic theory has long been an intellectual hobby. "We wanted to learn about it."

And indeed, as the axioms and theorems and proofs marched across the overhead projection screen, the physicists could only be awestruck at their counterparts' mathematical prowess—awestruck and appalled. They had the same objection that Arthur and many other economists had been voicing from within the field for years. "They were almost too good," says one young physicist, who remembers shaking his head in disbelief. "It seemed as though they were dazzling themselves with fancy mathematics, until they really couldn't see the forest for the trees. So much time was being spent on trying to absorb the mathematics that I thought they often weren't looking at what the models were for, and what they did, and whether the underlying assumptions were any good. In a lot of cases, what was required was just some common sense. Maybe if they all had lower IQs, they'd have been making some better models."

The physicists had no objections to the mathematics itself, of course. Physics is far and away the most thoroughly mathematized science in existence. But what most of the economists didn't know—and were startled to find out—was that physicists are comparatively casual about their math. "They use a little rigorous thinking, a little intuition, a little back-of-the-envelope calculation—so their style is really quite different," says Arrow, who remembers being pretty surprised himself. And the reason is that physical scientists are obsessive about founding their assumptions and their theories on empirical fact. "I don't know what it's like in fields like relativity theory, where the ratio of theory to observation is much greater," says Arrow, "but the general tendency is that you make a calculation, and then find some experimental data to test it. So the lack of rigor isn't so serious. The errors will be detected anyway. Well, we don't *have* data of that quality in economics. We can't generate data the way the physicists can. We have to go pretty far on a small basis. So we have to make sure every step of the way is correct."

Fair enough. But the physicists were nonetheless disconcerted at how seldom the economists seemed to pay attention to the empirical data that *did* exist. Again and again, for example, someone would ask a question like "What about noneconomic influences such as political motives in OPEC oil pricing, and mass psychology in the stock market? Have you consulted sociologists, or psychologists, or anthropologists, or social scientists in general?" And the economists—when they weren't curling their lips at the thought of these lesser social sciences, which they considered horribly mushy—would come back with answers like "Such noneconomic forces really aren't important"; "They *are* important, but they are too hard to treat"; "They aren't always too hard to treat, and in fact, we're doing so in specific cases"; and "We don't need to treat them because they're automatically satisfied through economic effects."

And then there was this business of "rational expectations." Arthur remembers someone asking him during his talk that first day, "Isn't economics a good deal simpler than physics?"

"Well," Arthur replied, "in one sense it is. We call our particles 'agents'—banks, firms, consumers, governments. And those agents react to other agents, just as particles react to other particles. Only we don't usually consider the spatial dimension in economics much, so that makes economics a lot simpler."

However, he added, there is one big difference: "Our particles in economics are smart, whereas yours in physics are dumb." In physics, an elementary particle has no past, no experience, no goals, no hopes or fears about the future. It just *is*. That's why physicists can talk so freely about "universal laws": their particles respond to forces blindly, with absolute obedience. But in economics, said Arthur, "Our particles have to think ahead, and try to figure out how other particles might react if they were to undertake certain actions. Our particles have to act on the basis of expectations and strategies. And regardless of how you model that, that's what makes economics truly difficult."

Immediately, he says, he could see all the physicists in the room sitting up: "Here was a subject that wasn't trivial. It was like their subject, but it has these two interesting quirks—strategy and expectations."

Unfortunately, the economists' standard solution to the problem of expectations—perfect rationality—drove the physicists nuts. Perfectly rational agents do have the virtue of being perfectly predictable. That is, they know everything that can be known about the choices they will face infinitely far into the future, and they use flawless reasoning to foresee all the possible implications of their actions. So you can safely say that they will always take *the* most advantageous action in any given situation, based on the available information. Of course, they may sometimes be caught short by oil shocks, technological revolutions, political decisions about interest rates, and other noneconomic surprises. But they are so smart and so fast in their adjustments that they will always keep the economy in a kind of rolling equilibrium, with supply precisely equal to demand.

The only problem, of course, is that real human beings are neither perfectly rational nor perfectly predictable—as the physicists pointed out at great length. Furthermore, as several of them also pointed out, there are real theoretical pitfalls in assuming perfect predictions, even if you do assume that people are perfectly rational. In nonlinear systems—and the economy is most certainly nonlinear—chaos theory tells you that the slightest uncertainty in your knowledge of the initial conditions will often grow inexorably. After a while, your predictions are nonsense.

"They kept pushing us and pushing us," says Arthur. "The physicists were shocked at the assumptions the economists were making—that the test was not a match against reality, but whether the assumptions were the common currency of the field. I can just see Phil Anderson, laid back with a smile on his face, saying, 'You guys really *believe* that?' "

The economists, backed into a corner, would reply, "Yeah, but this allows us to solve these problems. If you don't make these assumptions, then you can't do *anything*."

And the physicists would come right back, "Yeah, but where does that get you—you're solving the wrong problem if that's not reality."

Economists as a group are not exactly noted for their intellectual modesty, and the economists at Santa Fe would have been less than human if they hadn't felt a touch of resentment about all this. They were perfectly willing to complain about the failings of their field among themselves; Arrow, after all, had deliberately been looking for well-informed skeptics. But who wants to hear the same thing from a bunch of outsiders? Everyone was doing his damnedest to listen and to be polite and to make the meeting work. And yet there was a distinct undercurrent of "What has physics got to offer us? What makes you guys so damn smart?"

Of course, physicists aren't exactly noted for their intellectual modesty, either. For many nonphysicists, in fact, the phrase that comes to mind is "insufferable arrogance." It isn't a deliberate attitude or even personal. It's more like the unconscious superiority of the British aristocracy. Indeed, in their own minds, physicists *are* the aristocracy of science. From the day they sign up for Physics 101, they absorb the culture in a thousand subtle and not-so-subtle ways: they are the heirs of Newton, Maxwell, Einstein, and Bohr. Physics is the hardest, purest, toughest science there is. And physicists have the hardest, purest, toughest minds around. So if the economists started off the Santa Fe conference with a certain touchiness, their counterparts matched it with what Harvard economist Larry Summers dubbed the "Me Tarzan, You Jane" attitude: "Give us three weeks to master this subject and we'll show you how to do it right."

The potential clash of egos was a constant concern of Citicorp's Eugenia Singer, who was sitting in as John Reed's representative. "I was afraid that if the 'Tarzan' effect got started, then this whole project would be down the tubes before it ever got off the ground," she recalls. And in the beginning it looked as though it might. "Most of the economists sat on one side of the table, and most of the physical scientists sat on the other side," she says. "I was horrified by it." She would periodically pull Pines or Cowan aside and say, "Couldn't we get them to sit a little closer to each other?" But it never changed.

The potential for total miscommunication was likewise a nightmare for George Cowan—and not just because the institute might lose any hope of Citicorp funding if the conference failed to come off. The fact was that this meeting was the most stringent proof of concept yet for the Santa Fe Institute. In the founding workshops two years earlier they had brought people together for a weekend of talk; but now they were asking two very different and very proud groups to sit down with each other for ten days and do something substantive. "We were trying to create a community that didn't exist before," says Cowan. "The odds were that it wouldn't be successful, that they'd find nothing to talk about, that there would simply be polemics."

This was not an idle fear; later Santa Fe workshops have occasionally degenerated into shouting matches and sulking. But in September 1987, the gods of interdisciplinary research decided to smile once again. Anderson and Arrow had tried hard to recruit people who could listen as well as talk. And for all the initial bristling, the participants eventually began to discover that they had plenty to talk about. Looking back on it, in fact, the two sides began to find common ground in a remarkably short time.

Certainly that was true for Arthur. In his case, the discovery took about half a day.

According to the agenda, the second presentation of the economics workshop would begin after lunch on that first day and run the rest of the afternoon. "The Global Economy as an Adaptive Process," it was called, by John H. Holland of the University of Michigan.

Now that Brian Arthur had finished his own talk, he had the energy to feel curious about this one—and not just because the title sounded interesting. John Holland was another of the institute's visiting fellows that fall, and the two of them were supposed to be sharing a house. But Holland hadn't arrived in Santa Fe until late the night before, while Arthur was down at the convent going over and over his talk just one last time. Arthur had never laid eyes on the man. All he knew was that Holland was a computer scientist and, according to the institute, "a very nice guy."

Well, the institute appeared to be right about that. As people were filing back into the chapel and settling into their chairs around the long row of folding tables, Holland was up front getting ready to start. He proved to be a compact, sixtyish midwesterner with a broad, ruddy face that seemed fixed in a perpetual grin, and a high-pitched voice that made him sound like an enthusiastic graduate student. Arthur liked him instantly.

But then Holland began his talk. And within minutes, Arthur wasn't just sleepily following along. He was wide awake and listening hard.

Perpetual Novelty

Holland started by pointing out that the economy is an example par excellence of what the Santa Fe Institute had come to call "complex adaptive systems." In the natural world such systems included brains, immune systems, ecologies, cells, developing embryos, and ant colonies. In the human world they included cultural and social systems such as political parties or scientific communities. Once you learned how to recognize them, in fact, these systems were everywhere. But wherever you found them, said Holland, they all seemed to share certain crucial properties.

First, he said, each of these systems is a network of many "agents" acting in parallel. In a brain the agents are nerve cells, in an ecology the agents are species, in a cell the agents are organelles such as the nucleus and the mitochondria, in an embryo the agents are cells, and so on. In an economy, the agents might be individuals or households. Or if you were looking at business cycles, the agents might be firms. And if you were looking at international trade, the agents might even be whole nations. But regardless of how you define them, each agent finds itself in an environment produced by its interactions with the other agents in the system. It is constantly acting and reacting to what the other agents are doing. And because of that, essentially nothing in its environment is fixed.

Furthermore, said Holland, the control of a complex adaptive system tends to be highly dispersed. There is no master neuron in the brain, for example, nor is there any master cell within a developing embryo. If there is to be any coherent behavior in the system, it has to arise from competition and cooperation among the agents themselves. This is true even in an economy. Ask any president trying to cope with a stubborn recession: no matter what Washington does to fiddle with interest rates and tax policy and the money supply, the overall behavior of the economy is still the result of myriad economic decisions made every day by millions of individual people.

Second, said Holland, a complex adaptive system has many levels of organization, with agents at any one level serving as the building blocks for agents at a higher level. A group of proteins, lipids, and nucleic acids will form a cell, a group of cells will form a tissue, a collection of tissues will form an organ, an association of organs will form a whole organism, and a group of organisms will form an ecosystem. In the brain, one group of neurons will form the speech centers, another the motor cortex, and still another the visual cortex. And in precisely the same way, a group of individual workers will compose a department, a group of departments will compose a division, and so on through companies, economic sectors, national economies, and finally the world economy.

Furthermore, said Holland—and this was something he considered very important—complex adaptive systems are constantly revising and rearranging their building blocks as they gain experience. Succeeding generations of organisms will modify and rearrange their tissues through the process of evolution. The brain will continually strengthen or weaken myriad connections between its neurons as an individual learns from his or her encounters with the world. A firm will promote individuals who do well and (more rarely) will reshuffle its organizational chart for greater efficiency. Countries will make new trading agreements or realign themselves into whole new alliances.

At some deep, fundamental level, said Holland, all these processes of learning, evolution, and adaptation are the same. And one of the fundamental mechanisms of adaptation in any given system is this revision and recombination of the building blocks.

Third, he said, all complex adaptive systems anticipate the future. Obviously, this is no surprise to the economists. The anticipation of an extended recession, for example, may lead individuals to defer buying a new car or taking an expensive vacation—thereby helping guarantee that the recession will be extended. The anticipation of an oil shortage can likewise send shock waves of buying and selling through the oil markets—whether or not the shortage ever comes to pass.

But in fact, said Holland, this business of anticipation and prediction goes far beyond issues of human foresight, or even consciousness. From bacteria on up, every living creature has an implicit prediction encoded in its genes: "In such and such an environment, the organism specified by this genetic blueprint is likely to do well." Likewise, every creature with a brain has myriad implicit predictions encoded in what it has learned: "In situation ABC, action XYZ is likely to pay off."

More generally, said Holland, every complex adaptive system is constantly making predictions based on its various internal models of the world—its implicit or explicit assumptions about the way things are out there. Furthermore, these models are much more than passive blueprints. They are active. Like subroutines in a computer program, they can come to life in a given situation and "execute," producing behavior in the system. In fact, you can think of internal models as the building blocks of behavior. And like any other building blocks, they can be tested, refined, and rearranged as the system gains experience.

Finally, said Holland, complex adaptive systems typically have many *niches*, each one of which can be exploited by an agent adapted to fill that niche. Thus, the economic world has a place for computer programmers, plumbers, steel mills, and pet stores, just as the rain forest has a place for tree sloths and butterflies. Moreover, the very act of filling one niche opens up more niches—for new parasites, for new predators and prey, for new symbiotic partners. So new opportunities are always being created by the system. And that, in turn, means that it's essentially meaningless to talk about a complex adaptive system being in equilibrium: the system can never get there. It is always unfolding, always in transition. In fact, if the system ever does reach equilibrium, it isn't just stable. It's dead. And by the same token, said Holland, there's no point in imagining that the agents in the system can ever "optimi﹏﹏" their fitness, or their utility, or whatever. The space of possibilities is too vast; they have no practical way of finding the optimum. The most they can ever do is to change and improve themselves relative to what the other agents are doing. In short, complex adaptive systems are characterized by perpetual novelty.

Multiple agents, building blocks, internal models, perpetual novelty—taking all this together, said Holland, it's no wonder that complex adaptive systems were so hard to analyze with standard mathematics. Most of the conventional techniques like calculus or linear analysis are very well suited to describe unchanging particles moving in a fixed environment. But to really get a deep understanding of the economy, or complex adaptive systems in general, what you need are mathematics and computer simulation techniques that emphasize internal models, the emergence of new building blocks, and the rich web of interactions between multiple agents.

By this point in Holland's talk, Arthur was scribbling notes furiously. And as Holland went on to describe the various computer techniques he had developed over the past thirty years to make these ideas more precise and more useful, Arthur scribbled even faster. "It was incredible," he says. "I just sat there all afternoon with my mouth open." It wasn't simply that Holland's point about perpetual novelty was exactly what he'd been trying to say for the past eight years with his increasing-returns economics. Nor was it that Holland's point about niches was exactly what he and Stuart Kauffman had been thrashing out for the past two weeks in the context of autocatalytic sets. It was that Holland's whole way of looking at things had a unity, a clarity, a *rightness* that made you slap your forehead and say, "Of course! Why didn't I think of that?" Holland's ideas produced a shock of recognition, the kind that made more ideas start exploding in your own brain.

"Sentence by sentence," says Arthur, "Holland was answering all kinds of questions I'd been asking myself for years: What is adaptation? What is emergence? And many more questions that I never realized I'd been asking." Arthur had no idea yet how it would all fit into economics. In fact, as he looked around the room he could see that a number of his fellow economists were looking puzzled, too, if not downright skeptical. (At least one was taking a little midafternoon siesta.) "But I was convinced that Holland was onto something much, much more sophisticated than what we did." Somehow, he felt, Holland's ideas had to be incredibly important.

Certainly the Santa Fe Institute thought so. However new Holland's thinking might have seemed to Arthur and the other visitors at the economics meeting, he had long since become a familiar and profoundly influential figure among the Santa Fe regulars.

His first contact with the institute had come in 1985 during a conference entitled "Evolution, Games, and Learning," which had been organized at Los Alamos by Doyne Farmer and Norman Packard. (As it happens, this was the same meeting in which Farmer, Packard, and Kauffman first reported the results of their autocatalytic set simulation.) Holland's talk there was on the subject of emergence, and it seemed to go quite well. But he remembers being peppered with sharp-edged questions from this person out in the audience—a white-haired guy with an intent, slightly cynical face peering out from behind dark-rimmed glasses. "I was fairly flip in my answers," says Holland. "I didn't know him—and I'd probably have been scared to death if I had!"

Flip answers or not, however, Murray Gell-Mann clearly liked what Holland had to say. Shortly thereafter, Gell-Mann called him up and asked him to serve on what was then called the Santa Fe Institute's advisory board, which was just being formed.

Holland agreed. "And as soon as I saw the place, I *really* liked it," he says. "What they were talking about, the way they went at things—my immediate response was, 'I sure hope these guys like me, because this is for me!' "

The feeling was mutual. When Gell-Mann speaks of Holland he uses words like "brilliant"—not a term he throws around casually. But, then, it's not often that Gell-Mann has had his eyes opened quite so abruptly.

In the early days, Gell-Mann, Cowan, and most of the other founders of the institute had thought about their new science of complexity almost entirely in terms of the physical concepts they were already familiar with, such as emergence, collective behavior, and spontaneous organization. Moreover, these concepts already seemed to promise an exceptionally rich research agenda, if only as metaphors for studying the same ideas—emergence, collective behavior, and spontaneous organization—in realms such as economics and biology. But then Holland came along with his analysis of adaptation—not to mention his working computer models. And suddenly Gell-Mann and the others realized that they'd left a gaping hole in their agenda: What do these emergent structures actually *do*? How do they respond and adapt to their environment?

Within months they were talking about the institute's program being not just complex systems, but complex *adaptive* systems. And Holland's personal intellectual agenda—to understand the intertwining processes of emergence and adaptation—essentially became the agenda of the institute as a whole. He was accordingly given star billing at one of the institute's first attempts at a large-scale meeting, the Complex Adaptive Systems workshop organized in August 1986 by Jack Cowan and Stanford biologist Marc Feldman. (This was the same meeting that introduced Stuart Kauffman to Santa Fe.) David Pines likewise made a point of bringing Holland out to talk to John Reed and company during the encounter at Rancho Encantado, which overlapped the Complex Adaptive Systems workshop by a day. And Anderson saw to it that Holland was in attendance at the big economics meeting in September 1987.

Holland participated in all this cheerfully—as well he might. He had labored on his ideas about adaptation in relative obscurity for a quarter of a century. And only now, at age fifty-seven, was he being discovered. "The ability to talk one-on-one and be treated as an equal with the likes of Murray Gell-Mann and Phil Anderson—that's great! Incredible!" If there had been any way for his wife to leave Ann Arbor—Maurita Holland was head of the university's system of nine science libraries—he would have been spending a lot more time in New Mexico than he already was.

But, then, Holland was almost always cheerful. He possessed the guileless good humor of a genuinely happy man who was doing what he genuinely wanted with his life—and who still seemed amazed at his good fortune. It was essentially impossible not to like John Holland.

Arthur, for one, didn't even try to resist. After Holland's session was over that first afternoon, he eagerly introduced himself. And over the ensuing days, the two men quickly became fast friends. Holland found

Arthur a delight. "Few people could have picked up on these ideas about adaptation and then integrated them so quickly and so thoroughly into their own outlook," he says. "Brian was very interested in all parts of it, very quick to explore."

Arthur, meanwhile, found that Holland was easily the most complex and fascinating intellect that he had come across in Santa Fe. Indeed, Holland was one of the main reasons that he spent the remaining days of the economics conference in a state of chronic sleep deprivation. He and Holland spent many a late night sitting around the kitchen table at their shared house, drinking beer, and discussing the whichness of what.

He remembers one such conversation in particular. Holland had come to the conference eager to learn what the crucial issues were in economics. ("If you're going to do interdisciplinary studies and enter someone else's domain," Holland says, "the least you should do is take their questions very seriously. They've spent a long time formulating them.") And that night, as the two of them were sitting around the kitchen table, Holland put the question to him straight: "Brian, what is the real problem with economics?"

"Chess!" replied Arthur, without thinking.

Chess? Holland didn't understand.

Well, said Arthur, taking a sip of beer and stumbling for words. He didn't quite know himself what he meant. Economists were always talking about systems that were simple and closed, in that they would quickly settle down into one or two or three sets of behavior, and after that, nothing much happens. They were also tacitly assuming that economic agents are infinitely smart and can instantly perceive the best thing to do in any given situation. But think about what that means in terms of chess. In the mathematical theory of games there is a theorem telling you that any finite, two-person, zero-sum game—such as chess—has an optimal solution. That is, there is a way of choosing moves that will allow each player, black and white, to do better than he would with any other choice of moves.

In reality, of course, no one has the slightest idea what that solution is or how to find it. But one of these ideal economic agents that the economists talk about could find it instantaneously. Confronted with the chessboard at the start of the game, two such agents would just formulate all the possibilities in their mind, and work backward from all the possible ways you could force a checkmate. Then they would work backward again and again, until they had considered all possible moves and found the optimal move to make at the opening. Say, pawn to king 4. And at that point, there would be no need to actually play the game. Whichever player held the theoretical advantage—say, white—would immediately claim victory, knowing that he would always win. And the other player would immediately concede defeat, knowing that he would always lose.

"Now, John," said Arthur, "does anybody play chess like that?"

Holland just laughed. As it happens he knew precisely how absurd this was. Back in the 1940s, when computers were new and researchers were making their first attempts to design an "intelligent" program that could play chess, the father of modern information theory, Claude Shannon of Bell Labs, had estimated the total number of possible moves in chess. His answer, 10^{120}, was a number so vast as to defy all metaphor. There haven't been that many microseconds since the Big Bang. There aren't that many elementary particles in the observable universe. There is no conceivable computer that could examine all of those moves. And there is certainly no human being who could. We human players have to make do with rules of thumb—hard-learned heuristic guides that tell us what kind of strategies will work best in a given situation. Even the greatest chess masters are always exploring their way in chess, as if they were descending into a deep, deep set of caves with a tiny lantern. Of course, they do make progress. As a chess player himself, Holland knew that a grand master from the 1920s wouldn't stand a chance against a contemporary grand master such as Gary Kasparov. But even so, it's as if they had only gotten a few yards down into this immense unknown. That's why Holland would call chess a fundamentally "open" system: it is effectively infinite.

Right, said Arthur. "The kinds of patterns that people can actually perceive to work on are very limited compared with what's 'optimal.' You have to assume your agents are a lot smarter than the average economist." And yet, he said, "that's the way we carry out economic problems. Trade with Japan is at least as complicated as chess. But economists will start out by saying, 'Assume rational play.' "

So there you have the economic problem in a nutshell, he told Holland. How do we make a science out of imperfectly smart agents exploring their way into an essentially infinite space of possibilities?

"A *ha*!" said Holland, the way he always does when he finally sees the light. Chess! Now this was a metaphor he could understand.

The Immense Space of Possibilities

John Holland loved games. All kinds of games. He had been a regular at one monthly poker game in Ann Arbor for nearly thirty years. One of his earliest memories was of watching the grown-ups play pinochle at his grandparents' house and wishing he was big enough to sit at the table. He learned to play checkers in first grade from his mother, who was also an expert bridge player. Everyone in the family was a passionate sailor, and both Holland and his mother frequently competed in regattas. His father was a first-class gymnast—Holland himself spent several years at that in junior high school—and an avid outdoorsman. Holland's family was always playing something. Bridge, golf, croquet, checkers, chess, Go—you name it.

And yet somehow, very early on, this business of games began to be more than just fun for him. He began to notice that certain games held a peculiar fascination, a magic that went well beyond any question of winning or losing. Back when he was a freshman in high school, for example—it must have been about 1942 or 1943, when his family was living in Van Wert, Ohio—he and a couple of his buddies spent a lot of time down in the basement of Wally Purmort's house making up brand new games. Their masterpiece, inspired by the daily headlines, was a war game that covered most of the basement floor. They had tanks and artillery. They had firing tables and range tables. They even had ways of covering parts of the board to simulate smoke screens. "It got pretty intricate," says Holland. "I remember using the mimeograph machine at my dad's office to run off lots of pieces." (The elder Holland had prospered through the Depression by founding a string of soybean processing plants throughout the Ohio soybean belt.)

"We wouldn't have said it this way," he says. "But we did it because all three of us were interested in chess. Chess was a game with just a small number of rules. And yet the incredible thing to us was that you never played the same game twice. The possibilities were just infinite. So we would try to invent new games that had that quality."

In one way or another, he laughs, he's been making up games ever since. "I just love these things where the situation unfolds and I say, 'Gee whiz! Did that really come from these assumptions!?' Because if I do it right, if the underlying rules of evolution of the themes are in control and not me, then I'll be surprised. And if I'm not surprised, then I'm not very happy, because I know I've built everything in from the start."

Nowadays, of course, this sort of thing is called "emergence." But long before Holland had ever heard the word, his fascination with it had led him into a lifelong love affair with science and mathematics. He couldn't get enough of either one. All during school, he says, "I remember going to the library and getting hold of any book I could on science and technology. By the time I was a sophomore in high school, I was determined to become a physicist." What captivated him wasn't that science allowed you to reduce everything in the universe to a few simple laws. It was just the opposite: that science showed you how a few simple laws could produce the enormously rich behavior of the world. "It really delights me," he says. "Science and math are the ultimate in reduction in one sense. But if you turn them on their heads, and look at the synthetic aspects, the possibilities for surprise are just unending. It's a way of making the universe comprehensible at one end and forever incomprehensible at the other end."

At MIT, where Holland arrived as a freshman in the fall of 1946, it didn't take him long to discover that same quality of surprise in computers. "I really don't know where this came from, either," he says, "but even early on I was always fascinated by 'thought processes.' And the way you could put just a few bits of data into a computer and then have it do all these things like integration, and so on—it just seemed to me that you were getting an awful lot out for an awful little in."

At first, unfortunately, there was very little that Holland could learn about computers other than the secondhand bits and pieces he picked up in his electrical engineering classes. Electronic computers were still very new then, and mostly classified. There were certainly no computer science courses to take, even at MIT. One day, however, as he was browsing in the MIT library—he did a lot of that—Holland came across a set of loose-leaf lecture notes bound in a simple thesis cover. And as he thumbed through it, he discovered that the notes contained a detailed account of a 1946 conference at the Moore School of Electrical Engineering at the University of Pennsylvania, where a wartime effort to calculate artillery tables had led to the development of the first digital computer in the United States, the ENIAC. "Those notes are famous," says Holland. "This was the first series of really detailed lectures on digital computing. They went all the way from what we would now call computer architectures right up to software." Along the way, the lectures dealt with such brand-new concepts as *information* and *information processing*, and defined a whole new mathematical art: *programming*. Holland immediately bought his own copy of the lectures—in fact, he still has it—and read it cover to cover. Several times.

Then in the fall of 1949, as Holland was starting his senior year at MIT and casting about for a topic for his bachelor's dissertation, he found out about the Whirlwind Project: MIT's effort to build a "real-time" computer fast enough to keep track of air traffic. Funded by the Navy at the then-astounding rate of $1 million per year, Whirlwind employed some seventy engineers and technicians, and was by far the largest computer project of its day. It was also to be among the most innovative. Whirlwind would be the first computer to use magnetic core memory and interactive display screens. It would give birth to computer networks and multiprocessing (running more than one program at once). And as the first real-time computer, it would pave the way for the use of computers in air-traffic control, industrial process control, ticket reservation systems, and banking.

But when Holland first heard of it, Whirlwind was still very much an experiment. "I knew Whirlwind was there," he says. "It wasn't finished yet; it was still being built. But you could use it." Somehow, he just had to get in on this. He started pounding on doors. In the electrical engineering department he found a Czech astronomer named Zednek Kopal, who had been his instructor in numerical analysis. "I convinced him to chair my dissertation committee. I convinced the physics department to *let* somebody from electrical engineering chair it. And then I convinced the Whirlwind people to give me access to their manuals—which were classified!

"That was probably my happiest year at MIT," he says. Kopal suggested that, as a topic for his dissertation, he write a program that would allow Whirlwind to solve Laplace's equation, which describes a variety of physical phenomena ranging from the distribution of electric fields around any electrically charged object, to the vibration of a tightly stretched drumhead. Holland went right to it.

It wasn't exactly the easiest senior thesis ever undertaken at MIT. In those days no one had ever heard of programming languages such as Pascal, or C, or FORTRAN. Indeed, the very concept of a programming language wasn't invented until the mid-1950s. Holland had to write his program in something called machine language, in which commands to the computer were encoded as numbers—and not even ordinary decimal numbers, at that. These were numbers written in hexadecimal notation, base 16. The project took him longer than he'd thought; he eventually ended up asking MIT for twice the time usually allotted for a senior thesis.

And yet he loved it. "I liked the logical nature of the process," he recalls. "Programming had some of the same flavor as math: you take this step, and then that lets you take the next step, and so on." But more than that, writing his program for Whirlwind showed him that a computer needn't be just a fantastically fast adding machine. In his arcane columns of hexadecimal numbers, he could envision a vibrating drumhead, a convoluted electric field, or anything else he wanted. In a world of circulating bits he could create an imaginary universe. All he had to do was to encode the proper laws, and everything else would unfold.

Holland never actually ran his program on the Whirlwind, since it had been intended as a paper exercise from the beginning. But his senior thesis paid off handsomely in another way: it made him one of the very few people in the country who knew anything at all about programming. And as a result, he was snatched up by IBM as soon as he graduated in 1950.

The timing couldn't have been better. At its plant in Poughkeepsie, New York, the giant office equipment manufacturer was designing its first commercial computer: the Defense Calculator, eventually to be renamed the IBM 701. At the time the machine represented a major and rather dubious gamble for the company; many of the old-line executives considered computers to be a waste of money better spent on developing better punch-card machines. In fact, the product planning department spent the entire year of 1950 insisting that the market would never amount to more than about eighteen computers nationwide. IBM was going ahead with the Defense Calculator largely because it was the pet project of a Young Turk known as Tom Junior, son and heir apparent of the company's aging president, Thomas B. Watson, Sr.

But the twenty-one-year-old Holland knew little of that. What he did know was that he had been transported to the Land of Oz. "Here I was, a really young guy at a *prime* place. I was one of the few people to know what was even going on with the 701." The team leaders at IBM put Holland on its seven-man logical planning group, which was in charge of designing the instruction set and general organization of the new machine. This turned out to be yet another stroke of good luck, because it was an ideal spot to exercise his programming skills. "After we got through the initial stages and had a prototype machine, it had to be tested in various ways," he says. "So the engineers would work on it all day, tearing it apart, and then putting it back together again as best they could in the evening. Then a few of us would go in about 11 P.M. and run our stuff all night, just to see if it would work."

Indeed it did work, sort of. By today's standards, of course, the 701 was right out of the Stone Age. It had an enormous control panel full of dials and switches, with no sign of a video monitor. It performed input and output via an IBM-standard punch-card machine. It boasted a full 4 kilobytes of memory. (Personal computers sold today often have a thousand times as much.) And it could multiply two numbers in only 30 microseconds. (Almost any modern hand calculator can do better.) "It also had lots of quirks," says Holland. "The mean time between failures was about 30 minutes at best, so we ran everything twice." Moreover, the 701 stored its data by generating spots of light on the face of a special cathode-ray tube. So Holland and his fellow programmers had to tailor their algorithms to avoid writing data too often at the same location in memory; otherwise, charge would build up on the face of the tube and distort nearby bits. "It was amazing we could get the machine to work at all," he laughs. But the fact was, he didn't care. "So far as we were concerned, it seemed like a giant. We thought it was great to have time on a fast machine to try out our stuff."

There was no shortage of stuff to try out. Those heady, early days of computers were a ferment of new ideas about *information*, *cybernetics*, *automata*—concepts that hadn't even existed ten years earlier. Who knew where the limits were? Almost anything you tried was liable to break new ground. And more than that, for the more philosophically minded pioneers like Holland, these big, clumsy banks of wire and vacuum tubes were opening up whole new ways to think about thinking. Computers might not be the "Giant Brains" of the more lurid Sunday Supplements. In the details of their structure and operation, in fact, they weren't anything like brains at all. But it was very tempting to speculate that computers and brains might be alike in a deeper and much more important sense: they might both be information processing devices. Because if that were the case, then thought itself could be understood as a form of information processing.

At the time, of course, nobody knew to call this sort of thing "artificial intelligence" or "cognitive science." But even so, the very act of programming computers—itself a totally new kind of endeavor—was forcing people to think much more carefully than ever before about what it meant to solve a problem. A computer was the ultimate Martian: you had to tell it *everything*: What are the data? How are they transformed? What are the steps to get from here to there? Those questions, in turn, led very quickly to issues that had bedeviled philosophers for centuries: What is knowledge? How is it acquired through sense impressions? How is it represented in the mind? How is it modified through experience? How is it used in reasoning? How are decisions transformed into action?

The answers were far from clear at the time. (In fact, they are far from clear now.) But the questions were being asked with unprecedented clarity and precision. And IBM's development team in Poughkeepsie, having suddenly become one of the premier collections of computer talent in the country, was in the forefront. Holland fondly remembers how a "regular irregular" group would meet in the evenings every two weeks or so to thrash out the issues over a game of poker or Go. One attendee was a summer intern named John McCarthy, a young Caltech graduate who later became one of the founding gurus of artificial intelligence. (In fact, McCarthy was the one who invented the phrase in 1956 to advertise a summer conference on the subject at Dartmouth College.)

Another was Arthur Samuel, a soft-spoken, fortyish electrical engineer who had been brought in by IBM from the University of Illinois to help the company figure out how to make reliable vacuum tubes, and who was one of Holland's most frequent companions during his all-night programming marathons. (He also had a daughter in nearby Vassar, whom Holland dated several times.) Samuel frankly didn't give a damn about vacuum tubes anymore. For the past five years he had been trying to write a program that could play checkers—and not only play the game, but learn to play it better and better with experience. In retrospect, Samuel's checker player is considered one of the milestones of artificial intelligence research; by the time he finally finished revising and refining it in 1967, it was playing at a world championship level. But even in the 701 days, it was doing remarkably well. Holland remembers being very impressed with it, particularly with its ability to adapt its tactics to what the other player was doing. In effect, the program was making a simple model of "opponent" and using that model to make predictions about the best line of play. And somehow, without being able to articulate it very well at the time, Holland felt that this aspect of the checker player captured something essential and right about learning and adaptation.

He filed that thought away in his mind as something to mull over. For the moment he was more than busy enough with his own project: an attempt to simulate the inner workings of the brain itself. It started in the spring of 1952, he remembers, as he was listening to J.C.R. Licklider, a psychologist from MIT who had come for a visit to the Poughkeepsie lab and who had agreed to give the group there a lecture on what was then one of the hottest topics in the field: the new theories of learning and memory being advanced by neurophysiologist Donald O. Hebb of McGill University in Montreal.

The problem was this, Licklider had explained: Through a microscope, most of the brain appears to be a study in chaos, with each nerve cell sending out thousands of random filaments that connect it willy-nilly to thousands of other nerve cells. And yet this densely interconnected network is obviously *not* random. A healthy brain produces perception, thought, and action quite coherently. Moreover, the brain is obviously not static. It refines and adapts its behavior through experience. It learns. The question is, How?

Three years earlier, in 1949, Hebb had published his answer in a book entitled *The Organization of Behavior*. His fundamental idea was to assume that the brain is constantly making subtle changes in the "synapses," the points of connection where nerve impulses make the leap from one cell to the next. This assumption was a bold move on Hebb's part, since at the time he had no evidence for it whatsoever. But having made it, he argued that these synaptic changes were in fact the basis of all learning and memory. A sensory impulse coming in from the eyes, for example, would leave its trace on the neural network by strengthening all the synapses that lay along its path. Much the same thing would happen with impulses coming in from the ears or from mental activity elsewhere in the brain itself. And as a result, said Hebb, a network that started out at random would rapidly organize itself. Experience would accumulate through a kind of positive feedback: the strong, frequently used synapses would grow stronger, while the weak, seldom-used synapses would atrophy. The favored synapses would eventually become so strong that the memories would be locked in. These memories, in turn, would tend to be widely distributed over the brain, with each one corresponding to a complex pattern of synapses involving thousands or millions of neurons. (Hebb was one of the first to describe such distributed memories as "connectionist.")

But there was more. In his lecture, Licklider went on to explain Hebb's second assumption: that the selective strengthening of the synapses would cause the brain to organize itself into "cell assemblies"—subsets of several thousand neurons in which circulating nerve impulses would reinforce themselves and continue to circulate. Hebb considered these cell assemblies to be the brain's basic building blocks of information. Each one would correspond to a tone, a flash of light, or a fragment of an idea. And yet these assemblies would *not* be physically distinct. Indeed, they would overlap, with any given neuron belonging to several of them. And because of that, activating one assembly would inevitably lead to the activation of others, so that these fundamental building blocks would quickly organize themselves into larger concepts and more complex behaviors. The cell assemblies, in short, would be the fundamental quanta of thought.

Sitting there in the audience, Holland was transfixed by all this. This wasn't just the arid stimulus/response view of psychology being pushed at the time by behaviorists such as Harvard's B. F. Skinner. Hebb was talking about what was going on inside the mind. His connectionist theory had the richness, the perpetual surprise that Holland responded so strongly to. It felt *right*. And Holland couldn't wait to do something with it. Hebb's theory was a window onto the essence of thought, and he wanted to watch. He wanted to see cell assemblies organize themselves out of random chaos and grow. He wanted to see them interact. He wanted to see them incorporate experience and evolve. He wanted to see the emergence of the mind itself. And he wanted to see all of it happen spontaneously, without external guidance.

No sooner had Licklider finished his lecture on Hebb than Holland turned to his leader on the 701 team, Nathaniel Rochester, and said, "Well, we've got this prototype machine. Let's program a neural network simulator."

And that's exactly what they did. "He programmed one," says Holland, "and I programmed another, rather different in form. We called them the 'Conceptors.' No hubris there!"

In fact, the IBM Conceptors stand as an impressive accomplishment even forty years later, when neural network simulations have long since become a standard tool in artificial intelligence research. The basic idea would still look familiar enough. In their programs, Holland and Rochester modeled their artificial neurons as "nodes"—in effect, tiny computers that can remember certain things about their internal state. They modeled their artificial synapses as abstract connections between various nodes, with each connection having a certain "weight" corresponding to the strength of the synapse. And they modeled Hebb's learning rule by adjusting the strengths as the network gained experience. However, Holland, Rochester, and their collaborators also incorporated far more details about basic neurophysiology than most neural network simulations do today, including such factors as how fast each simulated neuron fired and how "tired" it got if it was fired too often.

Not surprisingly, they had a tough time getting it all to work. Not only were their programs among the first neural network simulations ever, they marked one of the first times a computer had been used to simulate anything (as opposed to calculating numbers or sorting data). Holland gives IBM a lot of credit for its corporate patience. He and his colleagues spent uncounted hours of computer time on their networks, and even took a trip up to Montreal to consult with Hebb himself—at company expense.

But in the end, by golly, the simulations worked. "There was a lot of emergence," says Holland, still sounding excited about it. "You could start with a uniform substrate of neurons and see the cell assemblies form." And when Holland, Rochester, and their colleagues finally published their results in 1956, several years after the bulk of the research was done, it was his first published paper.

Building Blocks

Looking back on it, says Holland, Hebb's theory and his own network simulation of it probably did more to shape his thinking over the next thirty years than did any other single thing. But at the time, the most immediate result was to goad him into leaving IBM.

The problem was that computer simulation had some definite limits, especially on the 701. Cell assemblies in a real nervous system have as many as 10,000 neurons distributed over a large part of the brain, with as many as 10,000 synapses per neuron. But the largest simulated network Holland and his coworkers ever ran on the 701 had only 1000 neurons and 16 connections per neuron—and they only managed to get that far by using every programming trick they could think of to speed things up. "The more I did that," says Holland, "the more I realized that the distance between what we could really test out and what I wanted to see was just too large."

The alternative was to try to analyze the networks mathematically. "But that proved to be really tough," he says. Everything he attempted ran up against a brick wall. A full-fledged Hebbian network was just too far beyond anything he could tackle with the math he had learned at MIT—and he had taken a lot more math courses than most physics majors. "It just seemed to me that the key to knowing more about networks was to know more mathematics," he says. So in the fall of 1952, with IBM's blessing and a nice little going away present—a contract to continue consulting for Big Blue at the rate of 100 hours per month—he entered the Ph.D. program in mathematics at the University of Michigan in Ann Arbor.

Again, his luck held. Of course, Michigan wouldn't have been a bad choice in any case. Not only did it have one of the best mathematics departments in the country then, but—a prime consideration for Holland—it had a football team. "A football weekend in the Big Ten, with 100,000 people coming into town—I still enjoy it!"

The real good fortune, however, was that at Michigan Holland encountered Arthur Burks, a philosopher who was no ordinary philosopher. A specialist on the pragmatist philosophy of Charles Peirce, Burks had gotten his Ph.D. in 1941 at a time when there was no hope whatsoever of finding a teaching job in his field. So the following year, he had taken a ten-week course at the Moore School of the University of Pennsylvania to turn himself into an engineer for the war effort. It proved a happy choice. Shortly thereafter, in 1943, he had been hired to work on the Moore School's top-secret ENIAC—the first electronic computer. And there he had met the legendary Hungarian mathematician John von Neumann, who was coming in frequently from the Institute of Advanced Study in Princeton to work as a consultant on the project. Under von Neumann, Burks had also worked on the design of ENIAC's successor, the EDVAC, the first computer to store its instructions electronically in the form of a program. Indeed, a 1946 paper by von Neumann, Burks, and the mathematician Herman Goldstine—"Preliminary Discussion of the Logical Design of an Electronic Computing Instrument"— is now regarded as one of the foundation stones of modern computer science. In it, the three men had defined the concept of a program in a precise logical form, and showed how a general-purpose computer could execute such a program by a continuous cycle of fetching each instruction from the computer's memory unit, executing that instruction in a central processing unit, and then storing the results back in memory. This "von Neumann architecture" is still the basis for almost all computers today.

When Holland met him at the University of Michigan in the mid-1950s, Burks was a slim, rather courtly man who looked very much like the minister he had once thought of becoming. (To this day, Burks never appears on the notoriously casual Michigan campus without a coat and tie.) But Burks also proved to be a warm friend and a superb mentor. He quickly brought Holland into his Logic of Computers group, a coterie of theorists studying computer languages, proving theorems about switching networks, and in general trying to understand these new machines at the most rigorous and fundamental level.

Burks also invited Holland into a new Ph.D. program that he was helping to organize and that would be dedicated to exploring the implications of computers and information processing in as broad a realm as possible. Soon known as Communication Sciences, this program would eventually evolve into a full-fledged computer department in 1967, when it became known as Computer and Communication Sciences. But at the time, Burks felt he was simply carrying on the legacy of von Neumann, who had died of cancer in 1954. "Von Neumann thought of using computers in two ways," he says. One was as a general-purpose computational device, the purpose for which they had been invented. "The other was as the basis for a general theory of automata, natural and artificial." Burks also felt that such a program would meet the needs of those students, Holland prominent among them, whose minds refused to flow in the normal channels.

Holland liked what he heard. "The idea was to develop some very tough courses in areas like biology, linguistics, and psychology, as well as a lot of the standard stuff such as information theory," he says. "The courses would be taught by professors from each subject, so that the students could get the linkages between these things and computer models. And the students who came out of these courses would come out with a very deep understanding of the fundamentals of the field—the problems, the questions, why the issues were difficult, and what computers could do to help. They wouldn't just have a surface understanding."

Holland liked the concept even more because he was getting totally disenchanted with mathematics. Like most mathematics groups in the post–World War II era, the department at Michigan was dominated by the ideals of the French Bourbaki school, which called for research of almost inhuman purity and abstraction. According to the Bourbaki standard, it was even considered gauche to illustrate the concepts behind your axioms and theorems with something so down to earth as a drawing. "The idea was to show that mathematics could be divorced from any interpretation," says Holland. And yet this wasn't what he had come for at all; he wanted to use mathematics to understand the world.

So when Burks suggested that Holland transfer into the Communication Sciences program, he didn't hesitate. He abandoned his nearly completed dissertation in mathematics and started all over. "It meant that I could do a dissertation in an area that was much closer to what I wanted to do," he says—namely, neural networks. (Ironically, the dissertation topic he finally decided on, "Cycles in Logical Nets," was devoted to an analysis of what happens in a network of on-off switches; in it he proved many of the same theorems that a young medical student named Stuart Kauffman independently struggled to prove in Berkeley four years later.) And when Holland finally earned his Ph.D. in 1959, it was the first that the Communication Sciences program had ever awarded.

None of this deflected Holland's attention from the broader issues that had brought him to Michigan in the first place. Quite the opposite. Burks' Communication Sciences program was the kind of environment where such questions could thrive. What *is* emergence? And what is thinking? How does it work? What are its laws? What does it really mean for a system to adapt? Holland jotted down reams of ideas on these questions, then systematically filed them away in manila folders labeled *Glasperlenspiel 1*, *Glasperlenspiel 2*, et cetera.

Glas—what? "*Das Glasperlenspiel*," he laughs. It was Herman Hesse's last novel, published in 1943 while the author was in exile in Switzerland. Holland discovered it one day in a stack of books that a roommate had brought home from the library. In German the title literally means *The Glass-Bead Game*, but in English translations the book is usually called *Master of the Game*, or its Latin equivalent, *Magister Ludi*. Laid in a society of the far future, the novel describes a game that was originally played by musicians; the idea was to set up a theme on a kind of abacus with glass beads, and then try to weave all kinds of counterpoint and variation on the theme by moving the beads back and forth. Over time, however, the game evolved from its simple origins into an instrument of profound sophistication, controlled by a cadre of powerful priest-intellectuals. "The great thing was that you could take any combination of themes," says Holland—"something from astrology, something from Chinese history, something from math—and then try to develop them like a music theme."

Of course, he says, Hesse was a little vague about exactly how this was done. But Holland didn't care; more than anything he'd ever seen or heard of, the Glass-Bead Game captured what had fascinated him about chess, about science, about computers, about the brain. In a metaphorical sense, the game was what he'd been after all his life: "I'd like to be able to take themes from all over and see what emerges when I put them together," he says.

A particularly fruitful source of ideas for the *Glasperlenspiel* files was another book that Holland came upon one day as he was browsing in the stacks of the math department library: R. A. Fisher's landmark 1929 tome on genetics, *The Genetical Theory of Natural Selection*.

At first, Holland was fascinated. "I'd always enjoyed reading about genetics and evolution, even in high school," he says. He loved the idea that genes from the parents are reshuffled in each new generation, and that you calculate how often specific traits such as blue eyes or dark hair will show up in their offspring. "I thought, 'Wow, this is really neat!' But reading Fisher's book was the first time I realized that you could do anything other than trivial algebra in this area." Indeed, Fisher used much more sophisticated ideas from differential and integral calculus, as well as probability theory. His book had provided biologists with the first really careful mathematical analysis of how the distribution of genes in a population will change as a result of natural selection. And as such it had laid the foundation for the modern, "neo-Darwinian" theory of evolutionary change. A quarter of a century later, it was still pretty much the state of the art in the theory of evolutionary dynamics.

So Holland devoured the book. "The fact that you could take calculus and differential equations and all the other things I had learned in my math classes to start a revolution in genetics—that was a real eye-opener. Once I saw that, I knew I could never let it go. I knew I had to do something with it. So I kept messing around with the ideas in the back of my mind, scribbling notes."

And yet, as much as Holland admired Fisher's math, there was something about the way Fisher used the math that began to bother him. As he thought about it more and more, in fact, it began to bother him a lot.

For one thing, Fisher's whole analysis of natural selection focused on the evolution of just one gene at a time, as if each gene's contribution to the organism's survival was totally independent of all the other genes. In effect, Fisher assumed that the action of genes was completely linear. "I *knew* that had to be wrong," says Holland. A single gene for green eyes isn't worth very much unless it's backed up by the dozens or hundreds of genes that specify the structure of the eye itself. Each gene had to work as part of a team, realized Holland. And any theory that didn't take that fact into account was missing a crucial part of the story. Come to think on it, that was also what Hebb had been saying in the mental realm. Hebb's cell assemblies were a bit like genes, in that they were supposed to be the fundamental units of thought. But in isolation the cell assemblies were almost nothing. A tone, a flash of light, a command for a muscle twitch— the only way they could mean anything was to link up into larger concepts and more complex behaviors.

For another thing, it bothered Holland that Fisher kept talking about evolution achieving a stable equilibrium—that state in which a given species has attained its optimum size, its optimum sharpness of tooth, its optimum *fitness* to survive and reproduce. Fisher's argument was essentially the same one that economists use to define economic equilibrium: once a species' fitness is at a maximum, he said, any mutation will lower the fitness. So natural selection can provide no further pressure for change. "An awful lot of Fisher is that way," says Holland: "He says, 'Well, the system will go to the Hardy-Weinberg equilibrium because of the following process. . . .' But that did not sound like evolution to me."

He went back and reread Darwin and Hebb. No, Fisher's concept of equilibrium didn't sound like evolution at all. Fisher seemed to be talking about the attainment of some pristine, eternal perfection. "But with Darwin, you see things getting broader and broader with time, more diverse," says Holland. "Fisher's math didn't touch on that." And with Hebb, who was talking about learning instead of evolution, you saw the same thing: minds getting richer, more subtle, more surprising as they gained experience with the world.

To Holland, evolution and learning seemed much more like—well, a game. In both cases, he thought, you have an agent playing against its environment, trying to win enough of what it needed to keep going. In evolution that payoff is literally survival, and a chance for the agent to pass, its genes on to the next generation. In learning, the payoff is a reward of some kind, such as food, a pleasant sensation, or emotional fulfillment. But either way, the payoff (or lack of it) gives agents the feedback they need to improve their performance: if they're going to be "adaptive" at all, they somehow have to keep the strategies that pay off well, and let the others die out.

Holland couldn't help thinking of Art Samuel's checker-playing program, which took advantage of exactly this kind of feedback: the program was constantly updating its tactics as it gained experience and learned more about the other player. But now Holland was beginning to realize just how prescient Samuel's focus on games had really been. This game analogy seemed to be true of *any* adaptive system. In economics the payoff is in money, in politics the payoff is in votes, and on and on. At some level, all these adaptive systems are fundamentally the same. And that meant, in turn, that all of them are fundamentally like checkers or chess: the space of possibilities is vast beyond imagining. An agent can learn to play the game better—that's what adaptation is, after all. But it has just about as much chance of finding the optimum, stable equilibrium point of the game as you or I have of solving chess.

No wonder "equilibrium" didn't sound like evolution to him; it didn't even sound like a war game that a trio of fourteen-year-old boys could cobble together in Wally Purmort's basement. Equilibrium implies an endpoint. But to Holland, the essence of evolution lay in the journey, the endlessly unfolding surprise: "It was becoming more and more clear to me that the things I wanted to understand, that I was curious about, that would please me if I found out about them—equilibrium wasn't an important part of any of them."

. . .

Holland had to keep all this on a back burner while he completed his Ph.D. dissertation. But once he had graduated in 1959—Burks had already invited him to stay on with the Logic of Computers group as a postdoc— he set himself the goal of turning his vision into a complete and rigorous theory of adaptation. "The belief was that if I looked at genetic adaptation as the longest-term adaptation, and the nervous system as the shortest term," he says, "then the general theoretical framework would be the same." To get the initial ideas straight in his own mind he even wrote a manifesto on the subject, a forty-eight-page technical report that he circulated in July 1961 under the title "A Logical Theory of Adaptive Systems Informally Described."

He also began to notice a lot of raised eyebrows among his colleagues in the Logic of Computers group. It wasn't a sense of hostility, exactly. It was just that a few people thought that this general theory of adaptation business sounded weird. Couldn't Holland be spending his time on something a little more—fruitful?

"The question was, is it crackpot?" recalls Holland, who cheerfully admits that he would have been skeptical, too, in his colleagues' place. "The stuff I was doing didn't fit very well in the nice, familiar categories. It wasn't hardware, exactly. It wasn't software, exactly. And at the time it certainly didn't fit into artificial intelligence. So you couldn't use any of the standard criteria and come up with a judgment."

One person who didn't need a lot of convincing was Burks. "I supported John," he says. "There was a clique of logicians who didn't think that what John was doing was what 'Logic of Computers' should be about. They were much more traditional. But I told them that this is what we needed to do, that it was as important for getting grants as their stuff." Burks won the day: as founder and guru of the program, his voice carried considerable weight. By and by, the skeptics drifted out of the program. And in 1964, with Burks' enthusiastic endorsement, Holland was awarded tenure. "An awful lot of those years I owe to Art Burks acting as a shield," he says.

Indeed, Burks' backing gave Holland the security he needed to bear down on the theory of adaptation as hard as he could. By 1962 he had put aside all his other research projects and was devoting himself to it essentially full time. In particular, he was determined to crack this problem of selection based on more than one gene—and not just because Fisher's independent-gene assumption had bugged him more than anything else about that book.

Moving to multiple genes was also the key to moving away from this obsession with equilibrium.

In fairness to Fisher, says Holland, equilibrium actually does make a lot of sense when you're talking about independent genes. For example, suppose you had a species with 1000 genes, which would make it roughly as complicated as seaweed. And suppose, for simplicity's sake, that each gene comes in just two varieties—green color versus brown color, wrinkled leaves versus smooth leaves, and so forth. How many trials does it take for natural selection to find the set of genes that gives the seaweed its highest fitness?

If you assume that all the genes are indeed independent, says Holland, then for each gene you just need two trials to find which variety is better. Then you have to perform those two trials on each of 1000 genes. So you need 2000 trials in all. And that's not very many, he says. In fact, it's such a comparatively small number that you can expect this seaweed to attain its maximum fitness fairly quickly, at which point the species will indeed be at an evolutionary equilibrium.

But now, says Holland, look what happens with that 1000-gene seaweed when you assume that the genes are *not* independent. To be sure of finding the highest level of fitness in this case, natural selection would now have to examine every conceivable combination of genes, because each combination potentially has a different fitness. And when you work out the total number of combinations, it isn't 2 multiplied by 1000. It's 2 multiplied by itself 1000 times. That's 2^{1000}, or about 10^{300}—a number so vast that it makes even the number of moves in chess seem infinitesimal. "Evolution can't even begin to try out that many things," says Holland. "And no matter how good we get with computers, we can't do it." Indeed, if every elementary particle in the observable universe were a supercomputer that had been number-crunching away since the Big Bang, they still wouldn't be close. And remember, that's just for seaweed. Humans and other mammals have roughly 100 times as many genes—and most of those genes come in many more than two varieties.

So once again, says Holland, you have a system exploring its way into an immense space of possibilities, with no realistic hope of ever finding the single "best" place to be. All evolution can do is look for improvements, not perfection. But that, of course, was precisely the question he had resolved to answer back in 1962: How? Understanding evolution with multiple genes obviously wasn't just a trivial matter of replacing Fisher's one-variable equations with many-variable equations. What Holland wanted to know was how evolution could explore this immense space of possibilities and find useful combinations of genes—without having to search over every square inch of territory.

As it happens, a similar explosion of possibilities was already well known to mainstream artificial intelligence researchers. At Carnegie Tech (now Carnegie Mellon University) in Pittsburgh, for example, Allen Newell and Herbert Simon had been conducting a landmark study of human problem-solving since the mid-1950s. By asking experimental subjects to verbalize their thoughts as they struggled through a wide variety of puzzles and games, including chess, Newell and Simon had concluded that problem-solving always involves a step-by-step mental search through a vast "problem space" of possibilities, with each step guided by a heuristic rule of thumb: "If *this* is the situation, then *that* step is worth taking." By building their theory into a program known as General Problem Solver, and by putting that program to work on those same puzzles and games, Newell and Simon had shown that the problem-space approach could reproduce human-style reasoning remarkably well. Indeed, their concept of heuristic search was already well on its way to becoming the dominant conventional wisdom in artificial intelligence. And General Problem Solver stood—as it still stands—as one of the most influential programs in the young field's history.

But Holland was dubious. It wasn't that he thought Newell and Simon were wrong about problem spaces or heuristics. Shortly after he got his Ph.D., in fact, he had made it a point to bring both of them to Michigan as part of a major seminar on artificial intelligence. He and Newell had been friends and intellectual sparring partners ever since. No, it was simply that the Newell-Simon approach didn't help him with biological evolution. The whole point of evolution is that there are *no* heuristic rules, no guidance of any sort; succeeding generations explore the space of possibilities by mutations and random reshuffling of genes among the sexes—in short, by trial and error. Furthermore, those succeeding generations don't conduct their search in a step-by-step manner. They explore it in parallel: each member of the population has a slightly different set of genes and explores a slightly different region of the space. And yet, despite these differences, evolution produces just as much creativity and surprise as mental activity does, even if it takes a little longer. To Holland, this meant that the real unifying principles in adaptation had to be found at a deeper level. But where?

Initially, all he had was this intuitive idea that certain sets of genes worked well together, forming coherent, self-reinforcing wholes. An example might be the cluster of genes that tells a cell how to extract energy from glucose molecules, or the cluster that controls cell division, or the cluster that governs how a cell combines with other cells to form a certain kind of tissue. Holland could also see analogs in Hebb's theory of the brain, where a set of resonating cell assemblies might form a coherent concept such as "car," or a coordinated motion such as lifting your arm.

But the more Holland thought about this idea of coherent, self-reinforcing clusters, the more subtle it began to seem. For one thing, you could find analogous examples almost anywhere you looked. Subroutines in a computer program. Departments in a bureaucracy. Gambits in the larger strategy of a chess game. Furthermore, you could find examples at every level of organization. If a cluster is coherent enough and stable enough, then it can usually serve as a building block for some larger cluster. Cells make tissues, tissues make organs, organs make organisms, organisms make ecosystems—on and on. Indeed, thought Holland, that's what this business of "emergence" was all about: building blocks at one level combining into new building blocks at a higher level. It seemed to be one of the fundamental organizing principles of the world. It certainly seemed to appear in every complex, adaptive system that you looked at.

But why? This hierarchical, building-block structure of things is as commonplace as air. It's so widespread that we never think much about it. But when you *do* think about it, it cries out for an explanation: Why is the world structured this way?

Well, there are actually any number of reasons. Computer programmers are taught to break things up into subroutines because small, simple problems are easier to solve than big, messy problems; it's simply the ancient principle of divide and conquer. Large creatures such as whales and redwoods are made of trillions of tiny cells because the cells came first; when large plants and animals first appeared on Earth some 570 million years ago, it was obviously easier for natural selection to bring together the single-celled creatures that already existed than to build big new blobs of protoplasm from scratch. General Motors is organized into several zillion divisions and subdivisions because the CEO doesn't want to have half a million employees reporting to him directly; there aren't enough hours in the day. In fact, as Herbert Simon had pointed out in the 1940s and 1950s in his studies of business organizations, a well-designed (emphasize *well*-designed) hierarchy is an excellent way of getting some work done without any one person being overwhelmed by meetings and memos.

As Holland thought about it, however, he became convinced that the most important reason lay deeper still, in the fact that a hierarchical, building-block structure utterly transforms a system's ability to learn, evolve, and adapt. Think of our cognitive building blocks, which include such concepts as *red*, *car*, and *road*. Once a set of building blocks like this has been tweaked and refined and thoroughly debugged through experience, says Holland, then it can generally be adapted and recombined to build a great many new concepts—say, "A red Saab by the side of the road." Certainly that's a much more efficient way to create something new than starting all over from scratch. And that fact, in turn, suggests a whole new mechanism for adaptation in general. Instead of moving through that immense space of possibilities step by step, so to speak, an adaptive system can reshuffle its building blocks and take giant leaps.

Holland's favorite illustration of this is the way police artists used to work in the days before computers, when they needed to make a drawing of a suspect to match a witness's description. The idea was to divide the face up into, say, 10 building blocks: hairline, forehead, eyes, nose, and so on down to the chin. Then the artist would have strips of paper with a variety of options for each: say, 10 different noses, 10 different hairlines, and so forth. That would make a total of 100 pieces of paper, says Holland. And armed with that, the artist could talk to the witness, assemble the appropriate pieces, and produce a sketch of the suspect very quickly. Of course, the artist couldn't reproduce every conceivable face that way. But he or she could almost always get pretty close: by shuffling those 100 pieces of paper, the artist could make a total of 10 billion different faces, enough to sample the space of possibilities quite widely. "So if I have a process that can discover building blocks," says Holland, "the combinatorics start working *for* me instead of against me. I can describe a great many complicated things with relatively few building blocks."

And that, he realized, was the key to the multiple-gene puzzle. "The cut and try of evolution isn't just to build a good animal, but to find good building blocks that can be put together to make many good animals." His challenge now was to show precisely and rigorously how that could happen. And the first step, he decided, was to make a computer model, a "genetic algorithm" that would both illustrate the process and help him clarify the issues in his own mind.

At one time or another, just about everyone in the Michigan computer science community had had the experience of seeing John Holland come running up with a fistful of fan-fold computer printout.

"Look at that!" he would say, eagerly pointing to something in the midst of a page full of hexadecimal gibberish.

"Oh. CCB1095E. That's—wonderful, John."

"No! No! Do you know what that means . . . !?"

Actually, there were quite a few people in the early 1960s who didn't know and who couldn't' quite figure it out. His skeptical colleagues had been right about one thing, at least: the genetic algorithm that Holland finally came up with was *weird*. Except in the most literal sense, in fact, it wasn't really a computer program at all. In its inner workings it was more like a simulated ecosystem—a kind of digital Serengeti in which whole populations of programs would compete and have sex and reproduce for generation after generation, always evolving their way toward the solution of whatever problem the programmer might set for them.

This wasn't the way programs were usually written, to put it mildly. So to explain to his colleagues why it made sense, Holland usually found it best to couch what he was doing in very practical terms. Normally, he would tell them, we think of a computer program as a string of instructions written in a special programming language such as FORTRAN or LISP. Indeed, the whole art of programming is to make sure that you've written precisely the right instructions in precisely the right order. And that's obviously the most effective way to do it—*if* you already know precisely what you want the computer to do. But suppose you don't know, said Holland. Suppose, for example, that you're trying to find the maximum value of some complicated mathematical function. The function could represent profit, or factory output, or vote counts, or almost anything else; the world is full of things that need to be maximized. Indeed, programmers have devised any number of sophisticated computer algorithms for doing so. And yet, not even the best of those algorithms is guaranteed to give you the correct maximum value in every situation. At some level, they always have to rely on old-fashioned trial and error—guessing.

But if that's the case, Holland told his colleagues, if you're going to be relying on trial and error anyway, maybe it's worth seeing what you can do with nature's method of trial and error—namely, natural selection. Instead of trying to *write* your programs to perform a task you don't quite know how to do, evolve them.

The genetic algorithm was a way of doing that. To see how it works, said Holland, forget about the FORTRAN code and go down into the guts of the computer, where the program is represented as a string of binary ones and zeros: 110100111100011001000101001110011 . . . , et cetera. In that form the program looks a heck of a lot like a chromosome, he said, with each binary digit being a single "gene." And once you start thinking of the binary code in biological terms, then you can use that same biological analogy to make it evolve.

First, said Holland, you have the computer generate a population of maybe 100 of these digital chromosomes, with lots of random variation from one to the next. Each chromosome corresponds to an individual zebra in a herd of zebras, so to speak. (For simplicity's sake, and because Holland was trying to get at the absolute essence of evolution, the genetic algorithm leaves aside such details as hoofs and stomachs and brains. It models the individual as a single piece of naked DNA. As a matter of practicality, moreover, Holland had to make his binary chromosomes no more than a few dozen binary digits long, so that they were actually not full-scale programs but fragments of programs. In his earliest work, in fact, the chromosome represented only a single variable. But none of that changed the basic principle of the algorithm.)

Second, said Holland, you test each individual chromosome on the problem at hand by running it as a computer program, and then giving it a score that measures how well it does. In biological terms, this score will determine the individual's "fitness"—its probability of reproductive success. The higher the fitness, the higher the individual's chances of being selected by the genetic algorithm to pass on its genes to the next generation.

Third, said Holland, you take those individuals you've selected as being fit enough to reproduce, and create a new generation of individuals through sexual reproduction. You allow the rest to die off. In practice, of course, the genetic algorithm leaves aside gender differences, courtship rituals, foreplay, the union of sperm and egg, and all the other intricacies of real sexual reproduction, and instead creates the new generation through the bare-bones exchange of genetic material. Schematically, the algorithm chooses a pair of individuals with chromosomes *ABCDEFG* and *abcdefg*, breaks each string at a random point in the middle, and then interchanges the pieces to form the chromosomes for a pair of offspring: *ABCDefg* and *abcdEFG*. (Holland got the idea from real chromosomes, where this sort of interchange, or "crossover," happens fairly frequently.)

Finally, said Holland, the offspring produced by this sexual exchange of genes go on to compete with each other and with their parents in a new generational cycle. And this is the crucial step, both in the genetic algorithm and in Darwinian natural selection. Without sexual exchange, the offspring would have been identical with their parents and the population would be well on its way to stagnation. The poor performers would gradually die off, but the good performers would never show any improvement. With sexual exchange, however, the offspring are similar to their parents, but different— and sometimes better. And when that happens, said Holland, those improvements stand a good chance of spreading through the population and improving the breed markedly. Natural selection provides a kind of upward ratchet.

In real organisms, of course, quite a bit of variation is also provided by mutations, typographical errors in the genetic code. And in fact, said Holland, the genetic algorithm does allow for an occasional mutation by deliberately turning a 1 into a 0, or vice versa. But for him, the heart of the genetic algorithm was sexual exchange. Not only does the exchange of genes through sex provide for variation in the population, but it turns out to be a very good mechanism for searching out clusters of genes that work well together and produce above-average fitness—in short, building blocks.

For example, said Holland, suppose that you've put the genetic algorithm to work on one of those optimization problems, where it's looking for a way to find the maximum value of some complicated function. And suppose that the digital chromosomes in the algorithm's internal population turn out to get very high scores when they have certain patterns of binary genes such as 11####11#10###10 or ##1001###11101##. (Holland used # to stand for "doesn't matter"; the digit in that position could be a 1 or a 0.) Such patterns will function as building blocks, he said. Maybe they happen to denote ranges of variables where the function does indeed have higher values than average. But whatever the reason, the chromosomes that contain such building blocks will tend to prosper and spread through the population, displacing chromosomes that don't have them.

Furthermore, he said, since sexual reproduction allows the digital chromosomes to shuffle their genetic material every generation, the population will constantly be coming up with new building blocks and new combinations of the existing building blocks. So the genetic algorithm will very quickly produce individuals that are doubly and triply blessed with good building blocks. And if those building blocks act together to confer extra advantages, Holland was able to show, then the individuals that have them will spread through the population even faster than before. The upshot is that the genetic algorithm will converge to the solution of the problem at hand quite rapidly—without ever having to know beforehand what the solution is.

Holland remembers being thrilled when he first realized this, back in the early 1960s. It never seemed to get his audiences very excited; at the time, most of his contemporaries in the still-young field of computer science felt that they had more than enough to do in laying the groundwork for conventional programming. In purely practical terms, the idea of evolving a program seemed a little off the wall. But Holland didn't care. This was exactly what he had been looking for since he set out to generalize Fisher's independent-genes assumption. Reproduction and crossover provided the mechanism for building blocks of genes to emerge and evolve together—and, not incidentally, provided a mechanism for a population of individuals to explore the space of possibilities with impressive efficiency. By the mid-1960s, in fact, Holland had proved what he called the schema theorem, the fundamental theorem of genetic algorithms: in the presence of reproduction, crossover, and mutation, almost any compact cluster of genes that provides above-average fitness will grow in the population exponentially. ("Schema" was his term for any specific pattern of genes.)

"It was when I finally got the schema theorem in a form that I liked that I started writing my book," he says.

Emergence of Mind

"The book"—a compilation of the schema theorem, the genetic algorithm, and his thinking on adaptation in general—was something that Holland thought he might be able to finish in a year or two. In fact, it took him a decade. Somehow, as the writing and the research continued in parallel, he was always finding a new idea to explore or a new aspect of the theory to analyze. He set several of his graduate students to work on computer experiments—demonstrations that the genetic algorithm really was a useful and efficient way to solve optimization problems. Holland felt that he was laying out the theory and practice of adaptation, and he wanted it done *right*—with detail, precision, and rigor.

He certainly had that. Published in 1975, *Adaptation in Natural and Artificial Systems* was dense with equations and analysis. It summarized two decades of Holland's thinking about the deep interrelationships of learning, evolution, and creativity. It laid out the genetic algorithm in exquisite detail.

And in the wider world of computer science outside Michigan, it was greeted with resounding silence. In a community of people who like their algorithms to be elegant, concise, and provably correct, this genetic algorithm stuff was still just too weird. The artificial intelligence community was a little more receptive, enough to keep the book selling at the rate of 100 to 200 copies per year. But even so, when there were any comments on the book at all, they were most often along the lines of, "John's a real bright guy, but . . ."

Of course, it has to be said that Holland didn't do much to make his case. Holland published his papers, although relatively few of them. He gave seminars when people invited him to. But that was just about it. He didn't make dramatic claims for genetic algorithms at the major conferences. He didn't apply genetic algorithms to flashy applications such as medical diagnosis, the kind that might make venture capitalists sit up and take notice. He didn't lobby for big grants to establish a "laboratory" for genetic algorithms. He didn't publish a popular book warning that massive federal funding of genetic algorithms was urgently needed to meet the Japanese threat.

In short, he simply didn't play the game of academic self-promotion. That seems to be the one game he doesn't like to play. More to the point, he really doesn't seem to care if he wins it or not. Metaphorically speaking, he still prefers to putter away with a few buddies in the basement. "It's like playing baseball," he says. "Just because you're playing on a sandlot team and not in the majors—it's the fun that matters. And the kind of science I do has always been a lot of fun for me."

"I think it would have bothered me if *no*body had been willing to listen," he adds. "But I've always been very lucky in having bright, interested graduate students to bounce ideas off of."

Indeed, that was the flip side of his buddies-in-the-basement attitude: Holland put a lot of energy into working with his immediate group at Michigan. At any one time he typically supervised six or seven graduate students—far above average. Starting in the mid-1960s, in fact, he managed to graduate them with Ph.D.'s at an average of more than one per year.

"Some of them have been really brilliant—and great fun for that reason," he says. Holland deliberately took a rather hands-off approach to guidance, having seen too many professors build up a huge publication list by publishing "joint" research papers that were in fact written entirely by their graduate students. "So they all followed their noses and did things they thought were interesting. Then we'd all meet around a table about once a week, one of them would tell where he stood on his dissertation, and we'd all critique it. That was usually a lot of fun for everybody involved."

In the mid-1970s, Holland also started meeting with a group of like-minded faculty members for a free-wheeling monthly seminar on—well, just about anything having to do with evolution or adaptation. In addition to Burks, the group included Robert Axelrod, a political scientist trying to understand why and when people will cooperate instead of stabbing each other in the back; Michael Cohen, another political scientist, specializing in the social dynamics of human organizations; and William Hamilton, an evolutionary biologist working with Axelrod to understand symbiosis, social behavior, and other forms of biological cooperation.

"Mike Cohen was the catalyst," recalls Holland. It was just after *Adaptation* came out. Cohen, who had been sitting in on one of Holland's courses, came up after class one day to introduce himself, and said, "You really ought to be talking to Bob Axelrod." Holland did, and through Axelrod soon met Hamilton. The BACH group—Burks, Axelrod, Cohen, Hamilton, and Holland—coalesced almost immediately. (They almost had to work in a "K"; very early in the group's existence they tried to recruit Stuart Kauffman, but lost out to the University of Pennsylvania.) "What tied us together was that we all had a very strong mathematical background," says Holland. "We also felt very strongly that the issues were wider than any one problem. We began meeting on a regular basis: someone would see a paper, and we'd all come in and discuss it. There was a lot of exploratory thinking."

Indeed there was—particularly on Holland's part. The book was done now, but his conversations with the BACH group only underscored what it had left undone. The genetic algorithm and the schema theorem had captured something essential and right about evolution; he was still convinced of that. But even so, he couldn't help but feel that the genetic algorithm's bare-bones version of evolution was just *too* bare. Something had to be missing in a theory in which "organisms" are just naked pieces of DNA that have been designed by a programmer. What could a theory like that tell you about complex organisms evolving in a complex environment? Nothing. The genetic algorithm was all very nice. But by itself, it simply wasn't an adaptive agent.

Nor, for that matter, was the genetic algorithm a model of adaptation in the human mind. Because it was so explicitly biological in its design, it couldn't tell you anything about how complex concepts grow, evolve, and recombine in the mind. And for Holland, that fact was becoming more and more frustrating. Nearly twenty-five years after he'd first heard about Donald Hebb's ideas, he was still convinced that adaptation in the mind and adaptation in nature were just two different aspects of the same thing. Moreover, he was still convinced that if they really were the same thing, they ought to be describable by the same theory.

So in the latter half of the 1970s, Holland set out to find that theory.

Back to basics. An adaptive agent is constantly playing a game with its environment. What exactly does that mean? Distilled to the essence, what actually has to happen for game-playing agents to survive and prosper?

Two things, Holland decided: prediction and feedback. It was an insight that he could trace all the way back to his IBM days and his conversations with Art Samuel about the checker player.

Prediction is just what it sounds like: thinking ahead. He can remember Samuel making the point again and again. "The very essence of playing a good game of checkers or chess is assigning value to the less-than-obvious stage-setting moves," says Holland—the moves that will put you in an advantageous position later on. Prediction is what helps you seize an opportunity or avoid getting suckered into a trap. An agent that can think ahead has an obvious advantage over one that can't.

But the concept of prediction also turns out to be at least as subtle as the concept of building blocks, says Holland. Ordinarily, for example, we think of prediction as being something that humans do consciously, based on some explicit model of the world. And there are certainly plenty of those explicit models around. A supercomputer's simulation of climate change is one example. A start-up company's business plan is another, as is an economic projection made by the Federal Reserve Board. Even Stonehenge is a model: its circular arrangement of stones provided the Druid priests with a rough but effective computer for predicting the arrival of the equinoxes. Very often, moreover, the models are literally inside our head, as when a shopper tries to imagine how a new couch might look in the living room, or when a timid employee tries to imagine the consequences of telling off his boss. We use these "mental models" so often, in fact, that many psychologists are convinced they are the basis of all conscious thought.

But to Holland, the concept of prediction and models actually ran far deeper than conscious thought—or for that matter, far deeper than the existence of a brain. "All complex, adaptive systems—economies, minds, organisms—build models that allow them to anticipate the world," he declares. Yes, even bacteria. As it turns out, says Holland, many bacteria have special enzyme systems that cause them to swim toward stronger concentrations of glucose. Implicitly, those enzymes model a crucial aspect of the bacterium's world: that chemicals diffuse outward from their source, growing less and less concentrated with distance. And the enzymes simultaneously encode an implicit prediction: If you swim toward higher concentrations, then you're likely to find something nutritious. "It's not a conscious model or anything of that sort," says Holland. "But it gives that organism an advantage over one that doesn't follow the gradient."

A similar story can be told about the viceroy butterfly, he says. The viceroy is a striking, orange-and-black insect that is apparently quite succulent to birds—if only they would eat it. But they rarely do, because the viceroy has evolved a wing pattern that closely resembles that of the vile-tasting monarch butterfly, which every young bird quickly learns to avoid. So in effect, said Holland, the DNA of the viceroy encodes a model of the world stating that birds exist, that the monarch exists, and that the monarch tastes horrible. And every day, the viceroy flutters from flower to flower, implicitly betting its life on the assumption that its model is correct.

The same story can be told yet again about a very different kind of organism, says Holland: the corporation. Imagine that a manufacturer receives a routine order for, say, 10,000 widgets. Since it's a routine order, the employees probably don't give any profound thought to the matter. Instead, they just set up the production run by invoking a "standard operating procedure"—a set of rules of the form, "If the situation is ABC, then take action XYZ." And just as with a bacterium or the viceroy, says Holland, those rules encode a model of the company's world and a prediction: "If the situation is ABC, then action XYZ *is a worthwhile thing to do and will lead to good results.*" The employees involved in carrying out the procedure may or may not know what that model is. After all, standard operating procedures are often taught by rote, without a lot of whys and wherefores. And if the company has been around for a while, there may not be anyone left who even remembers why things are done a certain way. Nonetheless, as the standard operating procedure collectively unfolds, the company as a whole will behave as if it understood that model perfectly.

In the cognitive realm, says Holland, anything we call a "skill" or "expertise" is an implicit model—or more precisely, a huge, interlocking set of standard operating procedures that have been inscribed on the nervous system and refined by years of experience. Show a textbook exercise to an experienced physics teacher and he won't waste any time scribbling every formula in sight, the way a novice will; his mental procedures will almost always show him a path to the solution instantly: "Aha! That's a conservation of energy problem." Lob a tennis ball across the net to Chris Evert and she won't spend any time debating how to respond: after years of experience and practice and coaching, her mental procedures will allow her to slam the ball back down your throat instinctively.

Holland's favorite example of implicit expertise is the skill of the medieval architects who created the great Gothic cathedrals. They had no way to calculate forces or load tolerances, or anything else that a modern architect might do. Modern physics and structural analysis didn't exist in the twelfth century. Instead, they built those high, vaulted ceilings and massive flying buttresses using standard operating procedures passed down from master to apprentice—rules of thumb that gave them a sense of which structures would stand up and which would collapse. Their model of physics was completely implicit and intuitive. And yet, these medieval craftsmen were able to create structures that are still standing nearly a thousand years later.

The examples could go on and on, says Holland. DNA itself is an implicit model: "Under the conditions we expect to find," say the genes, "the creature we specify has a chance of doing well." Human culture is an implicit model, a rich complex of myths and symbols that implicitly define a people's beliefs about their world and their rules for correct behavior. For that matter, Samuel's checker player contained an implicit model, which it created by changing the numerical value it assigned to various options as it gained experience with the opponent's playing style.

Indeed, models and predictions are everywhere, says Holland. But then, where do the models come from? How can *any* system, natural or artificial, learn enough about its universe to forecast future events? It doesn't do any good to talk about "consciousness," he says. Most models are quite obviously not conscious: witness the nutrient-seeking bacterium, which doesn't even have a brain. And in any case it simply begs the question. Where does the consciousness come from? Who programs the programmer?

Ultimately, says Holland, the answer has to be "no one." Because if there *is* a programmer lurking in the background—"the ghost in the machine"—then you haven't really explained anything. You've only pushed the mystery off someplace else. But fortunately, he says, there is an alternative: feedback from the environment. This was Darwin's great insight, that an agent can improve its internal models without any paranormal guidance whatsoever. It simply has to try the models out, see how well their predictions work in the real world, and—if it survives the experience—adjust the models to do better the next time. In biology, of course, the agents are individual organisms, the feedback is provided by natural selection, and the steady improvement of the models is called evolution. But in cognition, the process is essentially the same: the agents are individual minds, the feedback comes from teachers and direct experience, and the improvement is called learning. Indeed, that's exactly how it had worked in Samuel's checker player. Either way, says Holland, an adaptive agent has to be able to take advantage of what its world is trying to tell it.

The next question, of course, was how? Holland discussed the basic concept at length with his colleagues in the BACH group. But in the end, there was only one way to pin the ideas down: he would have to build a computer-simulated adaptive agent, just as he had done fifteen years earlier with genetic algorithms.

Unfortunately, he found that mainstream artificial intelligence was no more helpful in 1977 than it had been in 1962. The field had admittedly made some impressive progress in that time. Out at Stanford, for example, the artificial intelligence group was creating a series of startlingly effective programs known as expert systems, which modeled the expertise of, say, a doctor, by applying hundreds of rules: "If the patient has bacterial meningitis, and has been seriously burned, then the organism causing the infection may be *Pseudomonas aeruginosa*." Even then, the venture capitalists were starting to sit up and take notice.

But Holland wasn't interested in applications. What he wanted was a fundamental theory of adaptive agents. And so far as he could see, the past two decades of progress in artificial intelligence had been achieved at the price of leaving out almost everything important, starting with learning and feedback from the environment. To him, feedback was *the* fundamental issue. And yet, with a few exceptions such as Samuel, people in the field seemed to believe that learning was something that could be set aside for later, after they had gotten their programs working well with things like language understanding, or problem-solving, or some other form of abstract reasoning. The expert systems designers even seemed to take a certain macho pride in that fact. They talked about something called "knowledge engineering," in which they would create the hundreds of rules needed for a new expert system by sitting down with the relevant experts for months: "What would you do in *this* situation? What would you do in *that* situation?"

In fairness, even the knowledge engineers had to admit that things would go a lot more smoothly if the programs could only learn their expertise from teaching and experience, as people do—and if someone could only figure out how to implement learning without making the software far more complex and cumbersome than it already was. But to Holland, that was precisely the point. Rigging the software with some ad hoc "learning module" wasn't going to solve anything. Learning was as fundamental to cognition as evolution was to biology. And that meant that learning had to be built into the cognitive architecture from the beginning, not slapped on at the end. Holland's ideal was still the Hebbian neural network, where the neural impulses from every thought strengthen and reinforce the connections that make thinking possible in the first place. Thinking and learning were just two aspects of the same thing in the brain, Holland was convinced. And he wanted to capture that fundamental insight in his adaptive agent.

For all of that, however, Holland wasn't about to go back to doing neural network simulations. Even a quarter-century after the IBM 701, computers were still not powerful enough to do a full-fledged Hebbian simulation on the scale he wanted. True, neural networks had enjoyed a brief flurry of fame in the 1960s under the rubric of "perceptrons"—neural networks specialized to recognize features in a visual field. But perceptrons were highly, highly simplified versions of what Hebb had actually been talking about, and couldn't produce anything resembling a resonating cell assembly. (They also weren't very good at recognizing visual features, which is why they had fallen out of favor.) Nor was Holland much more impressed with the newer generation of neural networks, which were just coming back into fashion in the late 1970s and which have gotten a lot of attention in the years since then. These networks are somewhat more sophisticated than perceptrons, says Holland. But they still couldn't support cell assemblies. Indeed, most versions have no resonance at all; the signals cascade through the network in one direction only, front to back. "These connectionist networks are very good at stimulus-response behavior and pattern recognition," he says. "But by and large they ignore the need for internal feedback, which is what Hebb argued you needed for cell assemblies. And with few exceptions, they don't do much with internal models."

The upshot was that Holland decided to design his simulated adaptive agent as a hybrid, taking the best of both worlds. For computational efficiency he would go ahead and use the kind of if-then rules made famous by expert systems. But he would use them in the spirit of neural networks.

Actually, says Holland, there was a lot to like about if-then rules in any case. In the late 1960s, long before anyone had even heard of an expert system, rule-based systems had been introduced by Carnegie-Mellon's Allen Newell and Herbert Simon as a general-purpose computer model of human cognition. Newell and Simon saw each rule as corresponding to a single packet of knowledge or a single component of skill: "If Tweety is a bird, then Tweety has wings," for example, or "If there's a choice between taking your opponent's pawn and his queen, then take the queen." Moreover, they pointed out that when a program's knowledge is expressed in this way, it automatically acquires some of the wonderful flexibility of cognition. The condition-action structure of the rules—"If *this* is the case, then do *that*"—means that they don't execute in a fixed sequence like some subroutine written in FORTRAN or PASCAL. A given rule comes to life only when its conditions are met, so that its response is appropriate to the situation. Indeed, once a rule is activated it will very likely trigger a whole sequence of rules: "If A then B," "If B then C," "If C then D," and so on—in effect, a whole new program created on the fly and tailored to the problem at hand. And that, not the blind, rigid behavior of a wind-up toy, is exactly what you want from intelligence.

Furthermore, says Holland, rule-based systems turn out to make a lot of sense in terms of the neural architecture of the brain. A rule, for example, is just the computer equivalent of one of Hebb's resonating cell assemblies. "In Hebb's view," he says, "a cell assembly makes a simple statement: If such and such an event occurs, then I will fire for a while at a high rate." The interactions of the rules, with the activation of one rule setting off a whole cascade of others, are likewise a natural result of the dense inter-connectedness of the brain. "Each of Hebb's cell assemblies involves about one thousand to ten thousand neurons," says Holland. "And each of those neurons has about one thousand to ten thousand synapses connecting it to other neurons. So each cell assembly contacts *a lot* of other cell assemblies." In effect, he says, activating one cell assembly will post a message on a kind of internal bulletin board, where it can be seen by most or all of the other assemblies in the brain: "Cell Assembly 295834108 now active!" And when that message appears, those assemblies that are properly connected to the first one will also fire and post their own messages, causing the cycle to repeat again and again.

The internal architecture of a Newell-Simon type rule-based system actually follows this bulletin-board metaphor quite closely, says Holland. There is an internal data structure that corresponds to the bulletin board and contains a series of digital messages. And then there is a large population of rules, bits of computer code that number in the hundreds or even thousands. When the system is in operation, each of the rules constantly scans the bulletin board for the presence of a message that matches its "if" part. And whenever one of them finds such a message, it immediately posts a new digital message specified by its "then" part.

"Think of the system as a kind of office," says Holland. "The bulletin board contains the memos that are to be processed that day, and each rule corresponds to a desk in that office that has responsibility for memos of a given kind. At the beginning of the day, each desk collects the memos for which it is responsible. And at the end of the day, each desk posts the memos that result from its activities." In the morning, of course, the cycle repeats. In addition, he says, some of the memos may be posted by *detectors*, which keep the system up to date about what's going on in the outside world. And still other memos may activate *effectors*, subroutines that allow the system to affect the outside world. Detectors and effectors are the computer analog of eyes and muscles, says Holland. So, in principle, a rule-based system can easily get feedback from its environment—one of his prime requirements.

Holland accordingly used this same bulletin-board metaphor in the design of his own adaptive agent. Having done that, however, he went right back to being an iconoclast when it came to the details.

In the standard Newell-Simon approach, for example, both the rules and the memos on the bulletin board were supposed to be written in terms of symbols such as "Bird" or "Yellow," which were intended to be the analog of concepts in the human mind. And for most people in artificial intelligence research, this use of symbols to represent concepts was utterly noncontroversial. It had been standard doctrine in the field for decades—with Newell and Simon being among its most articulate champions. Moreover, it did seem to capture much of what actually goes on in our heads. Symbols in the computer could be linked into elaborate data structures to represent a complex situation, just as concepts are linked and merged to form the psychologists' mental models. And these data structures, in turn, could be manipulated by the program to emulate mental activities such as reasoning and problem-solving, just as mental models are remolded and changed by the mind during thinking. Indeed, if you took the Newell-Simon view literally, as many researchers did, this kind of symbol-processing *was* thinking.

Yet Holland just couldn't buy it. "Symbol-processing was a very good place to start," he says. "And it was a real advance in terms of understanding conscious thought processes." But symbols by themselves were far too rigid, and they left out far too much. How could a data register containing the characters *B-I-R-D* really capture all the subtle, shifting nuances of that concept? How could those characters really *mean* anything to the program if it had no way to interact with real birds in the outside world? And even leaving that issue aside, where do such symbolic concepts come from in the first place? How do they evolve and grow? How are they molded by feedback from the environment?

To Holland, it was all of a piece with the mainstream's lack of interest in learning. "You run into the same difficulties you do by classifying species without understanding how they evolved," says Holland. "You can learn a lot that way about comparative anatomy and such. But in the end, it just doesn't go far enough." He was still convinced that concepts had to be understood in Hebbian terms, as emergent structures growing from some deeper neural substrate that is constantly being adjusted and readjusted by input from the environment. Like clouds emerging from the physics and chemistry of water vapor, concepts are fuzzy, shifting, dynamic things. They are constantly recombining and changing shape. "The most crucial thing we've got to get at in understanding complex adaptive systems is how levels emerge," he says. "If you ignore the laws at the next level below, you'll never be able to understand this one."

To capture that sense of emergence in his adaptive agent, Holland decided that his rules and messages would *not* be written in terms of meaningful symbols. They would be arbitrary strings of binary 1's and 0's. A message might be a sequence such as 10010100, much like a chromosome in his genetic algorithm. And a rule, as paraphrased in English, might be something like, "If there is a message on the bulletin board with the pattern 1###0#00, where # stands for 'don't care,' then post the message 01110101."

This representation was so offbeat that Holland even took to calling his rules by a new name, "classifiers," because of the way their if-conditions classified different messages according to specific patterns of bits. But he considered this abstract representation essential, if only because he'd seen too many artificial intelligence researchers fool themselves about what their symbol-based programs "knew." In his classifier systems, the meaning of a message would have to emerge from the way it caused one classifier rule to trigger another, or from the fact that some of its bits were written directly by sensors looking at the real world. Concepts and mental models would likewise have to emerge as self-supporting clusters of classifiers, which would presumably organize and reorganize themselves in much the same way as autocatalytic sets.

Meanwhile, Holland was also taking exception to the standard ideas about centralized control in a rule-based system. According to the conventional wisdom, rule-based systems were *so* flexible that some form of centralized control was needed to prevent anarchy. With hundreds or thousands of rules watching a bulletin board crammed with messages, there was always the chance that several rules would suddenly hop up and start arguing over who got to post the next message. The presumption was that they couldn't all do so, because their messages might be utterly inconsistent. ("Take the queen." "Take the pawn.") Or their messages might lead to entirely different cascades of rules, and thus to entirely different behavior of the system as a whole. So to prevent the computer equivalent of schizophrenia, most systems implemented elaborate "conflict resolution" strategies to make sure that only one rule could be active at a time.

Holland, however, saw such top-down conflict resolution as precisely the wrong way to go. Is the world such a simple and predictable place that you always know the best rule in advance? Hardly. And if the system *has* been told what to do in advance, then it's a fraud to call the thing artificial intelligence: the intelligence isn't in the program but in the programmer. No, Holland wanted control to be *learned*. He wanted to see it emerging from the bottom up, just as it did from the neural substrate of the brain. Consistency be damned: if two of his classifier rules disagreed with one another, then let them fight it out on the basis of their performance, their proven contribution to the task at hand—*not* some preprogrammed choice made by a software designer.

"In contrast to mainstream artificial intelligence, I see competition as much more essential than consistency," he says. Consistency is a chimera, because in a complicated world there is no guarantee that experience will be consistent. But for agents playing a game against their environment, competition is forever. "Besides," says Holland, "despite all the work in economics and biology, we still haven't extracted what's central in competition." There's a richness there that we've only just begun to fathom. Consider the magical fact that competition can produce a very strong incentive for cooperation, as certain players spontaneously forge alliances and symbiotic relationships with each other for mutual support. It happens at every level and in every kind of complex, adaptive system, from biology to economics to politics. "Competition and cooperation may seem antithetical," he says, "but at some very deep level, they are two sides of the same coin."

To implement this competition, Holland decided to make the posting of messages into a kind of auction. His basic idea was to think of the classifiers not as computer commands but as hypotheses, conjectures about the best messages to post in any given situation. By assigning each hypothesis a numerical value measuring its plausibility, or strength, he then had a basis for bidding. In Holland's version of message posting, each cycle started just as before, with all the classifiers scanning the bulletin board in search of a match. And just as before, those that found a match would stand up and get ready to post their own messages. But instead of posting them immediately, each one would first shout out a bid proportional to its strength. A classifier as solidly grounded in experience as "The sun will rise in the east tomorrow morning" might bid 1000, while a classifier as well-grounded as "Elvis is alive and appearing nightly at the Walla Walla Motel 6" might bid only 1. The system would then collect all the bids and choose a set of winners by lottery, with the highest probability of winning going to the highest bidders. The chosen classifiers would post their messages, and the cycle would repeat.

Complex? Holland couldn't deny it. As things stood, moreover, the auction simply replaced arbitrary conflict resolution strategies with arbitrary plausibility values. But assuming for the moment that the system could somehow learn these plausibility values from experience, then the auction would eliminate the central arbiter and give Holland exactly what he wanted. Not every classifier could win: the bulletin board was big, not infinite. Nor would the race always go to the swift: even Elvis might get a chance to post his message if he got a lucky break. But on the average, control over the system's behavior would automatically be given to the strongest and most plausible hypotheses, with off-the-wall hypotheses appearing just often enough to give the system a little spontaneity. And if some of those hypotheses were inconsistent, well, that shouldn't be a crisis but an opportunity, a chance for the system to learn from experience which ones are *more* plausible.

So once again, it all came back to learning: How were the classifiers supposed to prove their worth and earn their plausibility values?

To Holland, the obvious answer was to implement a kind of Hebbian reinforcement. Whenever the agent does something right and gets a positive feedback from the environment, it should strengthen the classifiers responsible. Whenever it does something wrong, it should likewise weaken the classifiers responsible. And either way, it should ignore the classifiers that were irrelevant.

The trick, of course, was to figure out which classifiers were which. The agent couldn't just reward the classifiers that happen to be active at the moment of payoff. That would be like giving all the credit for a touchdown to the player who happened to carry the ball across the goal line—and none to the quarterback who called the play and passed him the ball, or to the linemen who blocked the other team and opened up a gap for him to run through, or to anyone who carried the ball in previous plays. It would be like giving all the credit for a victory in chess to the final move that trapped your opponent's king, and none to the crucial gambit many moves before that set up your whole endgame. And yet, what was the alternative? If the agent had to anticipate the payoff in order to reward the correct classifiers, how was it supposed to do so without being preprogrammed? How was it supposed to learn the value of these stage-setting moves without knowing about them already?

Good questions. Unfortunately, the general idea of Hebbian reinforcement was too broad-brush to provide any answers. Holland was at a loss—until one day he happened to think back on the basic economics course he'd taken at MIT from Paul Samuelson, author of the famous economics textbook, and realized that he'd almost solved the problem already. By auctioning off space on the bulletin board, he had created a kind of marketplace within the system. By allowing the classifiers to bid on the basis of their strength, he had created a currency. So why not take the next step? Why not create a full-fledged free-market economy, and allow the reinforcement to take place through the profit motive?

Why not, indeed? The analogy was obvious when you finally saw it. If you thought of the messages posted on the bulletin board as being goods and services up for sale, Holland realized, then you could think of the classifiers as being firms that produce those goods and services. And when a classifier sees a message satisfying its if-conditions and makes a bid, then you could think of it as a firm trying to purchase the supplies it needs to make its product. All he had to do to make the analogy perfect was to arrange for each classifier to pay for the supplies it used. When a classifier won the right to post its message, he decided, it would transfer some of its strength to its suppliers: namely, the classifiers responsible for posting the messages that triggered it. In the process, the classifier would then be weakened. But it would have a chance to recoup its strength and even make a profit during the next round of bidding, when its own message went on the market.

And where would the wealth ultimately come from? From the final consumer, of course: the environment, the source of all payoffs to the system. Except that now, Holland realized, it would be perfectly all right to reward the classifiers that happen to be active at the moment of payoff. Since each classifier pays its suppliers, the marketplace will see to it that the rewards propagate through the whole collection of classifiers and produce exactly the kind of automatic reward and punishment he was looking for. "If you produce the right intermediate product, then you'll make a profit," he says. "If not, then nobody will buy it and you'll go bankrupt." All the classifiers that lead to effective action will be strengthened, and yet none of the stage-setting classifiers will be neglected. Over time, in fact, as the system gains experience and gets feedback from the environment, the strength of each classifier will come to match its true value to the agent.

Holland dubbed this portion of his adaptive agent the "bucket-brigade" algorithm because of the way it passed reward from each classifier to the previous classifier. It was directly analogous to the strengthening of synapses in Hebb's theory of the brain—or, for that matter, to the kind of reinforcement used to train a simulated neural network in a computer. And when he had it, Holland knew he was almost home. Economic reinforcement via the profit motive was an enormously powerful organizing force, in much the same way that Adam Smith's Invisible Hand was enormously powerful in the real economy. In principle, Holland realized, you could start the system off with a set of totally random classifiers, so that the agent just thrashed around like the software equivalent of a newborn baby. And then, as the environment reinforced certain behaviors and as the bucket brigade did its work, you could watch the classifiers organize themselves into coherent sequences that would produce at least a semblance of the desired behavior. Learning, in short, would be built into the system from the beginning.

So, Holland was almost home—but not quite. By constructing the bucket-brigade algorithm on top of the basic rule-based system, Holland had given his adaptive agent one form of learning. But there was another form still missing. It was the difference between *exploitation* and *exploration*. The bucket-brigade algorithm could strengthen the classifiers that the agent already possessed. It could hone the skills that were already there. It could consolidate the gains that had already been made. But it couldn't create anything new. By itself, it could only lead the system into highly optimized mediocrity. It had no way to explore the immense space of possible new classifiers.

This, Holland decided, was a job for the genetic algorithm. When you thought about it, in fact, the Darwinian metaphor and the Adam Smith metaphor fit together quite nicely: Firms evolve over time, so why shouldn't classifiers?

Holland certainly wasn't surprised by this insight; he'd had the genetic algorithm in the back of his mind all along. He'd been thinking about it when he first set up the binary representation of classifiers. A classifier might be paraphrased in English as something like, "If there are two messages with the patterns 1###0#00 and 0#00####, then post the message 01110101." In the computer, however, its various parts would be concatenated together and written simply as a string of bits: "1###0#000#00####01110101." And to the genetic algorithm, that looked just like a digital chromosome. So the algorithm could be carried out in exactly the same way. Most of the time, the classifiers would merrily buy and sell in their digital marketplace as before. But every so often, the system would select a pair of the strongest classifiers for reproduction. These classifiers would reshuffle their digital building blocks by sexual exchange to produce a pair of offspring. The offspring would replace a pair of weak classifiers. And then the offspring would have a chance to prove their worth and grow stronger through the bucket-brigade algorithm.

The upshot was that the population of rules would change and evolve over time, constantly exploring new regions of the space of possibilities. And there you would have it: by adding the genetic algorithm as a third layer on top of the bucket brigade and the basic rule-based system, Holland could make an adaptive agent that not only learned from experience but could be spontaneous and creative.

And all he had to do was to turn it into a working program.

Holland started coding the first classifier system around 1977. And oddly enough, it didn't turn out to be as straightforward a job as he had hoped. "I really thought that in a couple of months I'd have something up and running that was useful to me," he says. "Actually, it was the better part of a year before I was fully satisfied."

On the other hand, he didn't exactly make things easy for himself. He coded that first classifier system in true Holland style: by himself. At home. In hexadecimal code, the same kind that he'd written for the Whirlwind thirty years earlier. On a Commodore home computer.

Holland's BACH colleagues still roll their eyes when they tell this story. The whole campus was crawling with computers: VAXs, mainframes, even high-powered graphics workstations. Why a *Commodore?* And hex! Almost nobody wrote in hex anymore. If you were really a hard-core computer jock trying to squeeze the last ounce of performance out of a machine, you might write in something called assembly language, which at least replaced the numbers with mnemonics like MOV, JMZ, and SUB. Otherwise, you went with a high-level language such as PASCAL, C, FORTRAN, or LISP—something that a human being could hope to understand. Cohen, in particular, remembers arguing long and hard with Holland: Who's going to believe that this thing works if it's written in alphanumeric gibberish? And even if anybody does believe you, who's going to use a classifier system if it only runs on a home computer?

Holland eventually had to concede the point—although it was well into the early 1980s before he agreed to hand over the classifier system code to a graduate student, Rick Riolo, who transformed it into a general-purpose package that would run on almost any type of computer. "This is just not my instinct," Holland admits. "My tendency is to do pieces of something until I see it can really be implemented. Then I tend to lose interest and go back to theory."

Be that as it may, he still maintains that the Commodore made a lot of sense at the time. The campus computers had to be shared, he explains, and that made them a pain: "I wanted to fuss with the program on-line, and nobody was likely to give me eight hours at a stretch." Holland saw the personal computer revolution as a godsend. "I realized that I could do my programming on *my* machine, that I could have it in my own home and be beholden to nobody."

Besides, having come of age programming Whirlwind and the IBM 701, Holland didn't find these little desktop machines primitive at all. When he finally got the Commodore, in fact, he considered it quite a step up. He had actually taken his first plunge into personal computing with something called the Micromind, which he bought in 1977 when it looked like a serious rival to the brand-new Apple II. "It was a very nice machine," he recalls. True, it wasn't much more than a bunch of circuit boards in a black box that could be hooked up to a teletype machine for input and output. It had no screen. But it did have 8 kilobytes of 8-bit memory. And it cost only $3000.

And as for hex—well, the Micromind didn't have any other programming languages available at that point, and Holland was not about to wait. "I was used to writing in assembler," he says, "and I could do hex almost as easily as I could assembler, so it wasn't hard."

All told, Holland says, it's really too bad that the Micromind company went bankrupt so quickly. He moved to the Commodore only when he began to feel the constraints of that tiny 8 kilobyte memory. At the time, it was the ideal choice, he says. It used the same microprocessor chip as the Micromind, which meant that it could run his hexadecimal code virtually unchanged. It had much more memory. It had a *screen*. And best of all, he says, "The Commodore would let me play games."

His colleagues' exasperation aside, Holland's first classifier system ran well enough to convince him that it really would work the way he intended it to—and, not incidentally, that it really did hold the seeds of a full-fledged theory of cognition. In tests of an early version of the system, which he published in 1978 in collaboration with Michigan psychology professor Judy Reitman, their agent learned how to run a simulated maze about ten times faster when it used the genetic algorithm than it did without the algorithm. The same tests also proved that a classifier system could exhibit what psychologists call *transfer*: it could apply rules learned in one maze to run other mazes later on.

These early results were impressive enough that word of classifier systems began to spread, even without Holland pushing them. In 1980, for example, Stephen Smith at the University of Pittsburgh built a classifier system that could play poker, and pitted it against an older poker-playing program that was also able to learn. It wasn't even a contest; the classifier system won in a walk. In 1982, Stewart Wilson of the Polaroid Corporation used a classifier system to coordinate the motion of a TV camera and a mechanical arm. He showed that the bucket-brigade and genetic algorithms caused a spontaneous organization of the classifier rules, so that they segregated themselves into groups that could function as control subroutines and produce specific, coordinated actions as needed. Also in 1982, Holland's student Lashon Booker completed a Ph.D. thesis in which he placed a classifier system in a simulated environment where it had to find "food" and avoid "poison." The system soon organized its rules into an internal model of that environment—in effect, a mental map.

For Holland, however, the most gratifying demonstration was the one produced in 1983 by David Goldberg, a Ph.D.-bound civil engineer who had enrolled in Holland's adaptive systems course several years before and had become a true believer. Persuading Holland to cochair his dissertation committee, Goldberg wrote a thesis demonstrating how genetic algorithms and classifier systems could be used to control a simulated gas pipeline. At the time it was by far the most complex problem that a classifier system had ever been presented with. The objective in any pipeline system is to meet demand at the end of the pipeline as economically as possible. But a pipeline consists of dozens or hundreds of compressors pumping gas through thousands of miles of large-diameter pipe. The customers' demand for gas changes on an hourly and seasonal basis. Compressors and pipes spring leaks, compromising the system's ability to deliver gas at the appropriate pressure. Safety constraints demand that pressure and flow rates have to be kept within proper bounds. And everything affects everything else. Optimizing even a very simple pipeline is far beyond the reach of mathematical analysis. Pipeline operators learn their craft through long apprenticeships—and then "drive" their system by instinct and feel, the way the rest of us drive the family car.

The pipeline problem seemed so intractable, in fact, that Holland fretted that Goldberg might have bitten off more than the classifier systems could chew. He needn't have worried. Goldberg's system learned to operate his simulated pipeline beautifully: starting from a set of totally random classifiers, it achieved expert-level performance in about 1000 days of simulated experience. Moreover, the system was incredibly simple for what it did. Its messages were only 16 binary digits long, its bulletin board held only 5 messages at a time, and it contained only 60 classifier rules, total. In fact, Goldberg ran the whole classifier system, plus the pipeline simulation, on his Apple II computer at home with just 64 kilobytes of memory. "He's a guy after my own heart," laughs Holland.

The pipeline simulation not only earned Goldberg a Ph.D. in 1983, it won him a Presidential Young Investigator Award in 1985. Holland himself considers his work a milestone for classifier systems. "It was very convincing," he says. "It really worked on a real problem—or, at least, a simulation of one." In a delicious irony, moreover, this most "practical" of the classifier systems devised until that time also turned out to have the most to say about basic cognitive theory.

You could see it most clearly in the way Goldberg's system organized its knowledge about leaks, says Holland. Starting from a random set of classifiers, it would first learn a series of broadly applicable rules that worked quite well for normal pipeline operations. An example that actually appeared in one run was a rule that could be paraphrased as "Always send a 'No leak' message." Clearly, this was an overgeneral rule that worked *only* if the pipeline was normal. But the system discovered that fact soon enough when Goldberg started punching simulated holes in various simulated compressors. Its performance immediately declined drastically. However, by means of the genetic algorithm and the bucket brigade, the system eventually recovered from its errors and started producing more specific rules such as "If the input pressure is low, the output pressure is low, and the rate of change of pressure is very negative, then send the 'Leak' message." Whenever this rule applied, moreover, it would give a much stronger bid than the first rule and knock it right off the bulletin board. So, in effect, the first rule governed the default behavior of the system under normal conditions, while the second rule and others like it would kick in to give the correct behavior under exceptional conditions.

Holland was tremendously excited when Goldberg told him about this. In psychology this kind of knowledge organization is known as a default hierarchy, and it happened to be a subject that was very much on Holland's mind at the time. Since 1980, he had been involved in an intense collaboration with three colleagues at Michigan—psychologists Keith Holyoak and Richard Nisbett and philosopher Paul Thagard—to build a general cognitive theory of learning, reasoning, and intellectual discovery. As they later recounted in their 1986 book, *Induction*, all four of them had independently come to believe that such a theory had to be founded on the three basic principles that happened to be the same three that underlay Holland's classifier system: namely, that knowledge can be expressed in terms of mental structures that behave very much like rules; that these rules are in competition, so that experience causes useful rules to grow stronger and unhelpful rules to grow weaker; and that plausible new rules are generated from combinations of old rules. Their argument, which they backed up with extensive observations and experiments, was that these principles could account for a wide variety of "Aha!" type insights, ranging from Newton's experience with the apple to such everyday abilities as understanding an analogy.

In particular, they argued that these three principles ought to cause the spontaneous emergence of default hierarchies as the basic organizational structure of all human knowledge—as indeed they appear to do. The cluster of rules forming a default hierarchy is essentially synonymous with what Holland calls an internal model. We use weak general rules with stronger exceptions to make predictions about how things should be assigned to categories: "If it's streamlined and has fins and lives in the water, then it's a fish"—but "If it also has hair and breathes air and is *big*, then it's a whale." We use the same structure to make predictions about how things should be done: "It's always 'i' before 'e' except after 'c' "—but "If it's a word like *neighbor, weigh,* or *weird,* then it's 'e' before 'i.' " And we use the same structure again to make predictions about causality: "If you whistle to a dog, then it will come to you"—but "If the dog is growling and raising its hackles, then it probably won't come."

The theory says that these default-hierarchy models ought to emerge whether the principles are implemented as a classifier system or in some other way, says Holland. (In fact, many of the computer simulations quoted in *Induction* were done with PI, a somewhat more conventional rule-based program devised by Thagard and Holyoak.) Nonetheless, he says, it was thrilling to see the hierarchies actually emerge in Goldberg's pipeline simulation. The classifier system had started with *nothing*. Its initial set of rules had been totally random, the computer equivalent of primordial chaos. And yet, here was this marvelous structure emerging out of the chaos to astonish and surprise them.

"We were elated," says Holland. "It was the first case of what someone could really call an emergent model."

A Place to Come Home To

The kitchen-table conversation between Holland and Arthur wandered on and on for hours, as kitchen-table conversations are wont to do. By the time they finally called it a night, their discussion had gone from chess to economics, from economics to checkers, and then on to internal models, genetic algorithms, and chess again. Arthur felt he was finally beginning to understand the full implications of learning and adaptation. And the two of them had rather sleepily begun to bat around an approach that might just crack this problem of rational expectations in economics: instead of assuming that your economic agents are perfectly rational, why not just model a bunch of them with Holland-style classifier systems and let them learn from experience like *real* economic agents?

Why not indeed? Before he turned in, Holland made a note to dig out an old set of overhead transparencies on Samuel's checker player that he happened to have with him. Arthur had been enchanted with the idea of a game-playing program that learned; he'd never heard of such a thing. Holland thought he might give the meeting participants an impromptu talk on the subject the next day.

The talk was a hit—especially when Holland pointed out to his audience that Samuel's program was still pretty much the state of the art in checkers playing some thirty years later. But, then, Holland's whole approach had been a hit at the meeting. Nor were such impromptu interchanges at all unusual by that point. Participants find it hard to pinpoint exactly when the mood of the economics meeting began to change. But somewhere about the third day, after they had cleared away the early barriers of jargon and mutual confusion, the meeting began to catch fire.

"I found it very exciting," says Stuart Kauffman, who felt primed for economics after two weeks of talking to Arthur. "In a funny way it was like kindergarten, when you get exposed to all sorts of new things like finger painting. Or it was like being a puppy, running around sniffing at things, with this wonderful sense of discovery, that the whole world was this wondrous place to explore. Everything was new. And somehow that's what this meeting was like to me. Wondering how these other guys think. What are the criteria? What are the questions in this new field? That's very much my style, personally. But I think it had that flavor for a fair number of people. We went on talking at one another long enough to *hear* one another."

Ironically, considering the physicists' early skepticism about mathematical abstractions, it was mathematics that provided the common language. "As I look back on it, I think Ken made the right decision," says Eugenia Singer, who had originally been disappointed at Arrow's failure to include sociologists and psychologists in the group. "He had the most highly, technically trained economists he could get. And as a result, there was a credibility that was built. The physical scientists were amazed at their technical background. They were familiar with a lot of the technical concepts, even some of the physical models. So they were able to start using common terms and building a language they could talk to each other in. But if they had gotten a lot of social scientists in there with no technical background, I'm not sure the gulf could have been crossed.

After most of the formal presentations were finished, the participants at the workshop started breaking up into informal working groups to focus on particular subjects. One of the most popular topics was chaos, the domain of one group that frequently gathered around David Ruelle in the small conference room. "All of us knew about chaos and had read articles," says Arthur. "Some of the economists had done considerable research in that area. But I remember that there was an awful lot of excitement in seeing some of the physicists' models."

Anderson and Arthur, meanwhile, were part of a group that met out on the terrace to discuss economic "patterns" such as technological lock-in or regional economic differences. "I was almost too tired to do an awful lot of talking or listening," says Arthur. "I used the working group to quiz Phil Anderson on various mathematical techniques."

Arthur actually found himself feeling very much in tune with Anderson and the other physicists. "I liked their emphasis on computer experiments," he says. Among economists, computer models had gotten a very bad odor back in the 1960s and 1970s because so many of the early ones had been rigged to give results supporting the programmers' favorite policy recommendations. "So it fascinated me to see computer modeling used properly in physics. And I think the openness of the field appealed to me. It was intellectually open, having a willingness to look at new ideas, and being nondogmatic about what was acceptable."

Arthur was also gratified to find that increasing returns was making quite an impact at the meeting. Quite aside from his own presentation, a number of the other economists had been thinking about it independently. One day, for example, the participants listened via telephone link to a lecture by Harvard emeritus professor Hollis Chenery, who had fallen too ill to travel. Chenery's lecture was about patterns of development—why countries show differences in how they grow, especially in the Third World. And during it, he mentioned increasing returns. "So after he hung up," says Arthur, "Arrow jumped up to the blackboard, and said, 'Hollis Chenery mentioned increasing returns. Let me tell you more about it'—and spontaneously gave an hour and a half lecture on the history of thought on increasing returns, along with what it had to say in trade theory, with no notes whatsoever. I would never have suspected that Arrow knew so much about the subject."

It was just a few days later that José Scheinkman, who had already done seminal work applying increasing returns to international trade, stayed up until three in the morning along with UCLA's Michele Boldrin to formulate a theory of economic development under increasing returns.

Inevitably, says Arthur, there was also a discussion about whether the stock market could get into a positive feedback loop, with stocks being bid higher and higher just because people see other investors coming in. Or, conversely, could there be an opposite effect, a crash, if people saw other investors getting out? "Given that the market was somewhat overheated at the time," says Arthur, "there was quite a lot of discussion of whether that was feasible, whether it did happen in reality—and whether it *might* happen soon."

The consensus was "Maybe." But the possibility seemed real enough to David Pines that he called his broker with an order to sell off some of his stock. The broker talked him out of it—and a month later, on October 19, 1987, the Dow fell 508 points in a day.

"That led to this rumor that the conference had predicted the stock market collapse a month before it happened," says Arthur. "We didn't. But the crash certainly had this positive feedback mechanism that we had discussed at length."

And so it went: a ten-day marathon with only one Saturday afternoon off. Everyone was exhausted—gloriously so. "By the end of the ten days I was on a huge scientific high," says Arthur. "I couldn't believe there were people willing to listen."

Indeed there were. Because of a prior commitment to deliver a paper in San Francisco on Friday, September 18, Arthur had had to miss the last day of the meeting, when the group had scheduled a wrap-up session and a press conference. (Reed, unable to get away from New York, sent a congratulatory message on video.) But as soon as he walked in the door of the convent that next Monday afternoon, Pines came up to him in the hallway with a smile on his face.

"Did the conference get over okay?" asked Arthur.

"Oh, we're very pleased," said Pines. Eugenia Singer had been particularly enthusiastic, and was preparing a glowing report for Reed. Meanwhile, he added, the science board had met right after the conference was over, and first off, they wanted to invite Arthur to join the science board.

Arthur was astonished. The science board was the institute's inner sanctum, the seat of all real policy-making power. "Certainly," he said.

"And there's been a further thought," said Pines. "We're very anxious not to let this opportunity slip. Everybody's so excited about the conference that we want to expand it into a full-scale research program. We'd been discussing that, and we were wondering if you and John Holland could come next year [meaning the next academic year, twelve months from then] and get the program up and running."

It took Arthur about two seconds to work that through. The science board was asking him and Holland to *run* the program. He stammered out something to the effect that he did have a sabbatical coming up, as a matter of fact, and it sounded like great fun. And—yes, he'd be delighted.

"I was enormously flattered," he says, "and I felt very humble indeed. But running throughout that—and still to this day—was this notion of 'Who, *me?*' I mean, this is Phil Anderson, or Ken Arrow, and here *I* am, and they're asking me what I think about this or that. So I had the reaction that—didn't they really mean somebody else? Certainly nothing like that had happened to me before in my academic life."

"You know," he adds, "it's perfectly possible for a scientist to feel that he has what it takes—but that he isn't accepted in the community. John Holland went through that for decades. I certainly felt like that—until I walked into the Santa Fe Institute, and all these incredibly smart people, people I'd only read about, were giving me the impression of 'What took you so long to get here?'"

For ten days, he had been talking and listening nonstop. His head was so full of ideas that it hurt. He was exhausted. He needed to catch up on about three weeks of sleep. And he felt as though he were in heaven.

"From then on," he says, "I stopped worrying about what other economists thought. The people I cared about sharing my work with were the people in Santa Fe. Santa Fe was a place to come home to."

On Tuesday, September 22, 1987, all too bright and early on the morning after he'd been offered the codirectorship of the Santa Fe Institute's new economics program, a sleepy Brian Arthur climbed into the car with John Holland and drove up to Los Alamos to visit the Artificial Life Workshop: a five-day happening that had started the day before.

Arthur was a little hazy about what "artificial life" actually meant. In fact, considering how exhausted he still felt after the economics meeting of the previous week, he was a little hazy about a lot of things. But as Holland explained it, artificial life was analogous to artificial intelligence. The difference was that, instead of using computers to model thought processes, you used computers to model the basic biological mechanisms of evolution and life itself. It was a lot like what he'd been trying to do with the genetic algorithm and classifier systems, said Holland, but even more broad-ranging and ambitious.

The whole thing was the brainchild of a postdoc up at the Los Alamos, Chris Langton, who had been a student of Holland's and Art Burks' at Michigan. Langton was something of a late bloomer, said Holland. At age thirty-nine, in fact, he was about ten years older than most postdocs. And he still hadn't quite put the finishing touches on his Ph.D. dissertation. But he'd been an extraordinary student. "A very fertile imagination," said Holland. "Very good at gathering in experience of all kinds." And he was putting tremendous energy into this workshop. Artificial life was Langton's baby. He'd invented the name. He'd spent most of the past decade trying to articulate the concept. He'd organized this workshop to try to turn artificial life into a real scientific discipline—without even knowing how many people would show up. He'd inspired enough confidence that the Los Alamos Center for Nonlinear Studies had put up $15,000 to pay for the workshop, while the Santa Fe Institute had put up another $5000 and agreed to publish the proceedings as part of its new book series on complexity. And from what Holland had seen of the workshop's kickoff yesterday, Langton was bringing it off beautifully. It was—well, Arthur would have to see for himself.

Indeed, Arthur did. When he and Holland walked into the auditorium building at Los Alamos, he formed two impressions very quickly. The first was that he'd badly underestimated his housemate. "It was like walking in with Gandhi," he says. "I'd thought I was rooming with a short, pleasant computer whiz. And here people were treating him like the great guru of this field: 'John *Holland!*' People would rush up to him in the hallway. What do you think of this? What do you think of that? Did you get the paper I sent you?"

Arthur's housemate tried to take it all in stride. But there was no getting around it: much to his own embarrassment, John Holland was becoming famous. Indeed, there wasn't much he could do to stop it. He'd been turning out one or two fresh Ph.D.s per year for twenty-five years, so that by now there were a lot of believers out there spreading the word. And in the meantime, the world had been catching up to him. Neural networks were very much back in vogue. And learning, by no coincidence, had now emerged as one of the hottest topics in mainstream artificial intelligence. The first international conference on genetic algorithms had been held in 1985, and more were in the offing. "It seemed to be the standard introduction to everyone's talk," says Arthur: "John Holland has such and such to say. Now here's my version."

Arthur's second impression was that artificial life was—strange. He never did get to meet Langton, who proved to be a tall, lanky guy with a mane of brown hair and a rumpled face that made him look strikingly like a young, amiable Walter Matthau. Langton was constantly on his way to somewhere else—coping, fixing, worrying, and frantically trying to make it all happen.

So, instead, Arthur spent a good part of the day wandering among the computer demonstrations that had been set up in the hallways around the auditorium. It was some of the damnedest stuff he'd ever seen: darting flocks of animated, electronic birds, strikingly realistic plants that grew and developed on screen before your eyes, bizarre, fractal-like creatures, patterns that undulated and sparkled. It was fascinating. But what did it mean?

And the talks! The ones that Arthur heard were a disconcerting mix of wild-eyed speculation and hard-nosed empiricism. It was as if no one knew what the speakers were going to say before they got up to say it. There were a lot of people there in ponytails and blue jeans. (One woman got up to give her talk in bare feet.) The word "emergence" seemed to crop up frequently. And most of all, there was this incredible energy and camaraderie in the air—a sense of barriers crumbling, a sense of new ideas let loose, a sense of spontaneous, unpredictable, open-ended freedom. In an odd, intellectual sort of way, the artificial life workshop felt like a throwback, like something right out of the Vietnam-era counterculture.

And, of course, in an odd, intellectual sort of way, it was.

Epiphany at Massachusetts General

Chris Langton can remember the precise moment when artificial life was born, if not the precise date. It was late 1971, early 1972—winter, anyway. And in classic hacker style, Langton was all alone up on the sixth floor of the Massachusetts General Hospital in Boston, sitting at the big, desklike console of the psychology department's PDP-9 computer and debugging code at three in the morning.

He liked working that way. "We didn't have to be there at any particular time," Langton explains. "The guy who ran this place, Frank Ervin, was a very creative, very hip kind of guy. He basically hired a whole bunch of bright kids to do the coding, and he gave them a pretty free hand. So the straight people, who were doing the real boring stuff, had the machine during the day. And we got into the habit of coming in at four or five in the afternoon and staying until three or four in the morning, when we could just play."

Indeed, so far as Langton was concerned, programming was the best game ever invented. It hadn't exactly been a deliberate career choice; he had just sort of drifted into Ervin's group about two years before, shortly after he arrived at Massachusetts General as a college dropout fulfilling his alternative service requirement as a conscientious objector to the Vietnam War. Except for a few summer courses back in high school, in fact, his programming skills were entirely self-taught. But once he really started messing around with computers, he'd started having so much fun that he stayed on, even after his requirement was finished.

"It was great," he says. "I'm a mechanic at heart. I like to construct things. I'd like to see this thing actually work." And with the kind of stuff he was doing on the PDP-9, he says, "You had to go knuckle-to-knuckle with the hardware. Your programs had to take into account what the machine was really doing, like 'load the accumulator from this specific address and then put it back.' It was logic, but it was also very mechanical."

But he also liked the weird kind of abstractions he was getting into. A good example was his very first project there, when he had gotten the experimental psychologists up and running on the PDP-9. For years they had been recording their data on an ancient and s-l-o-w PDP-8S, and they were getting sick of it. But the problem was that, in the process, they had created all manner of special-purpose software that nobody wanted to rewrite—and that wouldn't run on the PDP-9. So Langton's task had been to write a program that would trick the old software into thinking it was still running on the old machine. In effect, he was supposed to re-create the PDP-8S as a "virtual machine" inside the new one.

"I hadn't had any formal courses in computation theory," says Langton. "So I got my first, visceral exposure to the concept of a virtual machine by having to create one. And I just fell in love with the concept. The notion that you could take a real machine and abstract its laws of operation into a program meant that the program had captured everything that was *important* about the machine. You'd left the hardware behind."

Anyway, he says, on that particular night he was debugging code. And since he knew he wouldn't actually be running anything for a while, he pulled out one of the paper tapes that was always sitting in a box in front of the computer's big cathode-ray tube, and had run it through the tape reader to set the computer going with the Game of Life.

It was one of his favorites. "We'd gotten hold of the code from Bill Gosper and his group, who were hacking on the Game of Life over at MIT," says Langton, "and we were playing around with it, too." The thing was downright addictive. Developed the previous year by the English mathematician John Conway, the Game of Life wasn't actually a game that you played; it was more like a miniature universe that evolved as you watched. You started out with the computer screen showing a snapshot of this universe: a two-dimensional grid full of black squares that were "alive" and white squares that were "dead." The initial pattern could be anything you liked. But once you set the game going, the squares would live or die from then on according to a few simple rules. Each square in each generation would first look around at its immediate neighbors. If too many of those neighbors were already alive, then in the next generation the square would die of overcrowding. And if too few neighbors were alive, then the square would die of loneliness. But if the number of neighbors was just right, with either two living squares or three living squares, then in the next generation that central square would be alive—either by surviving if it were already alive or by being "born" if it weren't.

That was all. The rules were nothing but a kind of cartoon biology. But what made the Game of Life wonderful was that when you turned these simple rules into a program, they really did seem to make the screen come alive. Compared with what you would see on a present-day computer screen, the action was rather slow and jerky, as if it were being played back on a VCR in slow motion. In your mind's eye, however, the screen almost boiled with activity, as if you were looking through a microscope at the microbes in a drop of pond water. You could start up the game with a random scattering of live squares, and watch them instantly organize themselves into all manner of coherent structures. You could find structures that tumbled and structures that oscillated like beasts breathing in and out. You could find "gliders," little clusters of live cells that moved across the screen at constant velocity. You could find "glider guns" that fired off new gliders in a steady stream, and other structures that calmly ate the gliders. If you were lucky you might even find a "Cheshire Cat" that slowly faded away, leaving nothing behind but a smile and a paw print. Every run was different, and no one had ever exhausted the possibilities. "The first configuration I ever saw was a large, stable, diamond-shaped structure," says Langton. "But then you could introduce a glider from outside and it would interrupt the perfect crystalline beauty. And the structure would slowly decay into nothing, as if the glider was an infection from outside. It was like the Andromeda strain."

So that night, says Langton, the computer was humming, the computer screen was boiling with these little critters, and he was debugging code. "One time I glanced up," he says. "There's the Game of Life cranking away on the screen. Then I glanced back down at my computer code— and at the same time, the hairs on the back of my neck stood up. I sensed the presence of someone else in the room."

Langton looked around, sure that one of his fellow programmers was sneaking up on him. It was a crowded room, crammed with the big blue cabinets of the PDP-9, along with standing racks for electronic equipment, an old electroencephalograph machine, oscilloscopes, boxes pushed into corners trailing tubes and wires, and a lot of stuff that was never used anymore. It was the classic hacker's paradise. But no—no one was behind him; no one was hiding. He was definitely alone.

Langton looked back at the computer screen. "I realized that it must have been the Game of Life. There was something *alive* on that screen.

And at that moment, in a way I couldn't put into words at the time, I lost any distinction between the hardware and the process. I realized that at some deep level, there's really not that much difference between what could happen in the computer and what could happen in my own personal hardware—that it was really the same process that was going on up on the screen.

"I remember looking out the window in the middle of the night, with all this machinery humming away. It was one of those clear, frosty nights when the stars were sort of sparkling. Across the Charles River in Cambridge you could see the Science Museum and all the cars driving around. I thought about the patterns of activity, all the things going on out there. The city was sitting there, just *living*. And it seemed to be the same sort of thing as the Game of Life. It was certainly much more complex. But it was not necessarily different in kind."

The Self-Assembly of the Brain

Looking back on it with the perspective of twenty years, says Langton, that night of epiphany changed his life. But at the time it was little more than an intuition, a certain feeling he had. "It was one of those things where you have this flash of insight, and then it's gone. Like a thunderstorm, or a tornado, or a tidal wave that comes through and changes the landscape, and then it's past. The actual mental image itself was no longer really there, but it had set me up to feel certain ways about certain things. Things would come along that just smelled right, that would remind me of this pattern of activity. And for the rest of my career I've tried to follow that scent. Of course," he adds, "that scent has often led me somewhere and then just left me, not knowing where to go next."

That's actually an understatement. Not only was the Chris Langton of 1971 almost clueless as to what this feeling meant, he was a long way from being a systematic scholar. His idea of following the scent was to wander around the library or through bookstores, picking up articles here and there that somehow related to virtual machines, or to emergent, collective patterns, or to local rules making global dynamics. And every so often he would take a random course at Harvard, Boston University, or wherever. But basically he was content to take things as they came. There was just too much other stuff going on with his life. His real passion was his guitar; he and a friend were trying (unsuccessfully) to start a professional bluegrass band. He was still putting a lot of energy into draft resistance and protests against the Vietnam War. And so far as he was concerned, the whole counterculture scene around the periphery of the universities made Cambridge and Boston a great place to be. The fact was that Chris Langton was happier than he had been in a long while.

"High school was a disaster for me," he says. In 1962, when he was fourteen, Langton had gone from a very small elementary school in his hometown of Lincoln, Massachusetts, to Lincoln-Sudbury High, a big regional school in nearby Sudbury. "It was like going to jail, every day," he says. "This was an industrial-strength high school, where kids were treated like juvenile delinquents unless they somehow proved themselves otherwise and escaped into the special classes. And I was just not of the right mental demeanor to play along with that whole system. I had long hair. I played guitar and was into folk music. I was a hippie without there being any other hippies around. So I was very much a loner."

It probably didn't help that his parents, mystery writer Jane Langton and physicist William Langton, had been "radicals" from the earliest days of the civil rights movement and the Vietnam War. "During high school, my parents would occasionally take me out and we'd go into the city and participate in sit-ins and teach-ins for equality. We went to a lot of inner-city schools. We also took buses to Washington, and we'd protest this and that, and I got arrested for whatever excuse there was to arrest protesters."

Finally, Langton graduated in 1966. "This was the beginning of the hippie era," he says. "So a friend of mine and I hopped on a bus and went to California that summer, where things were a lot more advanced along that particular axis. We went straight to Haight-Ashbury. Listened to Janis Joplin, the Jefferson Airplane. It was a great summer."

In the fall, unfortunately, he had to report back for duty at Rockford College in Illinois. Personally, he didn't give a damn about college. And the feeling seemed to be mutual: with his high-school grade average hovering around a C, the Harvards and MITs of the world had given Langton's applications a decisive thumbs-down. But his parents had insisted that he go to college *somewhere*. And Rockford, having just converted itself from a finishing school for girls ("The Vassar of the Midwest") into a general liberal arts college, was eagerly recruiting.

To Langton, Rockford's brand-new campus out in the cornfields looked like a minimum-security farm prison. "It might as well have had barbed wire and razor wire along the top," he says. Because the school had done so much recruiting, however, it had managed to draw about ten East Coast hippie types that year out of a total student body of 500. "We got there and looked around, and there were these incredibly redneck, extreme right-wing—well, this basic area was the home of the Minutemen," he says. "At least on the East Coast, things were starting to happen. But out in the cornfields of Illinois, it was still somewhere back in the McCarthy era. And a hippie in the middle of Illinois in 1966 was basically dead meat. I got signed up for women's gym by the registrar after they saw me. One time several of us guys walked into a doughnut shop, and a couple of state cops walked in behind us, and one of them said, 'I don't know which one it is, but one of you guys has a pretty ugly girlfriend.' We got thrown out of every restaurant; nobody would serve us because we had long hair. The administration immediately started suspecting us of drugs and all kinds of other stuff."

The only thing to do, obviously, was to head north. Langton and his fellow "undesirables" started hitchhiking up to the University of Wisconsin in Madison, often staying weeks at a time. "This was where I belonged," he says. "The whole sociocultural upheaval that was the 1960s was happening in Madison, and there was zip going on in Rockford. There was a lot of antiwar activity going on in Madison. There were lots of hippies starting to experiment with drugs, so I did. I had an electric guitar, and a friend of mine had been exposed to Appalachian bluegrass, so we did some incredible jamming. There were lots of things going on—but nothing having to do with what you were supposed to be in college for."

By the start of his sophomore year at Rockford, not surprisingly, Langton was on academic probation. At the end of that fall semester, the school administration told him to leave and he told them he was quitting.

"I wanted to stay up in Madison," he says. "But I didn't have a job, and I didn't have any real way to support myself. So I ended up going back to Boston, where I got a lot more political and a lot more involved with antiwar activity." With no more student deferment, he filed for conscientious objector status. And after a long fight, he got his draft board to accept it. "Then I did alternative service at Massachusetts General Hospital starting in 1968."

Once there, of course, Langton was convinced he'd found his niche at last. He would have happily stayed in his programming job indefinitely. "It was a great job. I was learning a lot, I was having a great time with the people there. And there was no reason to leave that." But in 1972 he had no choice: the group leader, Frank Ervin, accepted a position at the University of California in Los Angeles and essentially packed up the lab to take it with him. Left at loose ends, Langton hooked up with another group of psychologists, who were studying social interactions among short-tailed macaque monkeys from Southeast Asia. And by Thanksgiving 1972 he found himself out in the jungles about forty miles from San Juan, Puerto Rico, at the Caribbean Primate Research Center.

This, as it turned out, was *not* a great job. Langton did like the monkeys: he spent eight to ten hours a day monitoring them during the experiments, becoming fascinated by their culture and how they passed that culture on to each new generation. The problem, unfortunately, was that the humans on the primate center staff were entirely too similar to their subjects. "One of the experiments there was to understand how the social system responded to stress," says Langton. "So they would slightly drug a monkey at some position in the hierarchy, and then see how the hierarchy responded when that monkey didn't do what it was supposed to do. The top male, for example, was supposed to threaten all the others, mate with all the females, settle the arguments, and chase certain ones around. So when he was a little out of it, the colony responded by breaking into factions. The sub-leaders might be very attentive to the chief monkey, but occasionally attack him, but then back off really quickly. You could see that they were trying to support him in his role, but also having to take on leadership responsibility. But he was still there, so there was this funny tension.

"Well, the head of the research center at the time was a complete, total alcoholic. He'd start off the morning with about a gallon of Bloody Marys, and then he'd be out of it for the rest of the day. He couldn't function in his role. So the rest of the staff weren't really empowered to do things, but *had* to do things. And there were all these fights about, 'You should have consulted with me!' I could have taken the data sheets I was using to observe the monkeys, lifted the roof off the research center, and seen exactly the same thing. It broke up into factions, there was a kind of revolution, and I was part of one faction that ultimately lost. I was asked to leave, and I was ready to leave."

At loose ends again after a year in Puerto Rico, Langton realized that it was time to start thinking a little more seriously about life. "I couldn't just keep jumping around and living for each day, without any kind of long-term idea of where I was going," he says. But where was that? He wondered if that mysterious scent might be trying to tell him something. He'd been following it the whole time he was in Puerto Rico, and he was beginning to think that maybe, just maybe he'd found the trail: cosmology and astrophysics.

"I didn't have any access to computers down there, so I wasn't doing any computer work to speak of. But I really did tons of reading," he says. The origin of the universe, the structure of the universe, the nature of time—it all seemed to have the right smell. "So when the situation deteriorated, I went back to Boston and started taking courses in mathematics and astronomy at Boston University."

He had taken a lot of the mathematics before, of course. But Langton thought it might be a good move to start all over from scratch. "I just wasn't paying attention before. I wasn't in school because I wanted to be. I went because that's what you did. You just got squeezed out of the tube of high school, onto the toothbrush of college." He could only afford to take a few courses at a time on an outstudent basis, while he worked at various odd jobs. Yet he threw himself into those courses wholeheartedly, and started doing remarkably well. Finally, one of his teachers, who had become a good friend, said, "Look, if you really want to do astronomy, go to the University of Arizona." Boston University was fine for a lot of things, he said. But Arizona was one of the astronomical capitals of the world. The campus in Tucson was right in the middle of the Sonora Desert, where you could find some of the clearest, driest, darkest skies on the planet. The mountaintops in the area sprouted telescope domes like mushrooms. Kitt Peak National Observatory was only forty miles away, and its headquarters was right there on campus. Arizona was the place to be.

That made sense to Langton. He applied to the University of Arizona, which accepted him for the fall of 1975.

When he was in the Caribbean, says Langton, he had learned to scuba dive. And there, among the corals and fishes, he had come to love moving in that third dimension. It was intoxicating. But once he was back in Boston, he'd soon discovered that scuba diving in the cold, brown waters of New England just wasn't the same. So as a substitute he'd tried hang-gliding. And he'd become hooked the first day. Sailing over the world, riding upward from thermal to thermal—this was the ultimate in three dimensions. He became a fanatic, buying his own hang-glider and spending every spare minute aloft.

All of which explains why, at the beginning of the summer of 1975, Langton set out for Tucson along with a couple of hang-gliding buddies who were moving to San Diego and who had a truck. Their plan was to spend the next few months making their way across the country at the slowest speed possible, while they went hang-gliding off any hill that looked halfway inviting. And that's exactly what they started to do, working their way down the Appalachians until they came to Grandfather Mountain, North Carolina.

As the highest peak in the Blue Ridge, Grandfather Mountain boasted a spectacular view; in fact, it was a privately owned tourist attraction. And it turned out to be just as spectacular a place to fly: "When the wind was right you could stay up for hours!" says Langton. Indeed, when the owner of the mountain realized how many hot dogs and hamburgers and souvenirs he was selling while the tourists stood around to watch these lunatics defying gravity, he offered them $25 a day if they would stay all summer.

"Well, it was quite unlikely we'd find a better place," says Langton. So they agreed. And as a tourist attraction, they were a tremendous success. Moreover, the owner got so interested in hang-gliding that he arranged for a national championship to be held at Grandfather Mountain at the end of the summer. Langton, figuring he would have a home court advantage if he competed, spent the rest of the summer practicing.

The accident was on August 5, he says. His friends with the truck had already left. And he was planning to head out the next day himself, figuring that he would go to Tucson, register, then come back to Grandfather Mountain for the championship before classes started. But in the meantime, he'd wanted to get in a few more practice runs for the spot-landing event, where you had to hit a bull's-eye on the ground.

So there he was, says Langton, coming in for his last attempt of the day. This particular spot landing was a tricky maneuver at best, since the target was in a small clearing in the trees; the only way to do it was to come in high and then spiral down almost at stall speed. But on that day the wind was so funky and uncooperative that it seemed just about impossible. Langton had had to abort his landing four times already, and he was really getting frustrated. This would be his last chance before the competition itself.

"I remember thinking, 'Damn, I'm too close and I'm too high. I'll just try for it anyway. What the hell.' And then as I settled down beneath the level of the trees, at about fifty feet, I sank into dead air. I was too slow, and I stalled out at just the wrong altitude. I remember thinking, 'Oh, shit.' I realized I was going to crash. It was going to be a bad crash. I remember thinking, 'God, I'm going to break a leg now. Shit.' " In a desperate attempt to pick up speed and regain control, he put the glider into a dive. No go. Then, as he'd been taught to do, he held out his legs to absorb some of the shock. "You know you're going to break your legs, but you don't pull them up," he says. "Because if you hit on your butt you'll break your back.

"I don't remember hitting. My memory is spotty after that. I do remember lying there, realizing that I was badly hurt and knowing that I should lie still. My friends ran over. People heard about it at the top of the mountain and came down. The owner of the mountain was taking pictures. Somebody was there with a two-way radio and called for an ambulance. A long time later I remember the medics showing up and saying, 'Where does it hurt?' and me saying, 'All over.' I remember them mumbling to themselves, rolling me onto the stretcher."

The ambulance took Langton down the mountain to the nearest emergency room, at Cannon Memorial Hospital in tiny Banner Elk, North Carolina. Much later, when he was lying semiconscious in the intensive-care unit, he remembers the nurses telling him, "Oh, you broke your legs. You'll be here a couple of weeks. Then we'll have you out of here and you'll be running around in no time."

"I was on morphine," he says, "so I believed them."

In fact, Langton was a mess. His crash helmet had saved his skull, and his legs had cushioned the impact enough to save his back and pelvis. But he had shattered thirty-five bones. The impact had broken both legs and both arms, almost ripping his right arm out of its socket. It had fractured most of his ribs and had collapsed one lung. And it had driven his knees into his face, smashing one knee, his jaw, and almost everything else. "Basically," says Langton, "my face was paste." His eyes wouldn't track: his cheekbones and the floors of his eye sockets were broken, and there was nothing solid for the eye muscles to pull against. And his brain wasn't working right: the crushing of his face had caused trauma deep inside. "They set a lot of bones and reinflated my lung in the emergency room," says Langton. "But I didn't come out of the anesthesia for a day longer than I should have. They were worried about a coma."

He did wake up, eventually. But it was a long time before he was coherent. "I had this weird experience of watching my mind come back," he says. "I could see myself as this passive observer back there somewhere. And there were all these things happening in my mind that were disconnected from my consciousness. It was very reminiscent of virtual machines, or like watching the Game of Life. I could see these disconnected patterns self-organize, come together, and merge with *me* in some way. I don't know how to describe it in any objectively verifiable way, and maybe it was just a figment of all these funny drugs they were giving me, but it was as if you took an ant colony and tore it up, and then watched the ants come back together, reorganize, and rebuild the colony.

"So my mind was rebuilding itself in this absolutely remarkable way.

And yet, still, there were a number of points along the way when I could tell I wasn't what I used to be, mentally. There were things missing—though I couldn't say what was missing. It was like a computer booting up: I could *feel* different levels of my operating system building up, each one with more capability than the last. I'd wake up one morning, and like an electric shock almost, I'd sort of shake my head and suddenly I'd be on some higher plateau. I'd think, 'Boy, I'm back!' Then I'd realize I wasn't really quite back. And then at some random point in the future, I'd go through another one of those, and—am I back yet or not? I still don't know until this day. A couple of years ago I went through another one of those episodes, a fairly major one. So who knows? When you're at one level, you don't know what's at a higher level."

Langton was one of the worst accident cases they'd ever seen in Banner Elk, where the hospital was much more used to gunshot wounds and ski accidents. Moreover, he was in traction from head to foot, and in no condition to be moved. Langton did, however, have one incredible piece of luck. Dr. Lawson Tate, the director of Cannon Memorial and the son of the founder, had practiced in a number of major medical schools before coming back to Banner Elk, and was an orthopedic surgeon of national caliber. Over the next several months he reconstructed Langton's crushed cheekbones and put in sheets of reinforcing plastic to rebuild the eye sockets. He pulled the sinus cavities back open and rebuilt the facial bones. He reconstructed the shattered knee from pieces of Langton's hip. And he rebuilt the dislocated right shoulder so that the nerves could grow back into the paralyzed arm. By Christmastime 1975, when Langton was finally flown to Emerson Hospital in Concord, Massachusetts, near his parents in Lincoln, Tate had performed fourteen operations on him. "The doctors there were amazed that one guy had done all these separate operations," says Langton.

In Concord, Langton was finally well enough to begin the long, slow process of learning to use his body again. "I'd been flat on my back for six months," he says, "a lot of that time in traction with my jaw wired shut. I went from 180 to 110 pounds. And I got no physical therapy that entire time. So a lot of things happen in that situation. You lose all your muscle; it just disappears. All of your ligaments and tendons tighten up. And you become very stiff, because if your joints aren't constantly being flexed to keep clear a certain range of motion, they fill up with this stuff that's secreted to replace worn-out cartilage, until there's no room for your joint to move at all.

"So I was this skeletal-looking anorexic," says Langton. "Of course, I'd had my jaws wired shut, so I'd lost a lot of the musculature that controls the jaws. It took me a long time to regain the ability to reopen my mouth more than just an inch or so. Eating was difficult; chewing was difficult. Talking—I talked almost through clenched teeth. And my face hung in funny ways. My cheekbones were way back instead of being pushed out. So I had this ghoulish expression. My eye sockets were very different shapes—they still are."

The physical therapists at Emerson Hospital got Langton up and walking. And they tried to get his right arm back to work. "A lot of the way I regained control was by playing guitar lying flat on my back," he says. "I forced myself. I didn't care what else happened, but I wasn't going to not play guitar anymore."

In the meantime, Langton was reading everything about science he could get his hands on. He'd started in Banner Elk, as soon as his eye sockets were back in place and he no longer had double vision. "I had people mail me books," he says. "I had books coming in by the truckload, and I was devouring them. Some of it was on cosmology. I read math books and did problems. But I also spent a lot more time on the history of ideas and biology in general. I read Lewis Thomas's *The Lives of a Cell*. I read a lot about the philosophy of science and the philosophy of evolution." He wasn't up to a truly concentrated effort, he says. The hospital in Banner Elk had put him on antidepressants and enough of the pain-killer Demerol to get him thoroughly addicted. Moreover, his mind was still in the midst of this funny process of reorganization. "But I was a sponge. I did a lot of non-specific, nondirected generic thinking about biology, physical science, and ideas of the universe, and about how those ideas changed with time. Then there was this scent I talk about. Through all of this, I was always following it, but without any direction. Cosmology and astronomy smelled good. But basically I still didn't understand it. I was still looking because I didn't know what was out there."

Artificial Life

When Langton finally made it to the University of Arizona campus in Tucson in the fall of 1976, he was able to hobble around with the aid of a cane, although there were still more operations to come on his knee and right shoulder. But he was also a twenty-eight-year-old sophomore, crippled and cadaverous-looking. He felt grotesque, like something out of a circus sideshow.

"It was bizarre," he says, "because the University of Arizona is a real Ken and Barbie kind of place, with frats and sororities and lots of beautiful people. Also, my mental state was such that I'd find myself rambling a lot. I'd get off on tangents of whatever the conversation was, and suddenly realize that I didn't have a clue of where this conversation had started from. My attention span was fairly narrow. So I felt mentally a freak, and physically a freak."

On the other hand, one of the really good things about Arizona was the university hospital and its first-class program in physical therapy and sports medicine. "I really benefited a lot from that program," says Langton. "They insisted that you keep plugging away, that you make progress. And I saw there that you had to pass a threshold; you had to go through a transition in your mind of accepting the way you were and work from there: not feeling bad about it, but feeling good about progress. So I just resolved to live with the ostracism and weirdness that I felt. And I still would answer questions in class—even though things would sometimes be a little strange, because I'd get off the topic. I just had to keep plugging away."

Unfortunately, however, even as his mind and body were slowly continuing to heal, Langton was discovering one thing that Arizona was *not* good for: astronomy. He had never thought to check whether the Astronomy Capital of the World offered an undergraduate major in the subject. It didn't. The university did have an astronomy Ph.D. program, which was superb. But to get there, undergraduates were supposed to major in physics and then switch. The only problem was that, so far as Langton was concerned, the Arizona physics department was abysmal. "It was completely disorganized," he says. "None of the people teaching the classes spoke English. The lab manuals were ancient. The equipment didn't match. Nobody knew what we were supposed to be learning."

This was not the kind of science that Langton had signed up for. Within a semester he was out of physics and out of astronomy. After all that, the elusive scent had led him straight into a dead end. (Langton wasn't the only person to feel this way about the physics department; in 1986, the university brought in a new chairman from Los Alamos to revitalize it: Peter Carruthers.)

The good news was that he had no regrets. Arizona did have an excellent department of philosophy, a subject that appealed to Langton because of his fascination with the history of ideas. And it had an equally fine department of anthropology, which appealed to him because of the affection he'd felt for the Puerto Rican monkeys. That first semester he took courses in both subjects to fill out his comprehensive program requirements. And by the time he left physics he was pursuing what eventually turned into a philosophy-anthropology double major.

It was an odd combination, to say the least. But to Langton they fit together perfectly. He'd sensed it the day he walked into Wesley Salmon's philosophy of science class. "Salmon had this very nice perspective," says Langton, who quickly asked him to be his adviser in the philosophy department. "He'd been a student of Hans Reichenbach, a philosopher of science from the Vienna Circle. Those guys had been doing very technical stuff—the philosophy of space and time, and quantum mechanics, and the curvature of space-time by gravity. And I very rapidly realized that I was much more interested, not in our specific, current understanding of the universe, but in how our world view had changed through time. What I was really interested in was the history of ideas. And cosmology just happened to be one of the most accessible arenas for studying that."

In the anthropology department, meanwhile, Langton was hearing about the rich variety of human mores, beliefs, and customs; about the rise and fall of civilizations; about the origins of humankind over three million years of hominid evolution. Indeed, his adviser there, physical anthropologist Stephen Zegura, was both a superb teacher and a man with a clear grasp of evolutionary theory.

So on every side, says Langton, "I was just immersed in this idea of the evolution of information. That quickly became my chief interest. It just smelled right." Indeed, the scent was overpowering. Somehow, he says, he knew he was getting very close.

One of Langton's favorite cartoons is a panel of Gary Larson's *The Far Side*, which shows a fully equipped mountaineer about to descend into an immense hole in the ground. As a reporter holds up a microphone, he proclaims, "Because it is not there!"

"That's how I felt," laughs Langton. The more he studied anthropology, he says, the more he sensed that the subject had a gaping hole. "It was a fundamental dichotomy. On the one hand, here was this nice, clear fossil record of biological evolution, together with a nice body of Darwinian theory that explained it. That theory involved the encoding of information, and the mechanisms by which that information was passed down from generation to generation. On the other hand, here was this nice, clear fossil record of cultural evolution, as discovered by the archeologists. And yet people in cultural anthropology wouldn't think about, or talk about, or even *listen* to you talk about a theory to explain that record. They seemed to be avoiding it."

Langton's impression was that evolutionary theories of culture still carried a stigma from the time of social Darwinism in the nineteenth century, when people were defending both war and gross social inequity on the grounds of "the survival of the fittest." But while he could certainly see the problem—after all, he'd been protesting war and social inequity most of his life—he just couldn't accept the gaping hole. If you could create a real theory of cultural evolution, as opposed to some pseudoscientific justification for the status quo, he reasoned, then you might be able to understand how cultures really worked—and among other things, actually do something about war and social inequity.

Now here was a goal worth pursuing. And most of all, here was something that smelled *right*. It wasn't just cultural evolution, Langton realized. It was biological evolution, intellectual evolution, cultural evolution, concepts combining and recombining and leaping from mind to mind over miles and generations—all wrapped together. Somehow, at the very deepest level, they were all just different aspects of the same thing. More than that, they were all just like the Game of Life—or for that matter like his own mind, still reassembling itself from fragments scattered by the fall. There was a unity here, a common story that involved elements coming together, structures evolving, and complicated systems acquiring the capacity to grow and be alive. And if he could only learn to look at that unity in the right way, if he could only abstract its laws of operation into the right kind of computer program, then he would have captured everything that was *important* about evolution.

"This was where things finally started to come together for me," says Langton. As a vision it was still almost impossible to articulate to anyone else. "But nothing else drove me. This is what I thought about all the time."

In the spring of 1978, Langton laid out his thinking in a twenty-six-page paper entitled "The Evolution of Belief." His basic argument was that biological and cultural evolution were simply two aspects of the same phenomenon, and that the "genes" of culture were beliefs—which in turn were recorded in the basic "DNA" of culture: language. In retrospect it was a pretty naive attempt, he says. But it was his manifesto—and not incidentally, his proposal for an interdisciplinary, self-designed Ph.D. program that would allow him to do research on this stuff. Moreover, the paper was enough to convince his anthropology adviser, Zegura. "He was a really good guy, a good teacher, a believer," says Langton. "He was the only one who really grokked what I was talking about. His attitude was, 'Go for it!' " But Zegura also warned him that for a special Ph.D. program, Langton would have to have advisers in other departments as well; as a physical anthropologist he simply was not competent to give Langton the guidance he would need in physics, biology, and computer science.

So Langton spent his senior year at Arizona making the rounds. "That's when I started calling it 'artificial life,' by analogy with artificial intelligence," he says. "I wanted a nice, short handle for it that would at least put people in the ballpark. Most people knew what artificial intelligence was, more or less. Well, artificial life meant trying to capture evolution in the same kind of way that artificial intelligence was trying to capture neuropsychology. I wasn't going to mimic exactly the evolution of the reptiles. I was going to capture an abstract model of evolution in the computer and experiment on that. So that phrase at least opened the door."

Unfortunately, the door usually slammed shut again as soon as he opened his mouth. "I talked to guys in computer science, and they didn't have a clue," he says. "They were into compilers and data structures and computer languages. They didn't even do artificial intelligence, so there was nobody there who could even come close to listening to me. They nodded their heads and said, 'This has nothing to do with computers.' "

Langton got exactly the same response from the biologists and the physicists. "I kept getting this look you get when they think you're a crackpot," he says. "It was very discouraging—especially coming as it did after the accident, when I felt unsure of what I was or who I was." Objectively, Langton had made enormous progress by this point; he could concentrate, he was strong, and he could run five miles at a stretch. But to himself, he still felt bizarre, grotesque, and mentally impaired. "I couldn't *tell*. Because of this neurological scrambling, I couldn't be sure of any of my thoughts anymore. So I couldn't be sure of this one. And it wasn't helping that nobody understood what I was trying to say."

And yet he kept plugging away. "I felt like it was the thing to do," he says. "I was willing to keep pushing because I knew this stuff had connections to what I'd thought about when I was sane and rational, before the accident. And I kept seeing things out there that related to it. I didn't know anything about nonlinear dynamics at the time, but there were all these intuitions for emergent properties, the interaction of lots of parts, the kinds of things that the group could do collectively that the individual couldn't."

Intuitions, unfortunately, weren't going to cut it. By the end of his senior year, for all his plugging away, Langton had to admit he was stuck. Zegura was behind him. But Zegura couldn't do it all by himself. It was time to fall back and regroup.

In the middle of all this, on December 22, 1979, Langton had gotten married to Elvira Segura, a feisty, blunt-talking master's in library science whom he had met in one of Steve Zegura's anthropology classes. "We started off as good friends, and it went from there," he says. And when he graduated with his double major in May 1980—largely because he'd accumulated so many credits that the university insisted on it—he and Elvira Segura-Langton moved into a little two-bedroom rental house just north of campus.

They were in a stable position for the time being. His wife had a good position at the university library, and Langton himself was working at hourly jobs, both as a carpenter with a home remodeling company—he thought of the exercise as good therapy—and as an assistant at a stained-glass shop. Indeed, there was a part of him that could have happily gone on doing that forever. "Good glass takes on a life of its own," he says. "You have lots of little pieces, but you're putting them all together to form a nice global effect." But Langton also knew that he had some serious decisions to make, and sooner rather than later. With Zegura's encouragement, he had already been accepted into the anthropology department's graduate program. But without an agreement to do an interdisciplinary Ph.D. on artificial life, that meant wasting a lot of time on courses he didn't want or need. So, should he just bag artificial life entirely, or what?

Not a chance. "By now I'd had the epiphany and I was a religious convert," he says. "This was clearly my life from now on. I *knew* I wanted to go on and do a Ph.D. in this general area. It's just that the path to take wasn't obvious."

The thing to do, he decided, was to get a computer and work some of these ideas out explicitly. That way, he could talk about artificial life and at least have something to show people. So with a loan from the proprietor of the stained-glass shop, he bought an Apple II home computer and set it up in the second bedroom. He also bought a little color television set to use as a monitor.

"I typically worked on it at night because I had to be on the job during the day," says Langton. "I'm almost always awake for two or three hours every night. For some reason, my mind is at its most active, most aware, and in its best mode for free, creative thinking then. I'll wake up with an idea on the tip of my mental tongue, and I'll just get up and pursue it."

His wife wasn't exactly happy about this. "Will you come back to bed and go to sleep!" he'd hear from the other bedroom. "You'll be exhausted tomorrow." Today, with 20/20 hindsight, Elvira Segura-Langton looks back on Chris Langton's nighttime hacking as being well worth the effort. But at the time she found it intensely annoying to have her husband treat the place like an office. To her the house was *home*, a place for family and an escape from the outside world. And yet—she could also see that this was very clearly what Chris Langton needed to do.

Langton's first attempt at artificial life was exceedingly simple: "organisms" that consisted of little more than a table of genes. "Each entry in the table was the genotype of the organism," he says. "It would have things like, How long is this organism supposed to live? How many years between producing offspring? What color is it? Where is it in space? And then there'd be some environment, like birds going through and picking things off that stood out too much from the background. So the creatures were evolving, because when they produced offspring there'd be a chance of mutation."

Once he got this program up and running, Langton was quite happy with it—at first. The organisms did indeed evolve; you could watch them do it. And yet he quickly became disillusioned. "It was all pretty damn linear," he says. The organisms were doing obvious things. They didn't take him beyond what he already understood. "There were no real organisms," he says. "What I had was this table of genes being manipulated by some external god—the program. Reproduction just happened magically. What I wanted was a little more closure—so that the process of reproduction would arise spontaneously and be part of the genotype itself."

Not knowing even how to begin, Langton decided it was time to go to the University of Arizona library, where he could do a computerized literature search. He tried the key words "self-reproduction."

"I got zillions of things back!" he says. There was one reference that leaped out immediately: *The Theory of Self-Reproducing Automata*, by John von Neumann (edited by A. W. Burks). Then there was another one, *Essays on Cellular Automata*, also edited by this guy Burks. And here was one called *Cellular Automata*, by Ted Codd, the guy who'd invented relational data bases. It just went on and on.

"Wow!" says Langton, "This was *right*. When I found all that, I said, 'Hey, I may be crazy, but these people are at least as crazy as I am!' " He checked out the books by von Neumann, Burks, and Codd, along with everything else on the list that he could find in the university library, and devoured them. Yes! It was all there: evolution, the Game of Life, self-assembly, emergent reproduction, everything.

Von Neumann, he discovered, had gotten interested in the issue of self-reproduction back in the late 1940s, in the aftermath of his work with Burks and Goldstine on the design of a programmable digital computer. At a time when the very concept of a programmable computer was still fresh and new, and when mathematicians and logicians were eager to understand what programmable machines could and couldn't do, the question was almost inevitable: Could a machine be programmed to make a copy of itself?

Von Neumann didn't have any doubt that the answer was yes, at least in principle. After all, plants and animals have been reproducing themselves for several billion years, and at the biochemical level they are just "machines" following the same natural laws as the stars and planets. But that fact didn't help him very much. Biological self-reproduction is immensely complicated, involving genetics, sex, the union of sperm and egg, cell divisions, and embryo development—to say nothing of the detailed molecular chemistry of proteins and DNA, which was still almost totally unknown in the 1940s. Machines obviously had none of that. So before von Neumann could answer the question about machine self-reproduction, he had to reduce that process to its essence, its abstract logical form. In effect, he had to operate in the same spirit that programmers would years later when they started to build virtual machines: he had to find out what was *important* about self-reproduction, independent of the detailed biochemical machinery.

To get a feel for the issues, von Neumann started out with a thought experiment. Imagine a machine that floats around on the surface of a pond, he said, together with lots of machine parts. Furthermore, imagine that this machine is a *universal constructor*: given a description of any machine, it will paddle around the pond until it locates the proper parts, and then construct that machine. In particular, given a description of itself, it will construct a copy of itself.

Now that sounds like self-reproduction, said von Neumann. But it isn't—at least, not quite. The newly created copy of the first machine will have all the right parts. But it won't have a description of itself, which means that it won't be able to make any further copies of itself. So von Neumann also postulated that the original machine should have a *description copier*: a device that will take the original description, duplicate it, and then attach the duplicate description to the offspring machine. Once that happens, he said, the offspring will have everything it needs to continue reproducing indefinitely. And then that *will* be self-reproduction.

As a thought experiment, von Neumann's analysis of self-reproduction was simplicity itself. To restate it in a slightly more formal way, he was saying that the genetic material of any self-reproducing system, natural or artificial, has to play two fundamentally different roles. On the one hand, it has to serve as a program, a kind of algorithm that can be executed during the construction of the offspring. On the other hand, it has to serve as passive data, a description that can be duplicated and given to the offspring.

But as a scientific prediction, that analysis turned out to be breathtaking: when Watson and Crick finally unraveled the molecular structure of DNA a few years later, in 1953, they discovered that it fulfilled von Neumann's two requirements precisely. As a genetic program, DNA encodes the instructions for making all the enzymes and structural proteins that the cell needs to function. And as a repository of genetic data, the DNA double helix unwinds and makes a copy of itself every time the cell divides in two. With admirable economy, evolution has built the dual nature of the genetic material into the structure of the DNA molecule itself.

That was still to come, however. In the meantime, von Neumann knew that a thought experiment alone wasn't enough. His image of a self-reproducing machine on a pond was still too concrete, too tied to the material details of the process. As a mathematician he wanted something that was completely formal and abstract. The solution, a formalism that eventually became known as the *cellular automaton*, was suggested by his colleague Stanislas Ulam, a Polish mathematician who had taken up residence at Los Alamos and who had been thinking about many of these issues himself.

What Ulam suggested was the same framework that John Conway was to use more than twenty years later when he invented the Game of Life; indeed, as Conway was well aware, the Game of Life was just one special case of a cellular automaton. Essentially, Ulam's suggestion to von Neumann was to imagine a programmable universe. "Time" in this universe would be defined by the ticking of a cosmic clock, and "space" would be defined to be a discrete lattice of cells, with each cell occupied by a very simple, abstractly defined computer—a *finite automaton*. At any given time and in any given cell, the automaton could be in only one of a finite number of states, which could be thought of as *red, white, blue, green, and yellow*, or *1,2,3,4*, or *living and dead*, or whatever. At each tick of the clock, moreover, the automaton would make a transition to a new state, which would be determined by its own current state and the current state of its neighbors. The "physical laws" of this universe would therefore be encoded in its *state transition table*: the rule that tells each automaton which state to change to for each possible configuration of states in its neighborhood.

Von Neumann loved the cellular automaton idea. Here was a system that was simple and abstract enough to analyze mathematically, yet rich enough to capture processes he was trying to understand. And, not incidentally, it was exactly the sort of system that you could simulate on a real computer—at least in principle. Von Neumann's work on the theory of cellular automata was left unfinished at the time of his death from cancer in 1954. But Art Burks, who had been asked to edit his papers on the subject, subsequently organized what was there, filled in the remaining details, and published the collection as *Theory of Self-Reproducing Automata* in 1966. One of the highlights was von Neumann's proof that there existed at least one cellular automaton pattern that could indeed reproduce itself. The pattern he'd found was immensely complicated, requiring a huge lattice and 29 different states per cell. It was far beyond the simulation capacity of any existing computer. But the very fact of its existence settled the essential question of principle: self-reproduction, once considered to be an exclusive characteristic of living things, could indeed be achieved by machines.

In reading all this, says Langton, "All of a sudden I felt very confident. I knew I was on the right track." He went back to his Apple II and quickly wrote a general-purpose cellular automaton program that would let him watch the cellular world as a grid of colored squares on the screen. The memory limitations of the Apple—it only had 64 kilobytes—meant that he could allow no more than eight states per cell. That ruled out von Neumann's 29-state self-reproducer. But it didn't rule out the possibility of finding *a* self-reproducing system within those limits. Langton had set up his program so that he could try out any set of states and any transition table he wanted. And with eight states per cell, that left him with only about $10^{30,000}$ different tables to explore. He went to it.

Langton already knew that his quest wasn't as hopeless as it seemed. In his reading he'd discovered that Ted Codd had found an eight-state self-reproducing pattern more than a decade earlier, when he was a grad student at Michigan working under some guy named John Holland. And while Codd's pattern was still too complex for the Apple II, Langton thought that by playing with the various components of it, he might be able to implement something simpler that would fit within his own constraints.

"All of Codd's components were like data paths," says Langton. That is, four of the eight automaton states in Codd's system acted as bits of data, while the other four played various auxiliary roles. In particular, there was one state that functioned as a conductor and another state that functioned as a kind of insulator, so that together they defined channels through which the data would flow from cell to cell as if the path were a copper wire. So Langton began by implementing Codd's "periodic emitter" structure: essentially a loop in which one bit of data circulated around and around like the second hand of a clock, along with a kind of arm that grew out from the side of the loop and periodically fired off a copy of the circulating bit. And then Langton started modifying the emitter, putting a cap on the arm so the signals wouldn't escape, adding a second circulating signal to make the cap, tweaking the rule table, ad infinitum. He knew he could do it, if only he could make that arm grow out, curl back on itself, and make a loop identical with the first.

It was slow going. Langton worked into the wee hours every night, while his wife Elvira did her best to be patient. "She cared that I was interested in it and that I thought something was happening," says Langton. "But she was more concerned with, What are we going to do? Where is all this work going to take us? How is it contributing to the progress of the domestic situation? Where are we going to be in two years? That was very hard to explain. So you've done this. Now you do *what* with it? And I didn't know. I just knew it was important."

Langton could only keep plugging away. "I kept getting this piece and that piece," he says. "I would start with a rule, then I'd modify it, and modify it again, and then I'd paint myself into a corner. I filled up fifteen floppy disks with preserved rule tables, so that I could back up and take off in a different direction. So I had to keep very careful records of what rule produced what behavior, and what changed, and what I'd backed up to, and what disk I'd stored it on."

All told, says Langton, it was about two months from the time he first read von Neumann until he finally got what he wanted. One night, he says, the pieces just finally came together. He sat staring at loops that extended their arms, curled those arms around to form new, identical loops, and went on to form still more loops ad infinitum. It looked like the growth of a coral reef. He had created the simplest self-reproducing cellular automaton ever discovered. "I had this incredible—volcano of emotion," he says. "This *is* possible. It *does* work. This *is* true. Evolution made sense now. This wasn't an external program that just manipulated a table. This had closure on itself, so that the organism *was* the program. It was complete.

And now all these things that I'd been thinking of that might be the case if I could do this—well, they were all possible, too. It was like a landslide of possibilities. The dominoes fell, and just kept falling and falling and falling."

The Edge of Chaos

"I'm in part a mechanic," says Langton. "I have to get my hands on something, put it together, see it work. And once I'd actually put something together, any doubts I had were gone. I could see where artificial life had to go from there." In his own mind it was all crystal clear: now that he had self-reproduction in the cellular automaton world, he would have to add on the requirement that these patterns perform some task before they can reproduce, like collect enough energy or enough of the right components. He would have to build whole populations of such patterns, so that they competed with each other for these resources. He would have to give them the ability to move around and sense each other. He would have to allow for the possibility of mutations and errors in reproduction. "These were all problems to be solved," he says. "But okay—in this von Neumann world, I knew I could embed evolution."

Armed with his new self-reproducing cellular automaton, Langton went back to campus and started making the rounds again, trying to drum up support for his interdisciplinary Ph.D. program. "This—," he would tell people, pointing to the unfolding structures on the screen, "*this* is what I want to do."

No go. If anything, the response was chillier than before. "There was so much to explain at this point," he says. "The people in the anthropology department didn't know about computers, period, let alone cellular automata. 'How is this any different from a video game?' The people in the computer science department didn't know about cellular automata, either. And they weren't interested in biology at all. 'What's self-reproduction got to do with computer science?' So when you tried to paint the whole picture—hey, you sounded like a complete, babbling idiot.

"Well, *I* knew I wasn't crazy," he says. "I felt very sane by then. I felt saner than everybody else. In fact, I worried about that—I'm sure that's what crackpots feel." But sane or not, he clearly wasn't getting anywhere at Arizona. It was time to start looking elsewhere.

. . .

Langton wrote to his former philosophy adviser, Wesley Salmon, who had moved to the University of Pittsburgh: "What do I do?" Salmon wrote back with a suggestion from his wife: "Study with Art Burks."

Burks? "I'd just assumed he was dead," says Langton. "Almost everybody else from that era was." But Burks turned out to be very much alive at the University of Michigan. More than that, once Langton started corresponding with him, Burks was very encouraging. He even arranged for Langton to get financial support as a teaching assistant and research assistant. Just apply, he wrote.

Langton wasted no time. By that point he'd learned that Michigan's Computer and Communication Sciences program was famous for exactly the kind of perspective that he was after. "To them," says Langton, "information processing was writ large across all of nature. However information is processed is worthwhile understanding. So I applied under that philosophy."

By and by he got a letter back from Professor Gideon Frieder, the chairman of the department. "Sorry," it said, "you don't have the proper background." Application denied.

Langton was enraged. He fired back a seven-page letter, the gist of which was, What the hell!? "Here is your whole philosophy, the purpose you claim to exist, live, and breathe for. This is exactly what I've been pursuing. And you're telling me no?"

A few weeks afterward Frieder wrote again, saying in effect, "Welcome aboard." As he told Langton later, "I just liked the idea of having somebody around who would say *that* to the chairman."

Actually, as Langton was to learn, there had been much more to it than that. Neither Burks nor Holland had ever seen his original application. For a variety of bureaucratic and financial reasons, the broad-based department they had spent thirty years building was on the verge of merging with the department of electrical engineering, where people had a much more hard-nosed, practical idea of what constituted worthwhile research. And in anticipation, Frieder and others were already trying to deemphasize such things as "adaptive computation." Burks and Holland were fighting a rear-guard action.

Fortunately or unfortunately, however, Langton didn't know this at the time. He was just glad to have gotten accepted. "I just couldn't pass this up," he says, "especially when I *knew* I was right." Elvira Segura-Langton was willing to give it a try. True, she would have to give up her job at the university. And they wouldn't be near her family in Arizona anymore. But considering that she was now pregnant with their first child, it wouldn't be such a bad idea to be covered by Chris's student health plan. And besides, as much as they both loved the Southwest, it might be fun to see a cloud every once and a while. So in the fall of 1982, they headed north.

Intellectually, at least, Langton had a grand time at Michigan. As a teaching assistant for Burks' history of computing class, he absorbed Burks' eyewitness account of those early days—and helped Burks put together an exhibit on ENIAC using some of the original hardware. He met John Holland. And for his class in integrated circuitry, Langton designed and built a chip for the ultra-fast execution of a part of Holland's classifier system.

But mostly, Langton studied like mad. Formal language theory, computational complexity theory, data structures, compiler construction—he was systematically learning material he'd only absorbed in bits and pieces before. And he loved it. Burks, Holland, and company were nothing if not demanding; once while Langton was at Michigan they flunked almost everybody taking the oral "qualifier" exams to get into the Ph.D. program. (The flunkees did get another chance.) "They could ask you stuff that wasn't even in the courses, and you'd have to say something intelligent about it," says Langton. "Well, I really enjoyed that kind of learning process. There's a big difference between having passed the courses and *knowing* the material."

On the academic politics front, however, things were not so grand. By late 1984, when Langton had finished his course work, gotten his master's degree, passed his qualifiers, and was ready to begin a Ph.D. dissertation, it had become painfully obvious that he wouldn't be allowed to do the thesis he wanted on the evolution of artificial life in a von Neumann universe. Burks' and Holland's rear-guard action was failing. The old Computer and Communication Sciences department had been absorbed into the engineering school in 1984. And in the dominant engineering culture of its new home, the Burks-Holland-style "natural systems" curriculum was effectively being phased out. (The situation was and is one of the few things that Holland gets visibly angry about; he had initially been one of the strongest voices in favor of the merger, having been convinced that the natural systems approach would be protected, and he now felt as though he had been suckered. Indeed, the situation gave Holland considerable extra incentive to start getting involved with the Santa Fe Institute about this time.) But as the better part of valor, Burks and Holland both urged Langton to do a thesis that was a little less biological and a little more computer-sciencelike. And Langton had to admit they had a point—if only as a matter of practicality. "I had enough savvy by then to know that this von Neumann universe was going to be an extremely difficult system to set up and get going," he says. So he started looking around for something that might be doable in the space of a year or two, instead of decades.

Well, he thought, instead of trying to build a complete von Neumann universe, why not start by trying to understand a little more about the "physics" of that universe? Why not try to understand why certain cellular automaton rule tables allowed you to build interesting structures, and others didn't? That would at least be a step in the right direction. It would probably contain enough hard-core computer science to satisfy the engineering types. And with any luck, it might produce some interesting connections to real physics. Indeed, this cellular automata-physics connection had become a certifiably hot topic of late. As physics whiz-kid Stephen Wolfram had pointed out in 1984, when he was still at Caltech, cellular automata not only have a rich mathematical structure but have deep similarities to nonlinear dynamical systems.

What Langton found particularly fascinating was Wolfram's contention that all cellular automata rules fall into one of four *universality classes*. Wolfram's Class I contained what you might call doomsday rules: no matter what pattern of living and dead cells you started them out with, everything would just die within one or two time steps. The grid on the computer screen would go monochrome. In the language of dynamical systems, such rules seemed to have a single "point attractor." That is, the system seemed to be mathematically like a marble rolling toward the bottom of a big cereal bowl: no matter where the marble started out on the sides of the bowl, it would always roll down very quickly to a point in the center—the dead state.

Wolfram's Class II rules were a little more lively, but not much. With these rules, an initial pattern that scattered living and dead cells over the screen at random would quickly coalesce into a set of static blobs, with perhaps a few other blobs that would sit there periodically oscillating. Such automata still gave the general impression of frozen stagnation and death. In the language of dynamical systems, these rules seemed to have fallen into a set of periodic attractors—that is, a series of hollows in the bottom of a bumpy cereal bowl where the marble could roll around and around the sides indefinitely.

Wolfram's Class III rules went to the opposite extreme: they were too lively. They produced so much activity that the screen seemed to be boiling. Nothing was stable and nothing was predictable: structures would break up almost as soon as they formed. In the language of dynamical systems, these rules corresponded to "strange" attractors—more commonly known as *chaos*. They were like a marble rolling around inside that cereal bowl so fast and so hard that it could never settle down.

Finally, there were Wolfram's Class IV rules. Included here were those rare, impossible-to-pigeonhole rules that didn't produce frozen blobs, but that didn't produce total chaos, either. What they did produce were coherent structures that propagated, grew, split apart, and recombined in a wonderfully complex way. They essentially never settled down. In that sense they were all very much like the most famous member of Class IV, the Game of Life. And in the language of dynamical systems, they were . . .

Well, that was just the problem. There was nothing in the conventional theory of dynamical systems that looked anything like a Class IV rule. Wolfram had conjectured that these rules represented a kind of behavior that was unique to cellular automata. But the fact was that no one had the slightest idea of what they represented. Nor, for that matter, did anyone have the slightest idea why one rule produced Class IV behavior and another didn't; the only way to find out which class a given rule belonged to was to try it out and see what happened.

For Langton, the situation not only was intriguing, but revived the old "Because it's not there" feeling he'd once had about anthropology. Here were the very rules that seemed essential to his vision of a von Neumann universe, that seemed to capture so much of what was important about the spontaneous emergence of life and self-reproduction. And yet they seemed to lie in some completely unknown realm of dynamics. So he decided to tackle the problem head on: How were Wolfram's classes related to one another, and what determined the class a given rule belonged to?

One idea struck him almost immediately. At about that same time, as it happens, he had been doing quite a bit of reading on dynamical systems and chaos. And in many real nonlinear systems, he knew, the equation of motion contains a numerical parameter that functions as a kind of tuning knob, controlling just how chaotic the system really is. If the system were a dripping water faucet, for example, the parameter would be the rate of water flow. Or if the system were a population of rabbits, the parameter would involve a ratio between the rabbits' birth rate and their death rate due to overcrowding. In general, a small value of the parameter would usually correspond to stable behavior: equal-sized drops, a constant population, and so forth. And that, in turn, seemed highly reminiscent of the static behavior seen in Wolfram's Class I and Class II. But as the parameter got progressively larger and larger, the behavior of the system would get progressively more complicated—different-sized drops, a fluctuating population, and so on—until ultimately it became completely chaotic. At that point it would resemble Wolfram's Class III.

Langton wasn't too clear where Class IV would fit into this picture. But the analogy was too good to ignore. If he could only find some way of assigning a similar kind of parameter to the cellular automaton rules, then the Wolfram classes might start to make some sense. Of course, he couldn't just assign the numbers arbitrarily; this parameter, whatever it turned out to be, would have to be derived from the rule itself. He might measure each rule's degree of reactivity—for example, how often it causes the central cell to change its state. But there were any number of things to try.

So Langton started programming his computer for every parameter that seemed halfway reasonable. (One of the first things he had done when he came to Michigan was to implement a more sophisticated version of his Apple II cellular automaton program on a high-powered, blazingly fast Apollo workstation.) And he got absolutely nowhere—until, one day, he tried one of the simplest parameters he could think of. The Greek letter lambda (λ), as he called it, was just the probability that any given cell would be "alive" in the next generation. So if a rule had a lambda value of precisely 0.0, for example, then nothing would be alive after the first time step and the rule would clearly be in Class I. If the rule had a lambda of 0.50, then the grid would be boiling with activity, with half the cells alive on the average and half dead. Presumably, such a rule would be in the chaotic Class III. The question was whether lambda would reveal anything interesting in between. (Beyond 0.50 the role of "alive" and "dead" would be reversed, and things would presumably get simpler again until you were back at Class I when you reached 1.0; it would be like looking at the same behavior in a photographic negative.)

To test the parameter, Langton wrote a little program telling his Apollo to automatically generate rules with a specific value of lambda, and then run a cellular automaton on the screen to show him what that rule would do. "Well, the first time I ran it," he says, "I set lambda equal to 0.50, thinking I was setting it to a totally random state—and I suddenly started getting all these Class IV rules, one after another! I thought, 'God, this is too good to be true!' So I discovered, sure enough, that there was a bug in the program that was actually setting lambda to a different value—which just happened to be the critical value for that class of automata."

Fixing the bug in his program, Langton started exploring various lambda values systematically. At very low values around 0.0 he found nothing but the dead, frozen Class I rules. As he increased the values a little bit, he started finding periodic Class II rules. As he increased the value a little more, he noticed that the Class II rules took longer and longer to settle down. Then if he jumped all the way to 0.50, he found himself in the total chaos of Class III, just as he expected. But right there in between Classes II and III, clustered tightly around this magic "critical" value of lambda (about 0.273), he found whole thickets of complex Class IV rules. And yes, the Game of Life was among them. He was flabbergasted. Somehow, this simpleminded lambda parameter had put the Wolfram classes into exactly the kind of sequence he'd wanted—and had found a place for the Class IV rules to boot, right at the transition point:

$$I \& II \rightarrow \text{"IV"} \rightarrow III$$

Moreover, that sequence suggested an equally provocative transition in dynamical systems:

$$\text{Order} \rightarrow \text{"Complexity"} \rightarrow \text{Chaos}$$

where "complexity" referred to the kind of eternally surprising dynamical behavior shown by the Class IV automata.

"It immediately brought to mind some kind of phase transition," he says. Suppose you thought of the parameter lambda as being like temperature. Then the Class I and II rules that you found at low values of lambda would correspond to a solid like ice, where the water molecules are rigidly locked into a crystal lattice. The Class III rules that you found at high values of lambda would correspond to a vapor like steam, where the molecules are flying around and slamming into each other in total chaos. And the Class IV rules you found in between would correspond to—what? Liquids?

"I knew very little about phase transitions at the time," says Langton, "but I dug into what was known about the molecular structure of liquids." At first it looked very promising: the molecules in a liquid, he discovered, are constantly tumbling over and around each other, bonding and clustering and breaking apart again billions of times per second—a lot like the structures in the Game of Life. "It seemed to me quite plausible that something like the Game of Life might be going on at the molecular level just in a glass of water," he says.

Langton loved that idea. And yet, as he thought about it some more, he began to realize that it wasn't quite right. Class IV rules typically produced "extended transients," such as the Game of Life's gliders: structures that could survive and propagate for an arbitrarily long time. Ordinary liquids didn't seem to have anything like that at a molecular level. So far as anyone could tell, they were almost as completely chaotic as gaseous matter. Indeed, Langton had learned that by increasing the temperature and pressure enough, you could go from steam to water without ever going through a phase transition at all; in general, gases and liquids are just two aspects of a single *fluid* phase of matter. So the distinction wasn't a fundamental one, and the resemblance of liquids to the Game of Life was only superficial.

Langton went back to his physics texts and kept reading. "Finally, I came across the basic distinction between first-order and second-order phase transitions," he says. First-order transitions are the kind we're all familiar with: sharp and precise. Raise the temperature of an ice cube past 32°F, for example, and the change from ice to water happens all at once. Basically, what's going on is that the molecules are forced to make an either-or choice between order and chaos. At temperatures below the transition, they are vibrating slowly enough that they can make the decision for crystalline order (ice). At temperatures above the transition, however, the molecules are vibrating so hard that the molecular bonds are breaking faster than they can reform, so they are forced to opt for chaos (water).

Second-order phase transitions are much less common in nature, Langton learned. (At least, they are at the temperatures and pressures humans are used to.) But they are much less abrupt, largely because the molecules in such a system don't have to make that either-or choice. They combine chaos *and* order. Above the transition temperature, for example, most of the molecules are tumbling over one another in a completely chaotic, fluid phase. Yet tumbling among them are myriads of submicroscopic islands of orderly, latticework solid, with molecules constantly dissolving and recrystallizing around the edges. These islands are neither very big nor very long-lasting, even on a molecular scale. So the system is still mostly chaos. But as the temperature is lowered, the largest islands start to get very big indeed, and they begin to live for a correspondingly long time. The balance between chaos and order has begun to shift. Of course, if the temperature were taken all the way past the transition, the roles would reverse: the material would go from being a sea of fluid dotted with islands of solid, to being a continent of solid dotted with lakes of fluid. But right *at* the transition, the balance is perfect: the ordered structures fill a volume precisely equal to that of the chaotic fluid. Order and chaos intertwine in a complex, ever-changing dance of submicroscopic arms and fractal filaments. The largest ordered structures propagate their fingers across the material for arbitrarily long distances and last for an arbitrarily long time. And nothing ever really settles down.

Langton was electrified when he found this: *"There* was the critical connection! *There* was the analog to Wolfram's Class IV!" It was all there. The propagating, gliderlike "extended transients," the dynamics that took forever to settle down, the intricate dance of structures that grew and split and recombined with eternally surprising complexity—it practically defined a second-order phase transition.

So now Langton had yet a third analogy:

Cellular Automata Classes:
    I & II → "IV" → III

Dynamical Systems:
    Order → "Complexity" → Chaos

Matter:
    Solid → "Phase transition" → Fluid

The question was, was it anything more than an analogy? Langton went right to work, adapting all manner of statistical tests from the physicists' world and applying them to the von Neumann world. And when he plotted his results as a function of lambda, the graphs looked like something right out of a textbook. To a physicist they would have screamed "second-order phase transition." Langton had no idea why his lambda parameter worked so well, or why it seemed so closely analogous to temperature. (Indeed, no one really understands it yet.) But there was no denying that it did. The phase transition wasn't just an analogy. It was real.

At one time or another, Langton used every name for this phase transition that he could think of: the "transition to chaos," the "boundary of chaos," the "onset of chaos." But the one that really captured the visceral feeling it gave him was "the edge of chaos."

"It reminded me of the feelings I experienced when I learned to scuba dive in Puerto Rico," he explains. "For most of our dives we were fairly close to shore, where the water was crystal clear and you could see the bottom perfectly about 60 feet down. However, one day our instructor took us to the edge of the continental shelf, where the 60-foot bottom gave way to an 80-degree slope that disappeared into the depths—I believe at that point the transition was to about 2000 feet. It made you realize that all the diving you had been doing, which had certainly seemed adventurous and daring, was really just playing around on the beach. The continental shelves are like puddles compared to 'The Ocean.'

"Well, life emerged in the oceans," he adds, "so there you are at the edge, alive and appreciating that enormous fluid nursery. And that's why 'the edge of chaos' carries for me a very similar feeling: because I believe life also originated at the edge of chaos. So here we are at the edge, alive and appreciating the fact that physics itself should yield up such a nursery. . . ."

That's a poetic way to say it, certainly. But for Langton, that belief was far more than just poetics. The more he thought about it, in fact, the more he became convinced that there was a deep connection between phase transitions and computation—and between computation and life itself.

The connection goes right back to—of course—the Game of Life. After the game was discovered in 1970, says Langton, one of the first things people noticed was that propagating structures like gliders could carry signals across the von Neumann universe from one point to another. Indeed, you could think of a flock of gliders going single file as being like a stream of bits: "glider present" = 1; "glider absent" = 0. Then, as people played around with the game still more, they discovered various structures that could store such data, or that could emit new signals that encoded new information. Very quickly, in fact, it became clear that Game of Life structures could be used to build a complete computer, with data storage, information-processing capability, and all the rest. This Game of Life computer would have nothing to do with the machine the game was actually running on, of course; whatever *that* machine was—a PDP-9, an Apple II, an Apollo workstation—it would just serve as the engine that made the cellular automaton go. No, the Game of Life computer would exist entirely within the von Neumann universe, in exactly the same way that Langton's self-reproducing pattern did. It would be a crude and inefficient computer, to be sure. But in principle, it would be right up there with Seymour Cray's finest. It would be a *universal* computer, with the power to compute anything computable.

Now, that's a pretty astonishing result, says Langton—especially when you consider that only comparatively few cellular automaton rules allow it to happen. You couldn't make a universal computer in a cellular automaton governed by Class I or II rules, because the structures they produce are too static; you could store data in such a universe, but you would have no way to propagate the information from place to place. Nor could you make a computer in a chaotic Class III automaton; the signals would get lost in the noise, and the storage structures would quickly get battered to pieces. Indeed, says Langton, the only rules that allow you to build a universal computer are those that are in Class IV, like the Game of Life. These are the only rules that provide enough stability to store information *and* enough fluidity to send signals over arbitrary distances—the two things that seem essential for computation. And, of course, these are also the rules that sit right in this phase transition at the edge of chaos.

So phase transitions, complexity, and computation were all wrapped together, Langton realized. Or, at least, they were in the von Neumann universe. But Langton was convinced that the connections held true in the real world as well—in everything from social systems to economies to living cells. Because once you got to computation, you were getting awfully close to the essence of life itself. "Life is based to an incredible degree on its ability to process information," he says. "It stores information. It maps sensory information. And it makes some complex transformations on that information to produce action. [The English biologist Richard] Dawkins has this really nice example. If you take a rock and toss it into the air, it traces out a nice parabola. It's at the mercy of the laws of physics. It can only make a simple response to the forces that are acting on it from outside. But now if you take a bird and throw it into the air, its behavior is nothing like that. It flies off into the trees somewhere. The same forces are certainly acting on this bird. But there's an awful lot of internal information processing going on that's responsible for its behavior. And that's true even if you go down to simple cells: they aren't just doing what inanimate matter does. They aren't just responding to simple forces. So one of the interesting questions we can ask about living things is, Under what conditions do systems whose dynamics are dominated by information processing arise from things that just respond to physical forces? When and where does the processing of information and the storage of information become important?"

In an attempt to answer that question, says Langton, "I took my phase transition glasses and looked at the phenomenology of computation. And there were enormous numbers of analogies." When you take a class in the theory of computation, for example, one of the first things you learn about is the distinction between programs that "halt"—that is, take in a string of data and produce an answer in a finite amount of time—and those that just keep churning away forever. But that, Langton says, is just like the distinction between the behavior of matter at temperatures below and above a phase transition. There is a sense in which the material is constantly trying to "compute" how to arrange itself at a molecular level: if it's cold, then it reaches an answer very fast—and crystallizes completely. But if it's hot it can't reach an answer at all, and remains fluid.

In much the same way, he says, the distinction is analogous to the one between Class I and II cellular automata that eventually halt by freezing into a stable configuration, and chaotic Class III cellular automata that boil along nonstop. For example, suppose you had a program that just printed out one message on the screen—"HELLO WORLD!"—and quit. Such a program would correspond to one of those Class I cellular automata down around lambda = 0.0, which go to quiescence almost immediately. Conversely, suppose you had a program with a serious bug in it, so that it printed out a steady stream of gibberish without ever repeating itself. Such a program would correspond to Class III cellular automata out around lambda = 0.50, where the chaos is maximal.

Next, says Langton, suppose you moved away from the extremes, toward the phase transition. In the material world, you would find longer and longer transients: that is, as the temperature gets closer to the phase transition, the molecules require more and more time to reach their decision. Likewise, as lambda increased from zero in the von Neumann universe, you would start to find cellular automata that would churn around a bit before they reached quiescence, with the amount of churning depending on just what their initial state was. These would correspond to what are known as *polynomial-time* algorithms in computer science—the kind that have to do a significant amount of work before they finish, but that tend to be relatively fast and efficient at it. (Polynomial-time algorithms often crop up when the problem involves chores such as sorting a list.) As you went further, however, and as lambda began to get very close to the phase transition, you would begin to find cellular automata that churned around for a very long time indeed. These would correspond to *nonpolynomial-time* algorithms—the kind that might not halt for the lifetime of the universe, or longer. Such algorithms are effectively useless. (An extreme example would be a program that tried to play chess by looking ahead at every possible move.)

And right *at* the phase transition? In the material world, a given molecule might wind up in the ordered phase or the fluid phase; there would be no way to tell in advance, because order and chaos are so intimately intertwined at the molecular level. In the von Neumann universe, likewise, the Class

IV rules might eventually produce a frozen configuration, or they might not. But either way, Langton says, the phase transition at the edge of chaos would correspond to what computer scientists call "undecidable" algorithms. These are the algorithms that might halt very quickly with certain inputs—equivalent to starting off the Game of Life with a known stable structure. But they might run on forever with other inputs. And the point is, you can't always tell ahead of time which it will be—even in principle. In fact, says Langton, there's even a theorem to that effect: the "undecidability theorem" proved by the British logician Alan Turing back in the 1930s. Paraphrased, the theorem essentially says that no matter how smart you think you are, there will always be algorithms that do things you can't predict in advance. The only way to find out what they will do is to run them.

And, of course, those are exactly the kind of algorithms you want for modeling life and intelligence. So it's no wonder the Game of Life and other Class IV cellular automata seem so lifelike. They exist in the only dynamical regime where complexity, computation, and life itself are possible: the edge of chaos.

Langton now had four very detailed analogies—

Cellular Automata Classes:
   I & II → "IV" → III

Dynamical Systems:
   Order → "Complexity" → Chaos

Matter:
   Solid → "Phase Transition" → Fluid

Computation:
   Halting → "Undecidable" → Nonhalting

—along with a fifth and far more hypothetical one:

Life:
   Too static → "Life/Intelligence" → Too noisy

But what did they all add up to? Just this, Langton decided: "solid" and "fluid" are not just two fundamental phases of matter, as in water versus ice. They are two fundamental classes of dynamical behavior in general—including dynamical behavior in such utterly nonmaterial realms as the space of cellular automaton rules or the space of abstract algorithms. Furthermore, he realized, the existence of these two fundamental classes of dynamical behavior implies the existence of a third fundamental class: "phase transition" behavior at the edge of chaos, where you would encounter complex computation and quite possibly life itself.

So did this mean that you might one day be able to write down general physical laws for the phase transition, laws that would somehow encompass both the freezing and thawing of water and the origin of life? Maybe. And maybe life began some four billion years ago as some kind of real phase transition in the primordial soup. Langton had no idea. But he did have this irresistible vision of life as eternally trying to keep its balance on the edge of chaos, always in danger of falling off into too much order on the one side, and too much chaos on the other. Maybe that's what evolution is, he thought: just a process of life's learning how to seize control of more and more of its own parameters, so that it has a better and better chance to stay balanced on the edge.

Who knew? Sorting it all out could be the work of a lifetime. Meanwhile, by 1986 Langton had finally gotten the engineering school to agree that, as a thesis topic, his ideas on computation, dynamical systems, and phase transitions in cellular automata were quite acceptable. But he still had plenty of work to do in getting his basic framework fleshed out enough to satisfy his thesis committee.

Go, Go, Go, Yes, Yes!

Two years earlier, in June 1984, Langton had gone to a conference on cellular automata at MIT and had happened to sit down at lunch one day next to a tall, rangy guy with a ponytail.

"What are you working on?" asked Doyne Farmer.

"I don't really know how to describe it," admitted Langton. "I've been calling it artificial life."

"Artificial life!" exclaimed Farmer. "Wow, we gotta talk!"

So they had talked. A lot. After the conference, moreover, they had kept on talking via electronic mail. And Farmer had made it a point to bring Langton out to Los Alamos on several occasions to give talks and seminars. (Indeed, it was at the "Evolution, Games, and Learning" conference in May 1985 that Langton gave his first public discussion of his lambda parameter and the phase transition work. Farmer, Wolfram, Norman Packard, and company were profoundly impressed.) This was the same period when Farmer was busy with Packard and Stuart Kauffman on the autocatalytic set simulation for the origin of life—not to mention helping to get the Santa Fe Institute up and running—and he was getting deeply involved with issues of complexity himself. Chris Langton, he figured, was precisely the kind of guy he wanted to have around. Furthermore, as a former antiwar activist himself, he was able to convince Langton that doing science at a nuclear weapons laboratory wasn't quite as spooky as it might seem: since the kind of research that Farmer and his group did was completely non-classified and nonmilitary, you could think of it as a way of diverting some of that "dirty" money to a good purpose.

The upshot was that in August 1986, Langton, his wife, and their two infant sons headed south for New Mexico and a postdoctoral appointment at Los Alamos' Center for Nonlinear Studies. The move was a big relief to Elvira Segura-Langton; after four years of Michigan snow and rain, she couldn't wait to get back to the sun. And it sounded wonderful to Langton himself; the Center for Nonlinear Studies was exactly the kind of place he wanted to be. True, he still had a few more computer runs to do before he was quite finished with his dissertation. But that wasn't unusual for people just starting their first postdoc. He should be able to wrap it all up and actually get his Ph.D. in just a few months.

It didn't quite work out that way. To finish his computer runs at Los Alamos, Langton needed the use of a workstation. That was no problem in principle. By the time he got there the Center for Nonlinear Studies had already received a whole shipment of workstations from Sun Micro-systems, along with all the cables and hardware needed to link them into a local area network. But it was a nightmare in practice. The machines were scattered around in various buildings and trailers, and the center was full of physicists who had not a clue how to make the system work. "Well, I was a computer scientist," says Langton, "so they just assumed I knew how. So I became the default system manager for the machines in my area."

John Holland, who was cochairman of Langton's dissertation committee along with Burks, and who arrived at Los Alamos shortly after Langton did to begin a year as a visiting scholar, was appalled. "Chris is just too good a guy," he says. "Any time anyone had a problem with the network, or with using a Sun, they'd come to him. And Chris, being Chris, would spend whatever time was necessary trying to get it straightened out. The first several months I was there, he spent more time on that than anything else. So here he is pulling wires through the wall. But all that time his dissertation is standing still.

"Art Burks and I were continually at Chris on this—as was Doyne," Holland adds. "The theme was always, 'Look, you've got to get your union card, or you're going to regret it later.' "

Langton understood that message completely; he wanted to get his dissertation finished just as badly as they wanted him to. But even when the network was up and running, of course, he still had to take all the computer codes he'd written for the Apollo workstation at Michigan and rewrite them to run on his Sun at Los Alamos. Nasty job. And then there were the preparations for his artificial life workshop in September 1987. (Part of the agreement when he'd come to Los Alamos was that he could organize such a workshop.) And—"Well, things just ran away from me," he admits. "I didn't get anything done on cellular automata that first year."

What Langton did get done was the workshop. Indeed, he threw himself into it with everything he had. "I was desperate to get back into artificial life," he says. "At Michigan I had done tons of computerized literature searches, and they were frustrating. If you used the keyword 'Self-Reproduction,' you'd get a flood of stuff. But if you tried 'Computers and Self-Reproduction,' you got nothing. And yet I kept stumbling across articles in weird, out-of-the-way places."

He could sense it. Somewhere out there were the authors of those weird, out-of-the-way articles: people just like he had been, lonely souls trying to follow this bizarre scent all by themselves without quite knowing what it was, or who else might be doing it too. Langton wanted to find these people and bring them together, so that they could begin to forge a real scientific discipline. The question was how.

In the end, says Langton, there was only one way to do it: "I just announced that there would be a conference on artificial life, and we'd see who showed up." Artificial life was still a good label, he decided. "I'd been using it since the University of Arizona, and people immediately grokked what I meant." On the other hand, he considered it crucial that people understand that term very, very clearly, or else he might start pulling in every flake in the country who wanted to show a whacked-out video game. "I spent a long time, about one man-month, wording the invitation," he says. "We didn't want the conference to be too far out, or too science fiction. But we didn't want to just limit it to DNA data bases, either. So I passed the invitation around here at Los Alamos. I refined it. I went over and over and over it again."

Then, once he had the invitation worded the way he wanted it, there was the question of how to broadcast it. Via nationwide electronic mail, maybe? In the UNIX operating system there was a utility program called SENDMAIL, which had a well-known bug that could be exploited to make an electronic message generate multiple copies of itself as it traveled. "I thought about using that bug to send out a self-reproducing message that would spread through the network to announce the conference—and then cancel itself," says Langton. "But I thought better of it. That wasn't the association I wanted."

In retrospect that was probably just as well. Two years later, in November 1989, a Cornell University graduate student named Robert Morris tried to exploit that same bug to write a computer virus—and ended up nearly crashing the whole nationwide research network when a programming error allowed the virus to propagate out of control. But even in 1987, says Langton, computer viruses were one of the few subjects that he actually wanted to discourage at the meeting. In one sense, they were a natural. Computer viruses could grow, reproduce, respond to their environment, and in general do almost everything carbon-based life-forms did; it was (and is) a fascinating philosophical question whether or not they are truly "alive." But computer viruses were also dangerous. "I didn't really want to stimulate people to go out and play with them," says Langton. "And frankly, at that time, it was pretty iffy whether, if we said anything about viruses in the workshop, the lab might actually step in and say, 'No you can't do this.' We didn't want to attract hackers to come to Los Alamos and try get into the secure computers."

In any case, says Langton, he finally just mailed the invitation to all the people he knew about who might be interested, and asked them to spread the word. "I had no idea how many people would come," he says. "Five to 500—I was clueless."

In fact, the turnout was about 150, including a handful of slightly baffled-looking reporters from places like *The New York Times* and *Nature* magazine. "In the end, we attracted just the right set of people," says Langton. "We had some on the lunatic fringe and some hard-nosed scoffers, but a lot of solid people in the middle." There were the usual suspects from Los Alamos and the Santa Fe Institute, of course—people like Holland, Kauffman, Packard, and Farmer. But the British biologist Richard Dawkins, author of *The Selfish Gene*, came from Oxford to talk about his "Biomorph" program for simulated evolution. Aristid Lindenmeyer came from Holland to talk about his computer simulations of embryonic development and plant growth. A. K. Dewdney, who had already promoted the conference in his "Computer Recreations" column in *Scientific American*, came and organized the computer demonstrations; he also ran the "Artificial 4-H" contest for the best computer creature. Graham Cairns-Smith came from Glasgow to talk about his theory of the origin of life on the surface of microscopic clay crystals. Hans Moravec came from Carnegie Mellon to talk about robots, and his conviction that they would one day be the heirs of humanity.

The list went on and on. Langton had no idea what most of the speakers would say until they got up to say it. "The meeting was a very emotional experience for me," he says. "You'll never re-create that feeling. Everybody had been doing artificial life on his own, on the side, often at home. And everybody had had this feeling of, 'There must be something here.' But they didn't know who to turn to. It was a whole collection of people who'd had the same uncertainties, the same doubts, who'd wondered if they were crazy. And at the meeting we almost embraced each other. There was this real camaraderie, this sense of 'I may be crazy—but so are all these other people.' "

There were no breakthroughs in any of the presentations, he says. But you could see the potential in almost all of them. The talks ranged from the collective behavior of a simulated ant colony, to the evolution of digital ecosystems made out of assembly-language computer code, to the power of sticky protein molecules to assemble themselves into a virus. "It was fascinating to see how far people had gotten on their own," says Langton. And more than that, it was fascinating to see how the same theme kept cropping up again and again: in virtually every case, the essence of fluid, natural, "lifelike" behavior seemed to lie in such principles as bottom-up rules, no central controller, and emergent phenomena. Already, you could sense the outlines of a new science taking shape. "That's why we told people not to turn their papers in until *after* the conference," he says. "Because it was only after having listened to all these other ideas that people could see more clearly what they had been thinking.

"It was hard to say exactly what was happening at the workshop," he says. "But 90 percent of it was to give people the confidence to keep pushing. By the time we all went away, it was as if we had each risen above all the things that were blocking us. Before, everything had been 'Stop,' 'Wait,' and 'No,' like my not being able to do a thesis on artificial life at the University of Michigan. But now, everything was "Go, Go, Go, Yes, Yes"!

"I was so hyped up, it was like an altered state of consciousness," he says. "I have this image of a sea of gray matter, with ideas swimming around, ideas recombining, ideas leaping from mind to mind."

For that space of five days, he says, "it was like being incredibly *alive*."

Some time after the meeting was over, Langton received an electronic mail message from Eiiti Wada, who had come to the meeting from the University of Tokyo. "The workshop was so intensive," wrote Wada, "that I had no time to confess to you that I was in Hiroshima when the first atomic bomb was dropped there."

He wished to thank Langton again for that most exciting week in Los Alamos, discussing the technology of life.

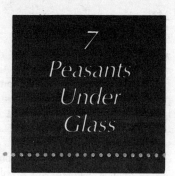

7

Peasants
Under
Glass

About five o'clock that same Tuesday, September 22, 1987, John Holland and Brian Arthur left the artificial life workshop at Los Alamos to drive back down the mesa toward Santa Fe. And aside from an occasional stop to savor the late afternoon vista to the east, where the Sangre de Cristo Mountains rose nearly 7000 feet above the valley of the Rio Grande, they spent the whole hour-long drive talking about "boids": a simulation presented at the workshop by Craig Reynolds of the Symbolics Corporation in Los Angeles.

Arthur was fascinated by the thing. Reynolds had billed the program as an attempt to capture the essence of flocking behavior in birds, or herding behavior in sheep, or schooling behavior in fish. And as far as Arthur could tell, he had succeeded beautifully. Reynolds' basic idea was to place a large collection of autonomous, birdlike agents—"boids"—into an on-screen environment full of walls and obstacles. Each boid followed three simple rules of behavior:

1. It tried to maintain a minimum distance from other objects in the environment, including other boids.
2. It tried to match velocities with boids in its neighborhood.
3. It tried to move toward the perceived center of mass of boids in its neighborhood.

What was striking about these rules was that none of them said, "Form a flock." Quite the opposite: the rules were entirely local, referring only to what an individual boid could see and do in its own vicinity. If a flock was going to form at all, it would have to do so from the bottom up, as an emergent phenomenon. And yet flocks *did* form, every time. Reynolds could start his simulation with boids scattered around the computer screen completely at random, and they would spontaneously collect themselves into a flock that could fly around obstacles in a very fluid and natural manner. Sometimes the flock would even break into subflocks that flowed around both sides of an obstacle, rejoining on the other side as if the boids had planned it all along. In one of the runs, in fact, a boid accidentally hit a pole, fluttered around for a moment as though stunned and lost—then darted forward to rejoin the flock as it moved on.

Reynolds had insisted that this last bit of business was proof that the behavior of the boids was truly emergent. There was nothing in the rules of behavior or in any of the other computer code that told that particular boid to act that way. So Arthur and Holland had started to chew the question over almost as soon as they got into the car: How much of the boids' behavior was built in, and how much of it was truly emergent behavior that was unexpected?

Holland remained to be convinced; he had seen all too many examples of "emergent" behavior that was tacitly built into the program from the start. "I was saying to Brian that you have to be cautious. Maybe everything that's happening in there, including the one bumping into the pole, is so obvious from the rules that you aren't learning anything new. At least, I'd want to have the ability to put other sorts of objects in, change the environment, and see if it still behaved in a reasonable way."

Arthur couldn't very well argue with that. "But for myself," he says, "I couldn't see how you could define 'truly' emergent behavior." In some sense, everything that happens in the universe, including life itself, is already built into the rules that govern the behavior of quarks. So what *is* emergence, anyway? And how do you recognize it when you see it? "That goes to the heart of the problem in artificial life," he says.

Since neither Holland nor anyone else had an answer to that, he and Arthur never did reach a firm conclusion. But in retrospect, says Arthur, that discussion did plant a seed in his sleep-deprived mind. In early October 1987, exhausted but happy, he ended his stint as a visiting scholar at the Santa Fe Institute and returned to Stanford. And there, once he had caught up on sleep, he continued to mull over what he had learned in Santa Fe. "I had been enormously impressed by Holland's genetic algorithms, and classifier systems, and the boids, and so on. I thought a good deal about them, and the possibilities they opened up. My instinct was that this was an answer. The problem was, What was the question in economics?

"Now, my earlier interests had been how economies change and develop in the Third World," he says. "So about November 1987 I phoned John and said that I'd had a vision of how these ideas might be applied to an economy. I had this notion that you could have within your office in the university a little peasant economy developing under a bubble of glass. Of course, it would really be in a computer. But it would have to be all these little agents, preprogrammed to get smart and interact with each other.

"Then in this dreamlike idea, you'd go in one morning and say, 'Hey, look at these guys! Two or three weeks ago all they were doing was bartering, and now they've got joint stock companies.' Then the next day you'd come in and say, 'Oh—they've discovered central banking.' Then a few days later you'd have all your colleagues clustered around and you're peering in: 'Wow! They've got labor unions! What'll they think of next?' Or half of them have gone Communist.

"At the time I still couldn't articulate it very well," says Arthur. But he knew that such an economy-under-glass would be profoundly different from conventional economic simulations, in which the computer just integrated a bunch of differential equations. His economic agents wouldn't be mathematical variables, but *agents*—entities caught up in a web of interaction and happenstance. They would make mistakes and learn. They would have histories. They would no more be governed by mathematical formulas than human beings are. As a practical matter, of course, they would have to be far simpler than real human beings. But if Reynolds could produce startlingly realistic flocking behavior with just three simple rules, then it was at least conceivable that a computer full of well-designed adaptive agents might produce startlingly realistic economic behavior.

"I thought vaguely we could cook up these agents via John's classifier system," says Arthur. "I couldn't see how to do it. John had no immediate suggestion of how to do it, either. But he was also enthused." The two men agreed that when the Santa Fe Institute's economics program got under way the next year, this would have to be a top priority.

The Fledgling Director

In the meantime, Arthur had plenty to keep him busy with the organization of that program. Indeed, he was just beginning to grasp the full implications of what he had committed himself to.

Holland, it soon developed, would not be able to share the directorship. He had already used up his sabbatical time the year before, when he had spent the 1986–87 academic year as a visiting scholar at Los Alamos. Back at Michigan he was still embroiled in the academic politics of his department's merger into the engineering school. And his wife, Maurita, was tied to her job as head of the science library system. At most, Holland would be able to come to Santa Fe for a month or so at a time.

So the job fell entirely to Arthur, who had never run a research program in his life, much less created one.

What does John Reed want us to do here? he asked Eugenia Singer, who was to be his liaison with the Citicorp chairman. "He says do anything you want," she replied after checking with Reed, "so long as it's not conventional."

What do *you* want us to do here? he asked Ken Arrow and Phil Anderson. They said they wanted him to create a completely rigorous new way of doing economics based on the complex adaptive systems point of view—whatever that was.

What does the institute want us to do here? he asked George Cowan and the other powers at Santa Fe. "The science board is hoping that you'll set radically new directions in economics," they told him. And by the way, your budget for the first year will be $560,000—partly from Citicorp, partly from the MacArthur Foundation, and partly from the National Science Foundation and the Department of Energy, which we're sure will be giving us a pair of hefty grants. Well, pretty sure, anyway. Of course, this will also be the first and the biggest major research program at the institute, so we'll all be watching closely to see how well you do.

"I left shaking my head," says Arthur. "Half a million dollars is about mid-sized on an academic scale. But this was an enormous challenge. It was like being told, 'Here's an ice ax and rope—go climb Mount Everest.' I was amazed. I was awestruck. I found it overwhelming."

In practice, of course, Arthur was far from alone. Both Arrow and Anderson were more than willing to give moral support, advice, and encouragement. "They were very much the bedrock, the gurus of the program," says Arthur. Indeed, he considered it *their* program. But they also made it very clear that Arthur was to be the chief executive officer. "They kept hands off," says Arthur. "It was up to me to direct the thing and get it going."

He made two key decisions early on, he says. The first had to do with topics. He was distinctly unenthused by the idea of applying chaos theory and nonlinear dynamics to economics, which seemed to be a big part of what Arrow had in mind. There were plenty of other groups doing that kind of thing already—and with very few worthwhile results, so far as he could tell. Nor was Arthur interested in having the program build some huge economic simulation of the whole world. "This may have been in John Reed's mind," he says, "and it seems to be the first thing engineers or physicists want to do. But it's as if I said to you, 'You're an astrophysicist, why don't you build a model of the universe?' " Such a model would be just about as hard to understand as the real universe, he says, which is why astrophysicists don't do it that way. Instead, they have one set of models for quasars, another set for spiral galaxies, another set for star formation, and so on. They go in with a computational scalpel to dissect specific phenomena.

And that's exactly what Arthur wanted to do in the Santa Fe program. He certainly didn't intend to back away from his economy-under-glass vision. But he also wanted people to learn how to walk before they tried to run. In particular, he says, he wanted to see the program take some of the classical problems in economics, the hoary old chestnuts of the field, and see how they changed when you looked at them in terms of adaptation, evolution, learning, multiple equilibria, emergence, and complexity—all the Santa Fe themes. Why, for example, are there speculative bubbles and crashes in the stock market? Or why is there money? (That is, how does one particular good such as gold or wampum become widely accepted as a medium of exchange?)

That emphasis on the old chestnuts got the program into hot water later, says Arthur, when a number of people on the institute's science board accused them of being insufficiently innovative. "But we thought it was just good science, good politics, and good procedure to approach the standard problems," he says. "These are problems that economists recognize. Above all, if we could prove that changing the theoretical assumptions to be more realistic made major differences to the insights you got, maybe getting a feeling of more realism in those insights, then we could show the field that we had really contributed something."

For much the same reason, he says, he resisted when Murray Gell-Mann urged him to come out with a manifesto for the economics program, something to nail to the Church Door. "Several times he pushed the idea," says Arthur. "He wanted something that said: 'The day has dawned for a different form of economics,' et cetera. But I thought about it and decided no. It would be far better to tackle it problem by problem—these old chestnuts in economics—and we'll be convincing."

The second key decision had to do with the kind of researchers Arthur recruited for the program. He needed people who were open-minded and sympathetic to the Santa Fe themes, of course. The ten-day economics workshop had proven to him just how fruitful and exciting such a group could be. "I realized early on that neither I nor Arrow, nor Anderson, nor anyone else was going to lay down the framework for the Santa Fe approach from on top," he says. "It would have to emerge from what we did, from the way we tackled problems, with everybody having his own ideas."

After his own fiasco trying to get that first increasing-returns paper published, however, Arthur also knew that it was critically important to establish the program's credibility among mainstream economists. So he wanted the participants to include economic theorists with impeccable, diamond-hard reputations—the likes of Arrow himself or Stanford's Tom Sargent. Not only would having them there help ensure that this still-to-be-defined Santa Fe approach met every existing standard of rigor, but if *they* went away talking Santa Fe-ism, then people were going to listen.

Unfortunately, assembling such a team turned out to be easier said than done. After consulting with Arrow, Anderson, Pines, and Holland to draw a list of candidates, Arthur was able to get pretty much everyone he wanted on the noneconomics side. Phil Anderson agreed to come for a short while, as did his former student, Richard Palmer of Duke University. Holland would come, of course. So would David Lane, a sharp-minded and argumentative probability theorist from the University of Minnesota. Arthur even got commitments from his Soviet coauthors, probability theorists Yuri Ermoliev and Yuri Kaniovski. And then there were Stuart Kauffman, Doyne Farmer, and all the rest of the crew from Los Alamos and Santa Fe. But when Arthur started calling the economists, he quickly discovered that his concerns about credibility were not misplaced. Almost everyone had heard rumors that *something* had happened in Santa Fe. Arrow was talking it up everywhere he went. But who or what was the Santa Fe Institute? "So when I called," says Arthur, "the economists were inclined to say, 'Well, yes, it's a bit late, and I've made plans.' I approached several who said that they'd wait and see how things worked out. Basically, it was very, very difficult to get economists who weren't at the meeting to be interested."

The good news was that the economists who had been at the economics meeting were a superb group—they'd been selected by Arrow, after all. Moreover, the response from outside that group was not entirely bleak. Arrow agreed to come for several months, as did Sargent. John Rust and William Brock agreed to come down from the University of Wisconsin. Ramon Marimon would likewise come down from Minnesota. John Miller would come from the University of Michigan, where he had just completed a Ph.D. dissertation that made heavy use of Holland's classifier systems. And in what Arthur regarded as a particularly satisfying coup, Frank Hahn agreed to come over from Cambridge, where he reigned as the leading economic theorist in England.

So all in all, says Arthur, about twenty people participated to a greater or lesser degree in the economics program that first year, with no more than seven or eight in residence at any one time. That was about the size of an economics department at a small college. And together they were supposed to reinvent the field.

The Santa Fe Approach

The economics program was due to start at the institute in September 1988, with a second week-long economics workshop to kick it off. So Arthur took up full-time residence there in June, to give himself the whole summer to get ready. He needed every minute of it. And he found that things only got more intense when the participants actually started arriving in the fall.

"People would come in to me daily," he says, "like one guy who didn't know how to change light bulbs. Could I do that? And the place was so small that occasionally I had to solve problems like, What office should you put someone in who smokes? Or should one person share an office with someone else who had hairy legs and wore shorts all the time? This person really objected to that. It also fell to me to organize the workshops. And part of it was traveling to recruit people, to talk to them, to get advice, and to spread the word."

When you're the boss, Arthur was discovering, you can't always go out and play with the other kids. You have to spend entirely too much of your time being the grown-up. Even with the yeoman help of the institute's permanent staff, Arthur found that about 80 percent of his time was being taken up by the nonscientific side of his job, and it wasn't a lot of fun. At one point, he says, he came home to their rented house in Santa Fe and started complaining to his wife, Susan, about how little time he had for research. "She finally said, 'Oh, knock it off. You've never been happier in your life.' And she was right."

She *was* right. Because for all the administrative nitty-gritty, says Arthur, that other 20 percent made up for everything. By the fall of 1988 the Santa Fe Institute was pulsing with energy—and not just because of the economics program. Late the previous autumn, the long-promised federal grants had indeed come through from the National Science Foundation and the Department of Energy. Cowan hadn't been able to talk those agencies into nearly as much money as he'd wanted—there was still no funding to hire a permanent faculty, for example—but he had gotten them to commit $1.7 million over the next three years, starting in January 1988. So the institute now had financial security until the beginning of 1991. And there was finally enough cash for people to get serious about doing what the institute had been created to do.

The science board, under Gell-Mann and Pines, had accordingly given the go-ahead for fifteen new workshops. Some workshops promised to come at the issue of complexity from a hard-core physics perspective, with a prime example being Physics of Information, Entropy, and Complexity, to be organized by a young Polish physicist from Los Alamos named Wojciech Zurek. Zurek's idea was to start from the concepts of information and computational complexity, as defined in computer science, and explore their deep connections with quantum mechanics, thermodynamics, the quantum radiance of black holes, and the (hypothetical) quantum origins of the universe.

Other workshops promised to approach complexity from the biological side, with two prime examples being a pair of workshops on the immune system to be organized by Los Alamos biologist Alan Perelson. Indeed, Perelson had already run a major Santa Fe workshop on immunology in June 1987, and was now leading a small ongoing research program there. The idea was that the body's immune system, which consists of billions of highly responsive cells that flood the bloodstream with antibodies to neutralize an invading virus or bacterium as soon as it appears, is a complex adaptive system in precisely the same sense that ecosystems and brains are. So the ideas and techniques floating around the institute ought to help illuminate immune-related problems such as AIDS, and autoimmune diseases such as multiple sclerosis or arthritis. And conversely, because so much was known about the molecular details of the immune system, a research program devoted to that system ought to help keep some of the more high-flying ideas around Santa Fe pinned down to reality.

Meanwhile, the science board also strongly endorsed the idea of bringing in visitors and postdocs not associated with any particular program or workshop. It was a continuation of the approach the institute had followed from the beginning: just get some very good, very smart people in here, and see what happens. The joke among the science board members was that the Santa Fe Institute was an emergent phenomenon all by itself. It was a joke they actually took quite seriously.

All this was fine by Cowan. He was always eager to find more people with that indefinable fire in their soul. It wasn't a matter of seeking out talent per se, he says. You could talk about the institute with an awful lot of excellent people who just didn't understand what you were driving at. Instead you had to look for a kind of resonance: "Either someone gets glassy-eyed or the communication begins," he says. "And if it does, then you're exercising a form of power that's extremely compelling: intellectual power. If you can get a person who understands the concept somewhere down in the bowels of the brain, where that same idea's been sitting forever, then you have a grasp on that person. You don't do it by physical coercion, but by a kind of intellectual appeal that amounts to a coercion. You grab them by the brains instead of by the balls."

It was no easier to find such people than it had ever been. But they were out there. And in ever-increasing numbers, they were beginning to come through Santa Fe—to the point where the tiny convent was often filled to overflowing. Indeed, with workshops and seminars constantly under way in the chapel, with as many as three or four people crowded into offices that would have been cramped for one, with office mates endlessly drawing on the blackboards and arguing, with bull sessions forming and reforming in the hallways or out on the patio under the trees, it was often almost impossible to think. And yet the energy and the camaraderie were electric. As Stuart Kauffman says, "I was learning a whole new way of looking at the world about twice a day."

They all were. "On a typical day there," says Arthur, "most people would disappear into their offices in the morning and you'd hear a lot of click-clacking of computer terminals and keyboards. But then someone would peer around your door. Have you done this? Have you thought of that? Have you got half an hour to talk to some visitor? Then we'd go to lunch, practically always together and usually to the Canyon Cafe, which we called the 'Faculty Club.' We became so well known there that the waitresses hardly even brought us menus anymore. We'd always say, 'Give me a number five,' so they didn't even have to ask."

The talk was endless and, for the most part, fascinating. Indeed, says Arthur, what he remembers more vividly than anything else were the impromptu seminar/bull sessions that were forever springing up in the late mornings or afternoons "We did that three, four, five times a week," he says. "Someone might go down the corridor and say, 'Hey, let's talk about X.' So about half a dozen of us would meet, either in the chapel or more often in the small meeting room by the kitchen. It was badly lit, but it was next to the coffee and the Coke machine. It had Navaho decorations dominating the room. And there was a photograph of Einstein in an Indian headdress, beaming down on us.

"So we would just sit around the table," he says, "maybe with Stuart leaning on the mantelpiece. Someone might put a problem on the board, and we'd start kicking around an awful lot of questions. These were actually very good arguments. They were never acrimonious. But they were quite hard-edged, because the issues that kept coming up were fundamental. They weren't the sort of technical problems that come up in academic economics, like how do you solve this or that fixed-point theorem, or in physics, like why does this material go superconducting at minus 253 degrees, or whatever. They were questions of where the science should go next. How do you deal with bounded rationality? How is economics supposed to proceed when the problems start to get truly complicated, as in chess? How would you think about an economy that is always evolving, that never settles down to an equilibrium? If you do computer experimentation in economics, how would that work?

"I think that's where Santa Fe came into its own," he says. "Because the answers we were starting to give and the techniques we were borrowing were, in my opinion, what began to define a Santa Fe approach to economics."

There was one series of discussions that Arthur remembers in particular, because it did so much to crystallize his own thinking. Arrow and Cambridge's Frank Hahn were both there, he says, so it must have been sometime during their visit in October or November 1988. "We would meet—myself and Holland, Arrow and Hahn, maybe Stuart and one or two others—and we would thrash out what economists could do about bounded rationality." That is, what would really happen to economic theory if they quit assuming that people could instantaneously compute their way to the solution of any economic problem—even if that problem were just as hard as chess?

They had almost daily meetings on the question in the small conference room. Arthur remembers Hahn pointing out once that the reason economists use perfect rationality is that it's a benchmark. If people are perfectly rational, then theorists can say exactly how they will react. But what would perfect *irrationality* be like? Hahn wondered.

"Brian!" he said. "You're Irish. You might know."

Seriously, Hahn continued while Arthur tried to laugh, there is only one way to be perfectly rational, while there are an infinity of ways to be partially rational. So which way is correct for human beings? "Where," he asked, "do you set the dial of rationality?"

Where do you set the dial of rationality? "That was Hahn's metaphor," says Arthur, "and it really stuck in my mind. I thought about it for a long time afterward. I chewed a lot of pencils. There was a lot of discussion." And slowly, like watching a photographic image emerging in a developer tray, he and the others began to see an answer: the way to set the dial of rationality was to leave it alone. Let the agents set it by themselves.

"You'd pull a John Holland on it," says Arthur. "You'd just model all these agents as classifier systems, or as neural nets, or as some other form of adaptive learning system, and then allow the dial to vary as the agents learn from experience. So all the agents could start off as perfectly stupid. That is, they would just make random, blundering decisions. But they would get smarter and smarter as they reacted to one another." Maybe they would get very smart indeed, and maybe they wouldn't; it all depended upon what they experienced. But either way, Arthur realized, these adaptive, artificially intelligent agents were exactly what you wanted for a real theory of dynamics in the economy. If you put them down in a stable, predictable economic situation, you might very well find them making exactly the kind of highly rational decisions that neoclassical theory predicts—not because they had perfect information and infinitely fast reasoning ability, but because the stability would give them time to learn the ropes.

However, if you put those same agents in the midst of simulated economic change and upheaval, they would still be able to function. Not very well, perhaps: they would stumble and fail and make any number of false starts, just as humans do. Nonetheless, under the influence of their built-in learning algorithms they would slowly grope their way toward some reasonable new course of action. And by the same token, if you put the agents into a competitive situation analogous to chess, where they had to choose moves against each other, you could watch them make their choices. If you put the agents into a simulated economy undergoing a simulated boom, you would watch them explore their immense space of possibilities. No matter where you put them, in fact, the agents would try to do *something*. So unlike the neoclassical theory, which has almost nothing at all to say about dynamics and change in the economy, a model full of adaptive agents would come with dynamics already built in.

This was obviously the same intuition as in his economy under glass, Arthur realized. Indeed, it was essentially the same vision he'd had almost a decade earlier, after reading *The Eighth Day of Creation*. Except that now he could see that vision with crystal clarity. Here was this elusive "Santa Fe approach": Instead of emphasizing decreasing returns, static equilibrium, and perfect rationality, as in the neoclassical view, the Santa Fe team would emphasize increasing returns, bounded rationality, and the dynamics of evolution and learning. Instead of basing their theory on assumptions that were mathematically convenient, they would try to make models that were psychologically realistic. Instead of viewing the economy as some kind of Newtonian machine, they would see it as something organic, adaptive, surprising, and alive. Instead of talking about the world as if it were a static thing buried deep in the frozen regime, as Chris Langton might have put it, they would learn how to think about the world as a dynamic, ever-changing system poised at the edge of chaos.

"Of course, that was not a totally new point of view in economics," says Arthur. The great economist Joseph Schumpeter may not have known words like "the edge of chaos," but he had pushed for an evolutionary approach to economics as far back as the 1930s. Richard Nelson and Sidney Winter of Yale University had been trying to foment an evolutionary movement in economics since the mid-1970s, with some success. And other researchers had even made some early attempts to model the effects of learning in economics. "But in these earlier learning models," says Arthur, "you assumed that the agents had already formed a more-or-less correct model of the situation they were in, and that learning was just a matter of sharpening up the model a bit by adjusting a few knobs. What we wanted was something much more realistic. We wanted these 'internal models' to *emerge*—to form inside the agents' minds, so to speak—as they learned. And we had a slew of methods we could use to analyze that process. There were Holland's classifier systems and genetic algorithms. Richard Palmer was just finishing a book on neural networks. David Lane and I knew how to mathematically analyze systems that learned on the basis of probability. Ermoliev and Kaniovski were experts on stochastic learning. And we had the whole literature in psychology. These approaches gave us a really fine-grained way to model adaptation, to make it algorithmically precise.

"In fact," Arthur adds, "the key intellectual influence that whole first year was machine learning in general and John Holland in particular—*not* condensed-matter physics, *not* increasing returns, *not* computer science, but learning and adaptation. And as we started to kick it around with Arrow and Hahn and the others, it was clear that what was exciting to all of us was an instinct that economics could be done in this very different way."

. . .

If the Santa Fe economists found the prospect exciting, however, they also found it vaguely disturbing. And the reason, says Arthur, was something that he didn't put his finger on until much later. "Economics, as it is usually practiced, operates in a purely deductive mode," he says. "Every economic situation is first translated into a mathematical exercise, which the economic agents are supposed to solve by rigorous, analytical reasoning. But then here were Holland, the neural net people, and the other machine-learning theorists. And they were all talking about agents that operate in an *in*ductive mode, in which they try to reason from fragmentary data to a useful internal model." Induction is what allows us to infer the existence of a cat from the glimpse of a tail vanishing around a corner. Induction is what allows us to walk through the zoo and classify some exotic feathered creature as a bird, even though we've never seen a scarlet-crested cockatoo before. Induction is what allows us to survive in a messy, unpredictable, and often incomprehensible world.

"It's as if you've parachuted into some negotiating session in Japan," says Arthur. "You've never been in Japan before, you don't know how the Japanese think, or act, or work. You can't quite understand what is going on. So most of the things you do are completely out of cultural context. And yet over time, you notice that some of the things you do are successful. So slowly, you and your company somehow learn to adapt and behave." (Of course, it's another story whether the Japanese actually buy your products.) Think of the situation as a competition like chess, he says. Players have fragmentary information about their opponents' intentions and abilities. And to fill in the gaps, they do indeed use logical, deductive reasoning. But they can only use it to look a few moves ahead, at most. Much more often the players operate in a world of *in*duction. They try to fill in the gaps on the fly by forming hypotheses, by making analogies, by drawing from past experience, by using heuristic rules of thumb. Whatever works, works—even if they don't understand why. And for that very reason, induction cannot depend upon precise, deductive logic.

At the time, Arthur admits, even he found this troubling. "Until I went to Santa Fe," he says, "I thought that an economic problem had to be well defined before you could even talk about it. And if it wasn't well defined, what the hell could you do with it? You certainly couldn't apply logic.

"But then John Holland taught us that this isn't so. When we talked to John and read his papers we began to realize that he was talking about cases where the problem context isn't well defined, and the environment isn't stationary over time. We said to him, 'John, how can you even *learn* in that environment?' "

Holland's answer was essentially that you learn in that environment because you have to: "Evolution doesn't care whether problems are well defined or not." Adaptive agents are just responding to a reward, he pointed out. They don't have to make assumptions about where the reward is coming from. In fact, that was the whole point of his classifier systems. Algorithmically speaking, these systems were defined with all the rigor you could ask for. And yet they could operate in an environment that was not well defined at all. Since the classifier rules were only hypotheses about the world, not "facts," they could be mutually contradictory. Moreover, because the system was always testing those hypotheses to find out which ones were useful and led to rewards, it could continue to learn even in the face of crummy, incomplete information—and even while the environment was changing in unexpected ways.

"But its behavior isn't optimal!" the economists complained, having convinced themselves that a rational agent is one who optimizes his "utility function."

"Optimal relative to what?" Holland replied. Talk about your ill-defined criterion: in any real environment, the space of possibilities is so huge that there is no way an agent can find the optimum—or even recognize it. And that's before you take into account the fact that the environment might be changing in unforeseen ways.

"This whole induction business fascinated me," says Arthur. "Here you could think about doing economics where the problem facing the economic agent was not even well defined, where the environment is not well defined, where the environment might be changing, where the changes were totally unknown. And, of course, you just had to think for about a tenth of a second to realize, that's what life is all about. People routinely make decisions in contexts that are not well defined, without even realizing it. You muddle through, you adapt ideas, you copy, you try what worked in the past, you try out things. And, in fact, economists had talked about this kind of behavior before. But we were finding ways to make it analytically precise, to build it into the heart of the theory."

Arthur remembers one key debate during that same period that went to the heart of the difficulty. "It was a long discussion in October, November 1988," he says. "Arrow, Hahn, Holland, myself, maybe half a dozen of us. We had just begun to realize that if you do economics this way—if there was this Santa Fe approach—then there might be no equilibrium in the economy at all. The economy would be like the biosphere: always evolving, always changing, always exploring new territory.

"Now, what worried us was that it didn't seem possible to do economics in that case," says Arthur. "Because economics had come to mean the investigation of equilibria. We'd gotten used to looking at problems as though they were butterflies, nailing them down to cardboard by holding them in equilibrium while we examined them, instead of letting them go past you and fly around. So Frankie Hahn said, 'If things are not repeating, if things are not in equilibrium, what can we, as economists, say? How could you predict anything? How could you have a science?' "

Holland took the question very seriously; he'd thought a lot about it. Look at meteorology, he told them. The weather never settles down. It never repeats itself exactly. It's essentially unpredictable more than a week or so in advance. And yet we can comprehend and explain almost everything that we see up there. We can identify important features such as weather fronts, jet streams, and high-pressure systems. We can understand their dynamics. We can understand how they interact to produce weather on a local and regional scale. In short, we have a real science of weather—without full prediction. And we can do it because prediction isn't the essence of science. The essence is comprehension and explanation. And that's precisely what Santa Fe could hope to do with economics and other social sciences, he said: they could look for the analog of weather fronts—dynamical social phenomena they could understand and explain.

"Well, Holland's answer was to me a revelation," says Arthur. "It left me almost gasping. I had been thinking for almost ten years that much of the economy would never be in equilibrium. But I couldn't see how to 'do' economics without equilibrium. John's comment cut through the knot for me. After that it seemed—straightforward."

Indeed, says Arthur, it was only during those conversations in the fall of 1988 that he really began to appreciate how profoundly this Santa Fe approach would change economics. "A lot of people, including myself, had naively assumed that what we'd get from the physicists and the machine-learning people like Holland would be new algorithms, new problem-solving techniques, new technical frameworks. But what we got was quite different—what we got was very often a new attitude, a new approach, a whole new world view."

The Darwinian Principle of Relativity

Holland, meanwhile, was having the time of his life at Santa Fe. He loved nothing better than to sit down with a bunch of very sharp people and kick ideas around. More important, however, these conversations were leading him to make a major course change in his own research—that is, the conversations plus the fact that he hadn't been able to figure out how to say no to Murray Gell-Mann.

"Murray's a great arm twister," laughs Holland. In the late summer of 1988, he says, Gell-Mann had telephoned him in Michigan: "John," he said, "you're doing all this stuff on genetic algorithms. Well, we need an example we can use against the creationists."

The fight against "creation science" was indeed one of Gell-Mann's many passions. He had gotten involved several years before, when the Louisiana Supreme Court was hearing arguments for and against a state law requiring that creation science be taught on an equal footing with Darwin's theory. Gell-Mann had persuaded almost all the U.S. recipients of what he calls "the Swedish prizes" in science to sign an amicus curiae brief urging repeal. And the court did vote 7 to 2 to throw out the law. But in the wake of that decision, as he read the newspaper correspondence, Gell-Mann had realized that the problem went well beyond the activities of a few religious fanatics: "People wrote in saying, 'Of course I'm not a fundamentalist, and I don't believe all that nonsense about creation science. However, the name brand of evolution they teach in our schools seems to have something wrong with it. Surely it couldn't be by blind chance that it all happened,' et cetera, et cetera. So these weren't creationists. But they couldn't be convinced, somehow, that just chance and selection have produced what we see."

So what he had in mind, he told Holland, was a series of computer programs, or even computer games, that would *show* people how it could happen. They would reveal how chance and selection pressure, operating over a vast number of generations, could produce a huge amount of evolutionary change. You would just set up the initial conditions—essentially a planet—and then let things rip. In fact, said Gell-Mann, he was thinking of organizing a workshop at the institute to talk about such games. Wouldn't Holland like to contribute something?

Well no, actually, Holland wouldn't. He was certainly sympathetic to what Gell-Mann was trying to do. But the fact was that he had a very full plate of research projects already—not the least of them being a classifier system he was writing for Arthur to apply to economic models. From his point of view, Gell-Mann's evolution simulator would be a distraction.

Besides, he'd already *done* the genetic algorithm, and he couldn't see that doing it all over again in another form would teach him anything new. So Holland said no as firmly as he could.

Okay then, said Gell-Mann, why don't you think about it. And not long after, he called again: John, this is really important. Wouldn't Holland change his mind?

Holland tried to say no once more—although he could already see that he was going to have a tough time making that answer stick. So in the end, after a long conversation, he abandoned all further resistance. "All right," he told Gell-Mann, "I'll try."

Actually, Holland admits, he wasn't putting up much of a fight by then anyway. In between the phone calls, as he'd thought about how to make Gell-Mann take no for an answer, he had started to think more and more about what he would do if he had to say yes. And he'd realized that there might actually be a rich opportunity here. Evolution, of course, was a lot more than just random mutation and natural selection. It was also emergence and self-organization. And that, despite the best efforts of Stuart Kauffman, Chris Langton, and a great many other people, was something that no one understood very well. Maybe this was a chance to do better. "I started looking at it," says Holland, "and realized that I could do a model that would satisfy Murray—or at least a piece of it might—and still do something interesting from the research viewpoint."

The model was actually a revival of something he'd done back in the early 1970s, Holland explains. At the time he was still working hard on genetic algorithms and his book *Adaptation*. But he was invited to give a talk at a conference in the Netherlands. And for the fun of it, he decided to tackle something completely different: the origin of life.

He called the talk and the paper based on it "Spontaneous Emergence," he says. And in retrospect, it took an approach that was quite similar in spirit to the autocatalytic models being independently pursued by Stuart Kauffman, Manfred Eigen, and Otto Rössler at about the same time. "My paper wasn't a computer model, as such," says Holland. "It was a formal model in which you could do mathematics. I was trying to show that you could design autocatalytic systems in which you could get a simple self-replicating entity, and that this would occur many orders of magnitude faster than the usual computations predicted."

Those usual calculations—still lovingly quoted by creationists—were first put forward by quite legitimate scientists back in the 1950s. The argument was that self-replicating life-forms could not possibly have originated from random chemical reactions in the prebiotic soup, because the time required would have been vastly greater than the age of the universe. It would have been like waiting for those fabled monkeys in the basement of the British Museum to produce the complete works of Shakespeare by banging away on typewriters: they will get there, but it will take them a *long* time.

Holland, however, wasn't any more discouraged by this argument than Kauffman and the others had been. Random reactions were all very well, he thought. But what about chemical catalysis, which is decidedly non-random? So in his mathematical model Holland postulated a soup of "molecules"—arbitrary symbols linked into strings of various lengths—which are acted upon by free-floating catalytic "enzymes": operators that did things to the strings. "These were very primitive operators like COPY, which could attach to any string whatsoever and make a copy of that string," says Holland. "I was actually able to prove a theorem. If you had a system with some of these operators floating around, and if you allowed recombination among arbitrary strings of different lengths—in effect, building blocks—then that system would produce a self-replicating entity much more rapidly than trying to do things purely at random."

That spontaneous emergence paper was what Holland calls "a singular point"; it was like nothing he had done before or since. And yet, the issues of emergence and self-organization were still very much in his mind. Just the year before, in fact, he had spent a lot of time batting them around with Doyne Farmer, Chris Langton, Stuart Kauffman, and others during his stay at Los Alamos. "So with Murray's arm-twisting, I thought maybe the time was ripe to do more along those lines," he says. "And maybe now I'd build a real computer model of these things."

Having spent all those intervening years on classifier systems, he says, the way to make a computer model seemed obvious. Since the free-floating operators in the original paper had had the effect of rules—"IF you encounter such and such a string, THEN do the following to it"—the thing to do was to write them that way in the program, and make the whole thing look as much like a classifier system as possible. And yet, as soon as he started thinking in those terms, Holland also realized that he was going to have to face up to the major philosophical flaw in classifier systems. In the spontaneous emergence paper, he says, the spontaneity had been real, and the emergence had been completely intrinsic. But in classifier systems, for all their learning ability and for all their power to discover emergent clusters of rules, there was still a *deus ex machina*; the systems still depended on the shadowy hand of the programmer. "A classifier system gets a payoff only because *I* assign winning or losing," says Holland.

It was something that had always bugged him. Leaving aside questions of religion, he says, the real world seems to get along just fine without a cosmic referee. Ecosystems, economies, societies—they all operate according to a kind of Darwinian principle of relativity: everyone is constantly adapting to everyone else. And because of that, there is no way to look at any one agent and say, "It's fitness is 1.375." Whatever "fitness" means—and biologists have been arguing about that since the time of Darwin—it cannot be a single, fixed number. That's like asking if a gymnast is a better or worse athlete than a sumo wrestler; the question is meaningless because there's no common scale to measure them. Any given organism's ability to survive and reproduce depends on what niche it is filling, what other organisms are around, what resources it can gather, even what its past history has been.

"That shift in viewpoint is *very* important," says Holland. Indeed, evolutionary biologists consider it so important that they've made up a special word for it: organisms in an ecosystem don't just evolve, they *coevolve*. Organisms don't change by climbing uphill to the highest peak of some abstract fitness landscape, the way biologists of R. A. Fisher's generation had it. (The fitness-maximizing organisms of classical population genetics actually look a lot like the utility-maximizing agents of neoclassical economics.) Real organisms constantly circle and chase one another in an infinitely complex dance of coevolution.

On the face of it, coevolution sounds like a recipe for chaos, says Holland. At the institute, Stuart Kauffman liked to compare it to climbing around in a fitness landscape made of rubber, so that the whole thing deforms every time you take a step. And yet somehow, says Holland, this dance of coevolution produces results that aren't chaotic at all. In the natural world it has produced flowers that evolved to be fertilized by bees, and bees that evolved to live off the nectar of flowers. It has produced cheetahs that evolved to chase down gazelles, and gazelles that evolved to escape from cheetahs. It has produced myriad creatures that are exquisitely adapted to each other and to the environment they live in. In the human world, moreover, the dance of coevolution has produced equally exquisite webs of economic and political dependencies—alliances, rivalries, customer-supplier relationships, and on and on. It is the dynamic that underlay Arthur's vision of an economy under glass, in which artificial economic agents would adapt to each other as you watched. It is the dynamic that underlay Arthur and Kauffman's analysis of autocatalytic technology change. It is the dynamic that underlies the affairs of nations in a world that has no central authority.

Indeed, says Holland, coevolution is a powerful force for emergence and self-organization in *any* complex adaptive system. And that's why he knew that, if he was ever going to understand these phenomena at the deepest level, he was going to have to start by eliminating this business of outside reward. Unfortunately, however, he also knew that the assumption of outside reward was intimately bound up in the classifier system's marketplace metaphor. Holland had set up a system in which each classifier rule was a very tiny, very simple agent that participated with the other rules in an internal economy where the currency was "strength," and where the only source of wealth was the payoff from the final consumer—that is, the programmer. And there was no way to get around that fact without changing the classifier system framework completely.

So that's what Holland did. What he needed, he decided, was a different and more elemental metaphor for interaction: combat. And what he came up with was Echo, a highly simplified biological community in which digital organisms roam the digital environment in search of the resources they need to stay alive and reproduce: the digital analogs of the water, grass, nuts, berries, et cetera. When the creatures meet, of course, they also try to make resources out of each other. ("Echo" is short for ecosystem.) "I compare it to a game that my daughter Manja has, called Mail Order Monsters," says Holland. "You have a bunch of possibilities of offense and defense, and how you put these together determines how well you do against other monsters."

More specifically, explains Holland, Echo represents the environment as a large flat plain dotted here and there with "fountains," which gush forth various kinds of resources represented by the symbols a, b, c, and d. Individual organisms randomly move about this environment in sheep mode, placidly grazing on whatever resources they come across and placing them in an internal reservoir. Whenever two organisms encounter one another, however, they instantly shift from sheep mode to wolf mode, and attack.

In the battle that follows, says Holland, the outcome is determined by each organism's pair of "chromosomes," which are just a set of resource symbols strung together into two sequences such as $aabc$ and $bbcd$. "If you're one of the organisms," he explains, "then you match your first string, which is called the 'offense,' with the other organism's second string, the 'defense.' And if they match, you get a high score. So it's much like the immune system: if your offense complements the other guy's defense, then you've made a breach. He does the same thing reciprocally back. His offense is a match against your defense. So the interaction here is awfully simple. Do your offensive and defensive capabilities overwhelm his?"

If the answer is yes, he says, then you get a meal: all the resource symbols in your opponent's reservoir and in both his chromosomes go into *your* reservoir. Furthermore, says Holland, if eating your erstwhile opponent means that you now have enough resource symbols in your reservoir to make a copy of your own chromosomes, then you can reproduce by creating a whole new organism—perhaps with a mutation or two. If not—well, back to grazing.

Echo wasn't exactly what Gell-Mann had had in mind, to put it mildly. It had nothing obvious for a user to play with, and nothing at all in the way of fancy graphics. Holland simply couldn't be bothered with any of that. To run the thing he would type in a string of cryptic numbers and symbols, and then watch as a cascade of even more cryptic output came pouring down the screen in columns of alphanumeric gibberish. (By this point he had graduated to a Macintosh II computer.) And yet Echo was Holland's kind of game. In it he had finally eliminated this business of explicit, outside reward. "It's the closure of the loop," he says. "You really are going clear back to the point of, If I don't gather enough resources to make a copy of myself, I don't survive." He had captured what he regarded as the essence of biological competition. And now he could use Echo as an intellectual playground, a place to explore and understand what co-evolution could really do. "I had a list of several phenomena that occur in ecological systems," he says. "And I wanted to show that even with this very simple structure, each of these phenomena would show up in one way or another."

At the head of that list was what the English biologist Richard Dawkins called the evolutionary arms race. This is where a plant, say, evolves ever tougher surfaces and ever more noxious chemical repellents to fend off hungry insects, even as the insects are evolving ever stronger jaws and ever more sophisticated chemical resistance mechanisms to press the attack. Also known as the Red Queen hypothesis, in honor of the Lewis Carroll character who told Alice that she had to run as fast as she could to stay in the same place, the evolutionary arms race seems to be a major impetus for ever-increasing complexity and specialization in the natural world—just as the real arms race was an impetus for ever more complex and specialized weaponry during the Cold War.

Holland certainly wasn't able to do much with the evolutionary arms race in the fall of 1988; at that point Echo was barely more than a design on paper. But within a year or so, it was all working beautifully. "If I started off with very simple organisms that used only one letter for their offense chromosome and one for their defense chromosome," says Holland, "then what I began to see were organisms that used multiple letters. [The organisms could lengthen their chromosomes through mutations.] They coevolved. One would add a little more offensive capability; the other would add defensive capability. So they got progressively more complex. And sometimes they would split, so I essentially got a new species.

"It was at that point," says Holland, "with the fact that even with such simple apparatus I could get arms races and speciation, that I began to be much more interested."

In particular, he says, he wanted to understand a deep paradox in evolution: the fact that the same relentless competition that gives rise to evolutionary arms races can also give rise to symbiosis and other forms of cooperation. Indeed, it was no accident that cooperation in its various guises actually underlay quite a few items on Holland's list. It was a fundamental problem in evolutionary biology—not to mention economics, political science, and all of human affairs. In a competitive world, why do organisms cooperate at all? Why do they leave themselves open to "allies" who could easily turn on them?

The essence of the problem is neatly captured by a scenario known as the Prisoners' Dilemma, which was originally developed in the branch of mathematics called game theory. Two prisoners are being held in separate rooms, goes the story, and the police are interrogating both of them about a crime they committed jointly. Each prisoner has a choice: he can inform on his partner ("defect") or else remain silent ("cooperate"—with his partner, not the police). Now, the prisoners know that if both of them remain silent, then both of them will be released; the police can't pin a thing on them without a confession. The police, however, are perfectly well aware of this. So they offer the prisoners a little incentive: if one of them defects and informs on his partner, then that prisoner will be granted immunity and go free—and will get a reward to boot. The partner, meanwhile, will be sentenced to the maximum—and to add insult to injury, will be assessed a fine to cover the first prisoner's reward. Of course, if *both* of the prisoners rat on each other, then both of them serve the maximum and neither gets a reward.

So what do the prisoners do—cooperate or defect? On the face of it, they ought to cooperate with each other and keep their mouths shut, because that way they both get the best result: freedom. But then they get to thinking. Prisoner A, being no fool, quickly realizes that there's no way he can trust his partner not to turn state's evidence and walk off with a fat reward, leaving him to pay for the privilege of sitting in a jail cell. The temptation is just too great. He also realizes that his partner, being no fool either, is thinking exactly the same thing about him. So prisoner A concludes that the only sane response is to defect and tell the police everything, because if his partner is crazy enough to keep his mouth shut, then prisoner A will be the one walking out with the cash. And if his partner does the logical thing and talks—well, since prisoner A has to serve time anyway, at least he won't be paying a fine on top of it. So the upshot is that both prisoners are led by ruthless logic to the least desirable outcome: jail.

In the real world, of course, the dilemma of trust and cooperation is rarely so stark. Negotiations, personal ties, enforceable contracts, and any number of other factors affect the players' decisions. Nonetheless, the Prisoners' Dilemma does capture a depressing amount of truth about mistrust and the need to guard against betrayal. Consider the Cold War, when the two superpowers locked themselves in to a forty-year arms race that was ultimately to the benefit of neither, or the seemingly endless Arab-Israeli deadlock, or the eternal temptation for nations to erect protectionist trade barriers. Or in the natural world, consider that an overly trusting creature might very well get eaten. So once again: Why should any organism ever dare to cooperate with another?

A big part of the answer came in the late 1970s with a computer tournament organized at Michigan by Holland's BACH colleague Robert Axelrod, a political scientist with a long-standing interest in the cooperation question. Axelrod's idea for the tournament was straightforward: anyone who liked could enter a computer program that would take the role of one of the prisoners. The programs would then be paired up in various combinations, and they would play the Prisoners' Dilemma game with each other by choosing whether to cooperate or defect. But there was a wrinkle: instead of playing the game just once, each pair of programs would play it over and over again for 200 moves. This would be what game theorists called an *iterated* Prisoners' Dilemma, arguably a much more realistic way of representing the kind of extended relationships we usually get into with each other. Moreover, this repetition would allow the programs to base their cooperate/defect decisions on what the other program had done in previous moves. If the programs meet only once, then defection is obviously the only rational choice. But when they meet many times, then each individual program will develop a history and a reputation. And it was far from obvious how the opposing program should deal with that. Indeed, that was one of the main things Axelrod wanted to learn from the tournament: What strategies would produce the highest payoff over the long run? Should a program always turn the other cheek and cooperate regardless of what the other player does? Should it always be a rat and defect? Should it respond to the other player's moves in some more complex manner? And if so, what?

In fact, the fourteen programs submitted in the first round of the tournament embodied a variety of complex strategies. But much to the astonishment of Axelrod and everyone else, the crown went to the simplest strategy of all: TIT FOR TAT. Submitted by psychologist Anatol Rapoport of the University of Toronto, TIT FOR TAT would start out by cooperating on the first move, and from there on out would do exactly what the other program had done on the move before. That is, the TIT FOR TAT strategy incorporated the essence of the carrot and the stick. It was "nice" in the sense that it would never defect first. It was "forgiving" in the sense that it would reward good behavior by cooperating the next time. And yet it was "tough" in the sense that it would punish uncooperative behavior by defecting the next time. Moreover, it was "clear" in the sense that its strategy was so simple that the opposing programs could easily figure out what they were dealing with.

Of course, with only a handful of programs entered in the tournament, there was always the possibility that TIT FOR TAT's success was a fluke. But maybe not. Of the fourteen programs submitted, eight were "nice" and would never defect first. And every one of them easily outperformed the six not-nice rules. So to settle the question Axelrod held a second round of the tournament, specifically inviting people to try to knock TIT FOR TAT off its throne. Sixty-two entrants tried—and TIT FOR TAT won again. The conclusion was almost inescapable. Nice guys—or more precisely, nice, forgiving, tough, and clear guys—can indeed finish first.

Holland and the other members of the BACH group were naturally enchanted by all this. "I'd always been tremendously bothered by the Prisoners' Dilemma," says Holland. "It was one of those things I just didn't *like*. So to see the resolution was a delight. Just invigorating. Great stuff."

It was lost on no one that TIT FOR TAT's success had profound implications both for biological evolution and for human affairs. In his 1984 book, *The Evolution of Cooperation*, Axelrod pointed out that TIT FOR TAT interaction can lead to cooperation in a wide variety of social settings—including some of the most unpromising situations imaginable. His favorite example was the "live-and-let-live" system that spontaneously developed during World War I, when units in the front-line trenches would refrain from shooting to kill, so long as the other side refrained as well. The troops on one side of no-man's-land had no chance to communicate with their counterparts on the other side, and they certainly weren't friends. But what made the system work was that the same units were bogged down on both sides for months at a time, giving them a chance to adapt to each other.

In a chapter of the book coauthored with biologist and BACH colleague William Hamilton (and adapted from a prize-winning 1981 paper in the journal *Science*), Axelrod also pointed out that TIT FOR TAT interactions lead to cooperation in the natural world even without the benefit of intelligence. Examples include lichens, in which a fungus extracts nutrients from the underlying rock while providing a home for algae that in turn provide the fungus with photosynthesis; the ant-acacia tree, which houses and feeds a type of ant that in turn protects the tree; and the fig tree, whose flowers serve as food for fig wasps that in turn pollinate the flowers and scatter the seeds.

More generally, Axelrod said, the process of coevolution should allow TIT FOR TAT–style cooperation to thrive even in a world full of treacherous sleazoids. Suppose a few TIT FOR TAT individuals arise in such a world by mutation, he argued. Then so long as those individuals meet one another often enough to have a stake in future encounters, they will start to form little pockets of cooperation. And once that happens, they will perform far better than the knife-in-the-back types around them. Their numbers will therefore increase. Rapidly. Indeed, said Axelrod, TIT FOR TAT–style cooperation will eventually take over. And once established, the cooperative individuals will be there to stay; if less-cooperative types try to invade and exploit their niceness, TIT FOR TAT's policy of toughness will punish them so severely that they cannot spread. "Thus," wrote Axelrod, "the gear wheels of social evolution have a ratchet."

Shortly after the book was published, Axelrod produced a computer simulation of this scenario in collaboration with Holland's then-graduate student Stephanie Forrest. The question was whether a population of individuals coevolving via the genetic algorithm could discover TIT FOR TAT. And the answer was yes: in the computer runs, either TIT FOR TAT or a strategy very much like it would appear and spread through the population very quickly. "When it did that," says Holland, "we all threw our hands up and said, Hooray!"

This TIT FOR TAT mechanism for the origin of cooperation was exactly the sort of thing Holland meant when he said that people at the institute ought to be looking for the analog of "weather fronts" in the social sciences. And the whole issue of cooperation was there at the back of his mind while he was developing Echo, he says. It was certainly not something that could arise in the first version of the program, because he had built in the assumption that individual organisms will always fight. But in a more recent version, he has tried to broaden the organisms' repertoire to include the possibility of cooperation. Indeed, he has been trying to make Echo into a kind of "unified" model of coevolution.

"At the institute we've now had three ongoing models outside of Echo," he explains. "We had a stock market model, the immune system model, and a model that [Stanford economist] Tom Sargent made that involved trading. I realized that these all had very similar features. They all had 'trade,' in that there were goods being exchanged in one way or another. They all had 'resource transformation,' such as might be produced by enzymes or production processes. And they all had 'mate selection,' which acted as a source of technological innovation. So I began on the unified model. I can remember Stephanie Forrest and John Miller and I sitting down and trying to figure out, What's the minimum apparatus that we could put into Echo to imitate all these things? It turned out you could do it by adding things to the offensive and defensive chromosomes without much change in the basic model. I added the possibility of trading by providing additional identifiers defined by the chromosomes; these would be analogous to trademarks, or molecular markers on the surface of the cell. And the minute I did that I had to add, for the first time, something that looked like a rule in Echo: 'If the other guy shows such and such an identifying tag, then I'm going to attempt to trade instead of fight.' That allowed the evolution of cooperation, and such 'aberrations' as lying and mimicry. With that, I sketched out how to do a version of Sargent's model. Then I started sketching out how I could make Echo look like an immune system model by stretching it in another direction, and so on. The current version of Echo stems from that."

This unified version of Echo has been quite successful, he says. With it he's been able to demonstrate both the evolution of cooperation and the evolution of predator-prey relationships simultaneously, in the same eco-system. And that success has inspired him to start work on still more sophisticated variations of Echo: "There's a later version that I'm program-ming right now that allows this thing to evolve multicelled organisms," he says. "So now, instead of talking about trading and so on, my hope is that I can talk about the emergence of individuals and organizations. There are nice things to be learned when each agent is trying to maximize its repro-ductive rate but is constrained by the necessity of the continuation of the overall organization. Cancers are a good example of failure in this dimen-sion—I'll not talk about the U.S. automotive industry!"

Practical applications of such models are still in the future, says Holland. But he's convinced that a few good computer simulations along these lines might do more for the world than almost anything else on the Santa Fe agenda. "If we do this right," he says, "then people who are not scientists—people in Washington, for example—will be able to create models that can give them some feeling for the implications of various policy options, without having to know all the details of how that model actually works." In effect, he says, such models would be like flight simulators for policy, and would allow politicians to practice crash-landing the economy without taking 250 million people along for the ride. The models wouldn't even have to be terribly complicated, so long as they gave people a realistic feel for the way situations develop and for how the most important variables interact.

Holland admits that listeners have been pretty underwhelmed when he talks about this flight simulator idea in Washington; most practicing politicians are too busy dodging the punches coming at them right *now* to think about strategy for their next fight. On the other hand, he is clearly not the only one thinking in simulator terms. In 1989 the Maxis Company of Orinda, California, brought out a simulation game called SimCity, in which the player takes the role of a mayor and tries to nurse his or her city to prosperity in the face of crime, pollution, traffic congestion, and tax revolts. The game quickly went to the top of the best-seller charts. Real city planners swore by it; as simple as the simulation was, they said, and as many details as it left out, SimCity *felt* right. Holland, of course, bought a copy immediately—and loved it. "SimCity is one of the best examples I know of this flight simulator idea," he says. The Santa Fe Institute is talking seriously with Maxis about adapting a SimCity-style interface to use with some of its own simulations. And Holland is now working with Maxis to develop a user-friendly version of Echo that anyone can use for computer experiments.

Wet Labs for the Mind

All through those early days of the Santa Fe economics program, meanwhile, Brian Arthur was taking a keen interest in computer experiments, too. "Mostly in the program we were doing mathematical analysis and proving theorems, just as in standard economics," he says. "But because we were studying increasing returns, learning, and this ill-defined world of adaptation and induction, the problems often got too complicated for the mathematics to handle. So then we had to resort to the computer to see how things would work out. The computer was like a wet lab where we could see our ideas play out in action."

Arthur's problem, however, was that the thought of computer modeling gave a lot of economists the willies, even in Santa Fe. "I guess we're going to have to do simulation in economics," Arrow glumly told him one day over lunch. "But I think I'm too old to change."

"Thank god, my boy, I'm retiring," said the sixtyish Hahn on another occasion. "If the era of theorems is passing, I don't want to be there."

Arthur had to admit that there was good reason for the economists to be leery; in many ways, he felt the same way himself. "The history of simulation in the field was absolutely dismal," he says. "Early in my own career, my colleague Geoff McNicoll and I spent a lot of time looking at simulation models in economics, and we came to two conclusions—which were widely shared. The first was that, by and large, only people who couldn't think analytically resorted to computer simulations. The whole culture of our discipline calls for deductive, logical analysis, and simulation runs counter to that. The second conclusion was that you could prove anything you wanted by tweaking the assumptions deep in your model. Often people would start from basically political positions—say, we need lower taxes—and then twiddle the assumptions to show that lower taxes would be wonderful. Geoff and I made a game of going into models and finding the one assumption you could tweak that would change the entire outcome. Other people did this as well. So computer simulation got a very bad name in social science and especially in economics. It was kind of the resort of the scoundrel."

Even after all these years, in fact, Arthur finds that he's still allergic to the word "simulation"; he much prefers to call what he and his colleagues did in the economics program "computer experiments"—a phrase that captures the kind of rigor and precision he saw being practiced by Holland and the Santa Fe physicists. At the time, he says, their approach to computer modeling was a revelation. "I thought it was wonderful," he says. "In the hands of someone who was being extremely careful, where all the assumptions were carefully laid out, where the entire algorithm was explicitly given, where the simulation was repeatable and rigorous, like a lab experiment—then I saw that computer experiments would be perfectly fine. In fact, the physicists were telling us that there were three ways now to proceed in science: mathematical theory, laboratory experiment, and computer modeling. You have to go back and forth. You discover something with a computer model that seems out of whack, and then you go and do theory to try to understand it. And then with the theory, you go back to the computer or to the laboratory for more experiments. To many of us, it seemed as though we could do the same thing in economics with great profit. We began to realize that we'd been restricting ourselves in economics unnaturally, by exploring only problems that might yield to mathematical analysis. But now that we were getting into this inductive world, where things started to get very complicated, we could extend ourselves to problems that might only be studied by computer experiment. I saw this as a necessary development—and a liberation."

The hope, of course, was that the Santa Fe economics program could come up with computer models compelling enough to convince the rest of the profession—or, at least, that would not turn them off any more than they already were. And, indeed, by the fall of 1988 Arthur and his team already had several such computer experiments under way.

Arthur's own effort, begun in collaboration with Holland, was a direct descendant of his original economy-under-glass vision. "By the time I got to Santa Fe in June 1988," he says, "I'd realized that we needed to start with a more modest problem than building a whole artificial economy. And that led to the artificial stock market."

Of all the hoary old chestnuts in economics, he explains, stock market behavior is one of the hoariest. And the reason is that neoclassical theory finds Wall Street utterly incomprehensible. Since all economic agents are perfectly rational, goes the argument, then all investors must be perfectly rational. Moreover, since these perfectly rational investors also have exactly the same information about the expected earnings of all stocks infinitely far into the future, they will always agree about what every stock is worth— namely, the "net present value" of its future earnings when they are discounted by the interest rate. So this perfectly rational market will never get caught up in speculative bubbles and crashes; at most it will go up or down a little bit as new information becomes available about the various stocks' future earnings. But either way, the logical conclusion is that the floor of the New York Stock Exchange must be a very quiet place.

In reality, of course, the floor of the New York Stock Exchange is a barely controlled riot. The place is wracked by bubbles and crashes all the time, not to mention fear, uncertainty, euphoria, and mob psychology in every conceivable combination. Indeed, says Arthur, a martian who subscribed to the interplanetary edition of the *Wall Street Journal* might very well end up thinking that the stock market was a living thing. "The stories refer to the market almost as if it has psychological moods," he says. "The market is jittery, the market is depressed, the market is confident." The place is a form of artificial life all by itself. So in 1988, says Arthur, it only seemed appropriate to try to model the stock market Santa Fe style: "Our idea was that we would go in with a scalpel, excise the perfectly rational agents from the standard neoclassical models, and in their place slide in artificially intelligent agents that could learn and adapt the way humans do. So the model would have one stock, which the agents could buy and sell. And as they learned rules for trading, you could watch what kinds of market behavior emerged."

The question, obviously, was what that emergent behavior would be. Would the agents just calmly settle down and start trading stock at the standard neoclassical price? Or would they go into a more realistic pattern of constant upheaval? Arthur and Holland had no doubt it would be the latter. But in fact there was quite a bit of skepticism on that score, even around the institute.

Arthur particularly remembers one meeting in March 1989, when Holland was back down from Ann Arbor for a time, and several other people had come in for an economics workshop in the convent's small conference room. When the subject of the stock market model came up, Tom Sargent and Minnesota's Ramon Marimon both argued very strongly that the prices bid by the adaptive agents would very quickly settle down to the stock's "fundamental value"—that is, the one predicted by neoclassical theory. The market might show a few random fluctuations up or down, they said. But the agents couldn't really do anything else; the fundamental value would draw them in like an immense gravitational field.

"Well, John and I looked at each other and just shook our heads," says Arthur. "We said no—it was our strong instinct that the stock market we were building had so much potential to self-organize its own behavior, to become complex, that rich new behavior would have to emerge."

It got to be quite a debate, Arthur recalls. He knew, of course, that Sargent had been an enthusiast for the Holland approach to learning since that first economics workshop in September 1987. Indeed, Sargent had started to study the impact of learning on economic behavior well before that. And Marimon, meanwhile, was as enthusiastic for computer experimentation as Arthur himself was. But to Arthur, it didn't seem that Marimon and Sargent saw learning as anything really *new* in economics. They seemed to see it as a way of strengthening the standard ideas—as a way of understanding how economic agents will grope their way toward neoclassical behavior even when they aren't perfectly rational.

Well, fair enough. Arthur had to admit that the two men had good reason to feel that way. Quite aside from theory—where Sargent's work on "rational expectations" was well known—they had quite a bit of experimental evidence on their side. In a number of laboratory simulations, with students playing the role of traders in simple stock markets, researchers had shown that the experimental subjects converged on the fundamental price very quickly. Moreover, Marimon and Sargent were well along on a Santa Fe–style computer model of their own: another old chestnut known as Wicksell's triangle. The scenario here is that three types of agents produce and consume three types of goods, one of which eventually emerges as a medium of exchange: money. And when Marimon and Sargent replaced the rational agents of the original model with classifier systems, they found that the system converged on the neoclassical solution every time. (That is, the medium of exchange was the good with the lowest storage cost— metal disks, say, instead of fresh milk.)

Nonetheless, Arthur and Holland stuck to their guns. "The question was," says Arthur, "does realistic adaptive behavior lead you to the rational expectations outcome? To my mind, the answer was yes—but only if the problem is simple enough, or if the conditions are repeated again and again. Basically, rational expectation is saying that people aren't stupid. Then it's like tic-tac-toe: after a few times I learn to anticipate my opponent, and we both play perfect games. But if it's an on-off situation that's never going to happen again, or if the situation is very complicated, so that your agents have to do an awful lot of computing, then you're asking for a hell of a lot. Because you're asking them to have knowledge of their own expectations, of the dynamics of the market, of other people's expectations, of other people's expectations *about* other people's expectations, et cetera. And pretty soon, economics is loading almost impossible conditions onto these hapless agents." Under those circumstances, Arthur and Holland argued, the agents are so far from equilibrium that the "gravitational pull" of the rational expectations outcome becomes very weak. Dynamics and surprise are everything.

The debate was both affable and intense, Arthur recalls, and it went on like that for some time. In the end, of course, neither side conceded. But Arthur definitely felt a challenge: If he and Holland believed that their stock market model could show realistic emergent behavior, then it was up to them to prove it.

Unfortunately, the programming of the stock market model had proceeded only in fits and starts by that point. Mostly fits. Arthur and Holland had roughed out an initial draft of the simulation over lunch one day in June 1988, when both of them were in Santa Fe to lecture at the institute's first complex systems summer school. During the summer, back in Ann Arbor, Holland had coded up a full-fledged classifier system and genetic algorithm in the only computer language that Arthur knew: BASIC. (This was what finally led Holland to give up writing his programs in hexadecimal notation; he had to teach himself the language, but he's written in BASIC ever since.) And during the fall, once Holland was back in Santa Fe for the first few months of the economics program, they had tried to develop the stock market model further. But with Holland getting deflected off into Echo, and with Arthur bogged down in administrative duties, nothing happened very fast.

Worse, Arthur was beginning to realize that classifier systems, for all their conceptual brilliance, could often be a bear to work with. "In the beginning," he says, "the atmosphere at Santa Fe was that classifier systems could do anything. They could crack the stock market problem. They could make your coffee in the morning. So I used to pull John's leg: 'Hey, John, is it true that classifier systems can produce cold fusion?'

"But then in early 1989 David Lane and Richard Palmer organized a study group on John Holland's ideas, where we would meet maybe four times per week before lunch. John had left by then, but for about a month we worked our way through his book *Induction*, took it apart. And as we got into classifier systems technically, I discovered that you had to have a very careful design to make sure the architecture worked in practice. You had to be very careful about how you hitched one rule to another. Also, you could have 'deep' classifier systems—that is, one where rules triggered rules triggered rules in long chains—or you could have 'wide' ones: stimulus-response–type systems where there would be 150 different ways to act under slightly different conditions, but where the rules weren't looking at what each other did. My experience was that wide systems learned very well, and deep systems didn't."

Arthur had a lot of talks about this problem with Holland's former student Stephanie Forrest, who was now at the University of New Mexico in Albuquerque and a frequent visitor to the institute. The problem, she told him, was in Holland's bucket-brigade algorithm for assigning credit to the rules. If a bucket brigade has to pass credit back through several generations of rules, it will usually have very little credit left to go around once it gets back to the ancestors. So it's no wonder that the shallow systems learn better. Indeed, coming up with refinements and alternatives to the bucket-brigade algorithm had become one of the most intensive areas of classifier system research.

"For those reasons I became skeptical of classifier systems," says Arthur.

"With familiarity, the drawbacks had become clear. And yet the more I looked at them, the more I admired the thinking behind them. I really loved the idea that you could have many conflicting hypotheses in your mind, and that these hypotheses could compete, so that you didn't have to preprogram the expertise. I began to conceive of Holland's systems slightly differently from John. I thought of them as being like an ordinary computer program with many modules and branching points, but where the program itself has to learn which module to trigger at any given time, rather than triggering them in a fixed sequence. And once I began to conceive of them as a self-adapting computer program, I became much more comfortable. That's what I thought John had achieved."

In any case, he says, they finally did get a version of the stock market model up and running. Sargent himself suggested a number of ways to simplify the original design, which helped a lot. And in the late spring of 1989 Duke University physicist Richard Palmer joined the project, bringing with him his inestimable programming skills.

Palmer, meanwhile, was intrigued by the model for much the same reasons that Holland and Arthur were. "It related to self-organization, which is an area that fascinates me a lot," he says. "How the brain is organized, the nature of self-awareness, how life spontaneously arose—a few big questions I keep in the back of my mind."

Besides, he says, he was getting a little restless with the project that had already devoured most of his time at Santa Fe: the Double Oral Auction Tournament, a joint effort with Carnegie-Mellon's John Miller and Wisconsin's John Rust. The tournament, which was ultimately held in early 1990, had been conceived during the first economics workshop in September 1987. It was very similar in spirit to Axelrod's tournament a decade before. But instead of playing the iterated Prisoners' Dilemma game, the programs would embody various strategies for brokers in a commodities market such as the stock exchange. Is it best to announce your bid at the start? Should you keep quiet and wait for a better price? What? Since the buyers and the sellers in such a market are both bidding simultaneously—thus the name "double auction"—the answer was not at all obvious.

Well, the tournament promised to be a lot of fun, says Palmer, and the programming that he and his colleagues were doing to get ready for it was certainly a challenge. But the agents in the model were essentially static. For him, the tournament just didn't have the magic of Arthur and Holland's model, where you could hope to see agents growing more and more complex, and developing a real economic life of their own.

So in the early spring, Palmer pitched in. By May 1989 he and Arthur had a preliminary version of the stock market up and running. As planned, they started their agents off from total stupidity—random rules—and let them learn how to bid. As expected, they saw the agents learning energetically.

And on every run, they watched the damn things doing exactly what Tom Sargent had said they would do. "We had a single stock with a $3 dividend in the model," says Arthur. "We had a 10 percent discount rate. So the fundamental value was $30. And the prices indeed settled down to fluctuations around $30. It proved the standard theory!"

Arthur was chagrined and disgusted. It seemed that the only thing left to do was to call Sargent back at Stanford and congratulate him. "But then Richard and I walked in one morning and ran it on my Macintosh. We kept looking at it, discussing how to improve it. And we noticed that every time the price hit 34, the agents would buy. We could graph that. It seemed anomalous behavior. We thought it was a bug in the model. But then after thinking hard about it for an hour or so, we realized that there was no mistake! The agents had discovered a primitive form of technical analysis. That is, they had come to the belief that if the price went up enough, then it would continue to go up. So buy. But, of course, that belief became a self-fulfilling prophecy: if enough agents tried to buy at price 34, that would *cause* the price to go up."

Furthermore, he says, exactly the inverse happened as the price fell to 25: the agents tried to sell, thereby creating a self-fulfilling prophecy for falling prices. Bubbles and crashes! Arthur was exuberant. And even Palmer, normally a most cautious man, found the enthusiasm infectious. The result would be confirmed again and again in later, more complete versions of the model, says Arthur. But that morning in May 1989 they knew they had it.

"Immediately," he says, "we realized that we had the first glimmer of an emergent property in the system. We had the first glimmer of *life*."

8
*Waiting
for
Carnot*

At the very end of November 1988, the secretaries at the Los Alamos Center for Nonlinear Studies handed Chris Langton a sealed, official-looking envelope, inside of which he found a memorandum signed by laboratory director Siegfried Hecker:

> It has recently come to our attention that you have entered the third year of your postdoctoral fellowship, at the same time that you have not yet completed your Ph.D. dissertation. According to DOE Rule 40-1130, we are not allowed to employ postdoctoral fellows beyond their third year unless they have obtained the Ph.D. degree. In your case, due to a clerical error, we neglected to warn you in advance of the possible violation of this rule. In view of this fact, we have obtained an extension from the DOE office, so that you will not be liable to return payment for the FY89 portion of your fellowship. However, until you have obtained your Ph.D. degree, we will not be able to extend your appointment beyond 12/1/88.

In short, "You're fired." Panic-stricken, Langton ran to Center associate director Gary Doolen, who gravely assured him that yes, there was such a rule. And yes, Hecker really could do this.

Langton still shudders at the memory. Those bastards left him in freak-out mode for a full two hours before they actually sprang the surprise party. "The DOE rule number should have given it away," says Doyne Farmer, who had written the memo and organized the entire charade. "Chris was turning forty, and his birthday was 11/30."

Happily, once Langton had recovered, the party turned out to be a pretty good one. After all, Ph.D. candidates don't turn forty every day. Farmer had even gotten Langton's colleagues at the Center and in the lab's theory division to chip in and buy him a new electric guitar. "But I was seriously trying to prod him to finish up," says Farmer, "because I was genuinely worried that the shit was going to hit the fan over his not having a degree. And I suspected that there really might be some kind of rule against it."

The A-Life Papers

Langton got the message loud and clear. He'd gotten it long ago. Nobody wanted to see the end of his dissertation more than he did. And in the year since the artificial life workshop, he had actually made quite a bit of progress on it. He had converted his old cellular automata code from Michigan to run on Los Alamos' Sun workstations. He had explored the edge-of-chaos phase transition with innumerable computer experiments. He had even gone deep into the physics literature, learning how to analyze the phase transition with hard-core statistical mechanics.

But as for actually writing the thing—well, the year had just gotten away from him. The fact was that most of the time since the artificial life workshop had been devoured by the workshop's aftermath. George Cowan and David Pines had invited him to collect written versions of the talks and publish them in the name of the Santa Fe Institute, as one of a series of books the institute was publishing on the sciences of complexity. But Pines and Cowan had also insisted that those papers be rigorously reviewed by outside scientists, in exactly the same way that they would have been in any other research publication. The institute couldn't afford to be associated with flakiness, they told him. This had to be *science*, not video games.

That was fine by Langton, who'd always felt that way himself. But the upshot was that he'd spent months playing editor—which meant reading forty-five papers an average of four times apiece, sending each of them out to several reviewers, sending the reviewers' comments back to the authors with demands for rewrites, and generally cajoling everybody to get it done before the sun grew cold. Then he'd spent more months writing a preface and introductory chapter to the book. "It just took enormous amounts of time," he sighs.

On the other hand, the whole process had been enormously educational. "It was like studying for your qualifiers," he says. "What's good? What's BS? It really made me a *master* of that material." And now that the volume was finally done—with all the rigor that Cowan and Pines could ask for—Langton felt that he had created something much more than a series of papers. His dissertation might be off in limbo somewhere, but the workshop volume promised to lay the foundation for artificial life as a serious science. Moreover, by taking all the ideas and insights that people had brought to the workshop, and distilling them into a preface and a forty-seven-page introductory article, Langton had written his clearest and most articulate manifesto yet for what artificial life was all about.

Artificial life, he wrote, is essentially just the inverse of conventional biology. Instead of being an effort to understand life by *analysis*—dissecting living communities into species, organisms, organs, tissues, cells, organelles, membranes, and finally molecules—artificial life is an effort to understand life by *synthesis:* putting simple pieces together to generate lifelike behavior in man-made systems. Its credo is that life is not a property of matter per se, but the organization of that matter. Its operating principle is that the laws of life must be laws of dynamical form, independent of the details of a particular carbon-based chemistry that happened to arise here on Earth four billion years ago. Its promise is that by exploring other possible biologies in a new medium—computers and perhaps robots—artificial life researchers can achieve what space scientists have achieved by sending probes to other planets: a new understanding of our own world through a cosmic perspective on what happened on other worlds. "Only when we are able to view *life-as-we-know-it* in the context of *life-as-it-could-be* will we really understand the nature of the beast," Langton declared.

The idea of viewing life in terms of its abstract organization is perhaps the single most compelling vision to come out of the workshop, he said. And it's no accident that this vision is so closely associated with computers: they share many of the same intellectual roots. Human beings have been searching for the secret of automata—machines that can generate their own behavior—at least since the time of the Pharaohs, when Egyptian craftsmen created clocks based on the steady drip of water through a small orifice. In the first century A.D., Hero of Alexandria produced his treatise *Pneumatics*, in which he described (among other things) how pressurized air could generate simple movements in various gadgets shaped like animals and humans. In Europe, during the great age of clockworks more than a thousand years later, medieval and Renaissance craftsmen devised increasingly elaborate figures known as "jacks," which would emerge from the interior of the clock to strike the hours; some of their public clocks eventually grew to include large numbers of figures that acted out entire plays. And during the Industrial Revolution, the technology of clockwork automata gave rise to the still more sophisticated technology of *process control*, in which factory machines were guided by intricate sets of rotating cams and interlinked mechanical arms. Moreover, by incorporating such refinements as movable cams, or rotating drums with movable pegs, nineteenth-century designers soon developed controllers that could be adjusted to generate many sequences of action from the same machine. Along with the development of calculating machines in the early twentieth century, noted Langton, "the introduction of such programmable controllers was one of the primary developments on the road to general-purpose computers."

Meanwhile, he said, the foundations for a general theory of computing were being laid by logicians who were trying to formalize the notion of a *procedure*, a sequence of logical steps. That effort culminated in the early decades of the twentieth century with the work of Alonzo Church, Kurt Gödel, Alan Turing, and others, who pointed out that the essence of a mechanical process—the "thing" responsible for its behavior—is not a thing at all. It is an abstract control structure, a program that can be expressed as a set of rules without regard to the material the machine is made of. Indeed, said Langton, this abstraction is what allows you to take a piece of software from one computer and run it on another computer: the "machineness" of the machine is in the software, not the hardware. And once you've accepted that, he said, echoing his own epiphany at Massachusetts General Hospital nearly eighteen years before, then it's a very small step to say that the "aliveness" of an organism is also in the software—in the organization of the molecules, not the molecules themselves.

Now admittedly, said Langton, that step doesn't always look so small, especially when you consider how fluid, spontaneous, and organic life is, and how *controlled* computers and other machines are. At first glance it seems ludicrous even to talk about living systems in those terms.

But the answer lies with a second great insight, which could be heard at the workshop again and again: living systems are machines, all right, but machines with a very different kind of organization from the ones we're used to. Instead of being designed from the top down, the way a human engineer would do it, living systems always seem to emerge from the bottom up, from a population of much simpler systems. A cell consists of proteins, DNA, and other biomolecules. A brain consists of neurons. An embryo consists of interacting cells. An ant colony consists of ants. And for that matter, an economy consists of firms and individuals.

Of course, this was exactly the point that John Holland and the rest of the Santa Fe crowd liked to make about complex adaptive systems in general. The difference was that Holland saw this population structure mainly as a collection of building blocks that could be reshuffled for very efficient evolution, whereas Langton saw it mainly as an opportunity for rich, lifelike dynamics. "The most surprising lesson we have learned from simulating complex physical systems on computers is that *complex behavior need not have complex roots*," he wrote, complete with italics. "Indeed, tremendously interesting and beguilingly complex behavior can emerge from collections of extremely simple components."

Langton was speaking from the heart here, since that statement so clearly reflects the experience of discovering his self-reproducing cellular automaton. But the statement applied equally well to one of the most vivid demonstrations at the artificial life workshop: Craig Reynolds' flock of "boids." Instead of writing global, top-down specifications for how the flock should behave, or telling his creatures to follow the lead of one Boss Boid, Reynolds had used only the three simple rules of local, boid-to-boid interaction. And it was precisely that locality that allowed his flock to adapt to changing conditions so organically. The rules always tended to pull the boids together, in somewhat the same way that Adam Smith's Invisible Hand tends to pull supply into balance with demand. But just as in the economy, the tendency to converge was only a tendency, the result of each boid reacting to what the other boids were doing in its immediate neighborhood. So when a flock encountered an obstacle such as a pillar, it had no trouble splitting apart and flowing to either side as each boid did its own thing.

Try doing that with a single set of top-level rules, said Langton. The system would be impossibly cumbersome and complicated, with the rules telling each boid precisely what to do in every conceivable situation. In fact, he had seen simulations like that; they usually ended up looking jerky and unnatural, more like an animated cartoon than like animated life. And besides, he said, since it's effectively impossible to cover *every* conceivable situation, top-down systems are forever running into combinations of events they don't know how to handle. They tend to be touchy and fragile, and they all too often grind to a halt in a dither of indecision.

The same kind of bottom-up, population thinking was responsible for the graphical plants presented by Aristid Lindenmayer of the University of Utrecht and Prezemyslaw Prusinkiewcz of the University of Regina in Saskatchewan. These plants weren't just drawn on the computer screen. They were *grown*. They started from a single stem, and then used a handful of simple rules to tell each branch how to make leaves, flowers, and more branches. Once again, the rules said nothing about the overall shape of the final plant. They were meant to model how a multitude of cells differentiate and interact with one another during the course of the plant's development. Nonetheless, they produced shrubs, or trees, or flowers that looked startlingly realistic. If the rules were chosen carefully enough, in fact, they could produce a computer plant that looked very much like a known species. (And if those rules were then changed even slightly, they might produce a radically different plant, thus illustrating how easy it is for evolution to make large leaps in outward appearances by making only tiny changes in the course of development.)

The theme was heard over and over again at the workshop, said Langton: the way to achieve lifelike behavior is to simulate populations of simple units instead of one big complex unit. Use local control instead of global control. Let the behavior emerge from the bottom up, instead of being specified from the top down. And while you're at it, focus on ongoing behavior instead of the final result. As Holland loved to point out, living systems never really settle down.

Indeed, said Langton, by taking this bottom-up idea to its logical conclusion, you could see it as a new and thoroughly scientific version of *vitalism*: the ancient idea that life involves some kind of energy, or force, or spirit that transcends mere matter. The fact is that life does transcend mere matter, he said—not because living systems are animated by some vital essence operating outside the laws of physics and chemistry, but because a population of simple things following simple rules of interaction can behave in eternally surprising ways. Life may indeed be a kind of biochemical machine, he said. But to animate such a machine "is not to bring life to a machine; rather, it is to organize a population of machines in such a way that their interacting dynamics are 'alive.' "

Finally, said Langton, there was a third great idea to be distilled from the workshop presentations: the possibility that life isn't just *like* a computation, in the sense of being a property of the organization rather than the molecules. Life literally *is* a computation.

To see why, said Langton, start with conventional, carbon-based biology. As biologists have been pointing out for more than a century, one of the most striking characteristics of any living organism is the distinction between its *genotype*—the genetic blueprint encoded in its DNA—and its *phenotype*—the structure that is created from those instructions. In practice, of course, the actual operation of a living cell is incredibly complicated, with each gene serving as a blueprint for a single type of protein molecule, and with myriad proteins interacting in the body of the cell in myriad ways.

But in effect, said Langton, you can think of the genotype as a collection of little computer programs executing in parallel, one program per gene. When activated, each of these programs enters into the logical fray by competing and cooperating with all the other active programs. And collectively, these interacting programs carry out an overall computation that is the phenotype: the structure that unfolds during an organism's development.

Next, move from carbon-based biology to the more general biology of artificial life. The same notions apply, said Langton. And to capture that fact, he coined the term *generalized genotype*, or GTYPE, to refer to any collection of low-level rules. He likewise coined the term *generalized phenotype*, or PTYPE, to refer to the structure and/or behavior that results when those rules are activated in some specific environment. In a conventional computer program, for example, the GTYPE is obviously just the computer code itself, and the PTYPE is what the program does in response to input from the user. In Langton's own self-reproducing cellular automaton, the GTYPE is the set of rules specifying how each cell interacted with its neighbors, and the PTYPE is the overall pattern. In Reynolds' boids program, the GTYPE is the set of three rules that guides the flight of each boid, and the PTYPE is the flocking behavior of the boids as a group.

More generally, said Langton, this concept of a GTYPE is essentially identical with John Holland's concept of an "internal model"; the only difference is that he placed even more emphasis than Holland did on its role as a computer program. And by no coincidence, the GTYPE concept applies perfectly well to Holland's classifier systems, where the GTYPE of a given system is just its set of classifier rules. It likewise applies to his Echo model, where a creature's GTYPE consists of its offense and defense chromosomes. It applies to Brian Arthur's economy-under-glass models, where an artificial agent's GTYPE is its hard-learned set of rules for economic behavior. And it applies, in principle, to any complex adaptive system whatsoever—anything that has agents interacting according to a set of rules. As their GTYPE unfolds into a PTYPE, they are all performing a computation.

Now, what's beautiful about all this, said Langton, is that once you've made the link between life and computation, you can bring an immense amount of theory to bear. For example, Why is life quite literally full of surprises? Because, in general, it is impossible to start from a given set of GTYPE rules and predict what their PTYPE behavior will be—even in principle. This is the undecidability theorem, one of the deepest results of computer science: unless a computer program is utterly trivial, the fastest way to find out what it will do is to run it and see. There is *no* general-purpose procedure that can scan the code and the input and give you the answer any faster than that. That's why the old saw about computers only doing what their programmers tell them to do is both perfectly true and virtually irrelevant; any piece of code that's complex enough to be interesting will always surprise its programmers. That's why any decent software package has to be endlessly tested and debugged before it is released—and that's why the users always discover very quickly that the debugging was never quite perfect. And, most important for artificial life purposes, that's why a living system can be a biochemical machine that is completely under the control of a program, a GTYPE, and yet still have a surprising, spontaneous behavior in the PTYPE.

Conversely, said Langton, there are other deep theorems from computer science stating that you can't go the other way, either. Given the specification for a certain desired behavior, a PTYPE, there is no general procedure for finding a set of GTYPE rules that will produce it. In practice, of course, those theorems don't stop human programmers from using well-tested algorithms to solve precisely specified problems in clearly defined environments. But in the poorly defined, constantly changing environments faced by living systems, said Langton, there seems to be only one way to proceed: trial and error, also known as Darwinian natural selection. The process may seem terribly cruel and wasteful, he pointed out. In effect, nature does its programming by building a lot of different machines with a lot of randomly differing GTYPES, and then smashing the ones that don't work very well. But, in fact, that messy, wasteful process may be the best that nature can do. And by the same token, John Holland's genetic algorithm approach may be the only realistic way of programming computers to cope with messy, ill-defined problems. "It is quite likely that this is the *only* efficient, *general* procedure that could find GTYPEs with specific PTYPE traits," Langton wrote.

In writing his introductory chapter, Langton very carefully avoided making any claim that the entities being studied by the artificial lifers were "really" alive. Obviously, they weren't. Boids, plants, self-reproducing cellular automata—none of them were anything more than a simulation, a highly simplified model of life having no existence outside of a computer. Nonetheless, since the whole point of artificial life research was to grapple with the most fundamental principles of life, there was no avoiding the question: Could human beings ultimately create artificial life for real?

Langton found that question to be a tough one, not least because neither he nor anyone else had a clear idea of what "real" artificial life would be like. Some kind of genetically engineered superorganism, perhaps? A self-reproducing robot? An overeducated computer virus? What *is* life, precisely? How do you know for sure when you've got it and when you haven't?

Not surprisingly, there had been a good deal of discussion on this point during the workshop, not only during the sessions, but in the hallways and in loud, lively debates over dinner. Computer viruses were a particularly hot topic: many of the participants felt that viruses had come uncomfortably close to crossing the line already. The pesky things met almost every criterion for life that anyone could think of. Computer viruses could reproduce and propagate by copying themselves into another computer or to a floppy disk. They could store a representation of themselves in computer code, analogous to DNA. They could commandeer the metabolism of their host (a computer) to carry out their own functions, much as real viruses commandeer the molecular metabolism of infected cells. They could respond to stimuli in their environment (the computer again). And—courtesy of certain hackers with a warped sense of humor—they could even mutate and evolve. True, computer viruses lived their lives entirely within the cyberspace of computers and computer networks. They didn't have any independent existence out in the material world. But that didn't necessarily rule them out as living things. If life was really just a question of organization, as Langton claimed, then a properly organized entity would literally *be* alive, no matter what it was made of.

Whatever the status of computer viruses, however, Langton had no doubt that "real" artificial life would one day come into being—and sooner rather than later. What with biotechnology, robotics, and advanced software development, moreover, it was going to happen for commercial and/or military reasons whether he and his colleagues studied the subject or not. But that just made the research all the more important, he argued: if we really are headed into the Brave New World of artificial life, then at least we ought to be doing it with our eyes open.

"By the middle of this century," he wrote, "mankind had acquired the power to extinguish life on Earth. By the middle of the next century, he will be able to create it. Of the two, it is hard to say which places the larger burden of responsibility on our shoulders. Not only the specific kinds of living things that will exist, but the very course of evolution itself will come more and more under our control."

Given that prospect, he said, he felt that everyone involved in the field should go right out and read *Frankenstein*: it's clear in the book (although

not in the movie) that the doctor disavowed any responsibility for his creation. That could not be allowed to happen here. The future effect of the
changes we make now are unpredictable, he pointed out, even in principle.
Yet we are responsible for the consequences, nonetheless. And that, in
turn, meant that the implications of artificial life had to be debated in the
open, with public input.

Furthermore, he said, suppose that you *could* create life. Then suddenly
you would be involved in something a lot bigger than some technical
definition of living versus nonliving. Very quickly, in fact, you would find
yourself engaged in a kind of empirical theology. Having created a living
creature, for example, would you then have the right to demand that it
worship you and make sacrifices to you? Would you have the right to act
as its god? Would you have the right to destroy it if it didn't behave the
way you wanted it to?

Good questions, said Langton. "Whether they have correct answers or
not, they must be addressed, honestly and openly. Artificial life is more
than just a scientific or technical challenge; it is a challenge to our most
fundamental social, moral, philosophical, and religious beliefs. Like the
Copernican model of the solar system, it will force us to reexamine our
place in the universe and our role in nature."

The New Second Law

If Langton's rhetoric sometimes seems to soar a bit higher than most
scientific prose—well, that wasn't at all unusual in his corner of Los Alamos. Doyne Farmer, for one, was famous for his cruises in the conceptual
stratosphere. A prime example was "Artificial Life: The Coming Evolution," a nontechnical paper that he coauthored in 1989 with his wife,
environmental lawyer Alletta Belin, and then delivered at a Caltech symposium celebrating Murray Gell-Mann's sixtieth birthday: "With the advent
of artificial life," they wrote, *"we may be the first creatures to create our
own successors. . . . If we fail in our task as creators, they may indeed be
cold and malevolent. However, if we succeed, they may be glorious, en-
lightened creatures that far surpass us in their intelligence and wisdom. It
is quite possible that, when the conscious beings of the future look back
on this era, we will be most noteworthy not in and of ourselves but rather
for what we gave rise to. Artificial life is potentially the most beautiful
creation of humanity."*

Rhetoric aside, however, Farmer was perfectly serious about artificial life as a new kind of science. (Most of the "Coming Evolution" paper was in fact a reasonably sober assessment of what the field might hope to accomplish.) By no coincidence, he was equally serious about supporting Chris Langton. It was Farmer, after all, who brought Langton to Los Alamos in the first place. And despite his exasperation over Langton's much-delayed dissertation, he found no reason to regret having done so. "Chris was definitely worth it," he says. "People like him, who have a real dream, a vision of what they want to do, are rare. Chris hadn't learned to be very efficient. But I think he had a good vision, one that was really needed. And I think he was doing a really good job carrying it out. He wasn't afraid to tackle the details."

Indeed, Farmer was wholeheartedly serving as a mentor to Langton—even though Langton happened to be five years older than he was. Down the hill, where Farmer was one of the very few young scientists included in the inner circle of the Santa Fe Institute, he had persuaded Cowan to contribute $5000 toward Langton's artificial life workshop in 1987. Farmer had made sure that Langton was invited to speak at institute meetings. He had served as an advocate on the institute's science board for bringing in visiting scientists to work on artificial life. He had likewise encouraged Langton to set up an ongoing series of artificial life seminars up at Los Alamos, with occasional sessions down in Santa Fe. And perhaps most important, when Farmer had agreed in 1987 to head the new Complex Systems group within the Los Alamos theory division, he had made artificial life one of the group's three major research efforts, along with machine learning and dynamical systems theory.

Farmer wasn't exactly a natural-born administrator type. At age thirty-five, he was a tall, angular New Mexican who still wore a graduate student's ponytail and tee-shirts that said things like, "Question Authority!" He found bureaucratic busywork to be a pain, and he found the writing of proposals begging money from "some bonehead back in Washington" to be even more of a pain. Yet Farmer had an undeniable gift for generating both funding and intellectual excitement. In the field of mathematical prediction, where he had originally made his reputation and where he still spent most of his research time, he was at the forefront of finding ways to project the future behavior of systems that seemed hopelessly random and chaotic—including certain systems, such as the stock market, where people had *incentive* to project the future. Moreover, Farmer had no compunction about channeling most of the group's "general-purpose" money toward Langton and the tiny cadre of artificial life researchers, while making his own nonlinear forecasting work and other efforts pay for themselves. "Forecasting produces practical results, so that I could promise the funding agencies a payoff within a year," he says, "whereas practical results from artificial life are farther in the future. With the current funding climate, this makes A-life almost unfundable. This was driven home to me when one of the agencies that funded my prediction work called to ask about a proposal they had received to study artificial life. From their attitude it was quite clear that they viewed artificial life on a par with flying saucers or astrology. They were upset to see my name appearing in the list of references."

This was not Farmer's idea of the ideal situation, by a long shot. He genuinely loved the forecasting work. But between that and the administrative BS, he had very little time left over to work on artificial life. And artificial life somehow struck a chord in him that nothing else did. Artificial life, he says, was where you could get right down into the deep questions of emergence and self-organization, questions that had haunted him all his life.

"I was already thinking about self-organization in nature when I was in high school," says Farmer, "although initially it was on a vague level, from reading science fiction stories." He particularly remembers one story by Isaac Asimov, "The Final Question," in which humans of the far future consult a cosmic supercomputer about how to repeal the second law of thermodynamics: the inexorable tendency for everything in the universe to cool off, decay, and run down as atoms try to randomize themselves. How can we reverse the increase in entropy, they ask, referring to the physicists' name for molecular-scale disorder. Eventually, long after the human race has vanished and all the stars have grown cold, the computer learns how to accomplish this feat—whereupon it declares, "Let there be light!" and gives rise to a fresh, new, low-entropy universe.

Farmer was fourteen when he read Asimov's story, and it seemed to him even then to point the way toward a profound question. If entropy is always increasing, he asked himself, and if atomic-scale randomness and disorder are inexorable, then why is the universe still able to bring forth stars and planets and clouds and trees? Why is matter constantly becoming more and more organized on a large scale, at the same time that it is becoming more and more *dis*organized on a small scale? Why hasn't everything in the universe long since dissolved into a formless miasma? "Frankly," says Farmer, "my interest in those questions was one of my driving concerns in becoming a physicist. Bill Wootters [physicist William Wootters, now at

Williams College in Massachusetts] and I spent a lot of time at Stanford sitting around on the lawn after physics class, talking about these questions. Ideas just seemed to spring into our heads. It was only years later that I discovered that other people had thought about this, too, and that there was a literature out there—Norbert Wiener and cybernetics, Ilya Prigogine and self-organization, Hermann Haken and synergetics." In fact, he says, you could even find the same issues latent in the work of Herbert Spencer, the English philosopher who helped popularize Darwin's theory back in the 1860s by coining such phrases as "survival of the fittest," and who saw Darwinian evolution as just a special case of a broader force driving the spontaneous origin of structure in the universe.

So these were questions that sprang forth in many heads independently, says Farmer. But at the time he felt frustrated: "I couldn't see a forum for pursuing them. Biologists weren't doing it—they were mired in the nitty-gritty of which protein interacted with which, missing the general principles. Yet as far as I could tell, physicists didn't seem to be doing anything like this either. That's one of the reasons I jumped into chaos."

The story of that jump rated a whole chapter in James Gleick's best-selling book *Chaos*: how Farmer and his lifelong friend Norman Packard became fascinated with roulette in the late 1970s while they were graduate students in physics at the University of California's campus in Santa Cruz; how the effort of calculating the moving ball's trajectory on the fly, so to speak, gave them an exquisite feel for the way a tiny change of the initial conditions in a physical system can produce a dramatic change in the outcome; how they and two other graduate students—Robert Shaw and James Crutchfield—came to realize that this sensitivity to initial conditions could be described by the emerging science of "chaos," more generally known as "dynamical systems theory"; and how the four of them were so determined to pursue research in this field that they became known as the Dynamical Systems Collective.

"After a while, though, I got pretty bored with chaos," says Farmer. "I felt 'So what?' The basic theory had already been fleshed out. So there wasn't that excitement of being on the frontier, where things *aren't* understood." Besides, he says, chaos theory by itself didn't go far enough. It told you a lot about how certain simple rules of behavior could give rise to astonishingly complicated dynamics. But despite all the beautiful pictures of fractals and such, chaos theory actually had very little to say about the fundamental principles of living systems or of evolution. It didn't explain how systems starting out in a state of random nothingness could then organize themselves into complex wholes. Most important, it didn't answer his old question about the inexorable growth of order and structure in the universe.

Somehow, Farmer was convinced, there was a whole new level of understanding yet to be reached. Thus his work with Stuart Kauffman and Norm Packard on autocatalytic sets and the origin of life, and his enthusiastic support of Langton's artificial life. Like so many of the other people around Los Alamos and Santa Fe, Farmer could *feel* it—an understanding, an answer, a principle, a law hovering almost within reach.

"I'm of the school of thought that life and organization are inexorable," he says, "just as inexorable as the increase in entropy. They just *seem* more fluky because they proceed in fits and starts, and they build on themselves. Life is a reflection of a much more general phenomenon that I'd like to believe is described by some counterpart of the second law of thermodynamics—some law that would describe the tendency of matter to organize itself, and that would predict the general properties of organization we'd expect to see in the universe."

Farmer has no clear idea of what this new second law would look like. "If we knew that," he says, "we'd have a big clue how to get there. At this point it's purely speculative, something that intuition suggests when you stand back and stroke your beard and contemplate." In fact, he has no idea whether it would be one law, or several. What he does know, however, is that people have recently been finding so many hints about things like emergence, adaptation, and the edge of chaos that they can begin to sketch at least a broad outline of what this hypothetical new second law might be like.

Emergence

First, says Farmer, this putative law would have to give a rigorous account of emergence: What does it really mean to say that the whole is greater than the sum of its parts? "It's not magic," he says. "But to us humans, with our crude little human brains, it *feels* like magic." Flying boids (and real birds) adapt to the actions of their neighbors, thereby becoming a flock. Organisms cooperate and compete in a dance of coevolution, thereby becoming an exquisitely tuned ecosystem. Atoms search for a minimum energy state by forming chemical bonds with each other, thereby becoming the emergent structures known as molecules. Human beings try to satisfy their material needs by buying, selling, and trading with each other, thereby creating an emergent structure known as a market. Humans likewise interact with each other to satisfy less quantifiable goals, thereby forming families, religions, and cultures. Somehow, by constantly seeking mutual accommodation and self-consistency, groups of agents manage to transcend themselves and become something more. The trick is to figure out how, without falling back into sterile philosophizing or New Age mysticism.

And that, says Farmer, is the beauty of computer simulation in general and artificial life in particular: by experimenting with a simple model that you can run on your desktop, you can try out ideas and see how well they really work. You can try to pin down vague notions with more and more precision. And you can try to extract the essence of how emergence really works in nature. These days, moreover, there is a wide variety of models to choose from. One that Farmer has given particular attention to is *connectionism:* the idea of representing a population of interacting agents as a network of "nodes" linked by "connections." And in that, he has plenty of company. Connectionist models have been popping up everywhere in the past decade or so. Exhibit A has to be the neural network movement, in which researchers use webs of artificial neurons to model such things as perception and memory retrieval—and, not incidentally, to mount a radical attack on the symbol-processing methods of mainstream artificial intelligence. But close behind are many of the models that have found a home at the Santa Fe Institute, including John Holland's classifier systems, Stuart Kauffman's genetic networks, the autocatalytic set model for the origin of life, and the immune system model that he and Packard did in the mid-1980s with Los Alamos' Alan Perelson. Admittedly, says Farmer, some of these models don't *look* very connectionist, and a lot of people are surprised the first time they hear the things being described that way. But that's only because the models were created by different people at different times to solve different problems—and then described in different language. "When you peel everything back," he says, "they end up looking the same. You can literally map one model into another."

In a neural network, of course, the node-and-connection structure is obvious. The nodes correspond to neurons, and the connections correspond to synapses linking the neurons. If a programmer has a neural network model of vision, for example, he or she can simulate the pattern of light and dark falling on the retina by activating certain input nodes, and then letting the activation spread through the connections into the rest of the network. The effect is a bit like sending shiploads of goods into a few port cities along the seacoast, and then letting a zillion trucks cart the stuff along the highways among the inland cities. But if the connections have been properly arranged, the network will soon settle into a self-consistent pattern of activation that corresponds to a classification of the scene: "That's a cat!" Moreover, it will do so even if the input data are noisy and incomplete—or, for that matter, even if some of its nodes are burned out.

In John Holland's classifier system the node-and-connection structure is considerably less obvious, says Farmer, but it's there. The set of nodes is just the set of all possible internal messages, such as 1001001110111110. And the connections are just the classifier rules, each of which looks for a certain message on the system's internal bulletin board, and then responds to it by posting another. By activating certain input nodes—that is, by posting the corresponding input messages on the bulletin board—the programmer can cause the classifiers to activate more messages, and then still more. The result will be a cascade of messages analogous to the spreading activation in a neural network. And, just as the neural net eventually settles down into a self-consistent state, the classifier system will eventually settle down into a stable set of active messages and classifiers that solves the problem at hand—or, in Holland's picture, that represents an emergent mental model.

The network structure is also there in the model he did with Kauffman and Packard on autocatalysis and the origin of life, says Farmer. In this case the set of nodes is the set of all possible polymer species, such as *abbcaad*. And the connections are the simulated chemical reactions among those polymers: polymer A catalyzes the formation of polymer B, and so on. By activating certain input nodes—that is, by seeding the system with a steady flow of small "food" polymers from the simulated environment—the three of them could set off a cascade of reactions, which eventually settled down into a pattern of active polymers and catalytic reactions that could sustain itself: an "autocatalytic set" that presumably corresponds to some sort of protoorganism emerging from the primordial soup.

The analysis is much the same in Kauffman's network models of the genome and in any number of other models, says Farmer. Underlying them all is this same node-and-connection framework. Indeed, when he first recognized the parallels several years ago, he was so delighted that he wrote it all up for publication in a paper entitled "A Rosetta Stone for Connectionism." If nothing else, he says, the very existence of a common framework is reassuring, in the sense that most of the blind men at least seem to have their hands on the same elephant. But more than that, a common framework should help the people working on these models to communicate a lot more easily than they usually do, without the babel of different jargons. "The thing I considered important in that paper was that I hammered out the actual translation machinery for going from one model to another. I could take a model of the immune system and say, 'If that were a neural net, here's what it would look like.' "

But perhaps the most important reason for having a common framework, says Farmer, is that it helps you distill out the essence of the models, so that you can focus on what they actually have to say about emergence. And in this case, the lesson is clear: the power really does lie in the connections. That's what gets so many people so excited about connectionism. You can start with very, very simple nodes—linear "polymers," "messages" that are just binary numbers, "neurons" that are essentially just on-off switches—and still generate surprising and sophisticated outcomes just from the way they interact.

Take learning and evolution, for example. Since the nodes are very simple, the behavior of the network as a whole is determined almost entirely by the connections. Or in Chris Langton's language, the connections encode the GTYPE of the network. So to modify the system's PTYPE behavior, you simply have to change those connections. In fact, says Farmer, you can change them in two different ways. The first way is to leave the connections in place but modify their "strength." This corresponds to what Holland calls *exploitation* learning: improving what you already have. In Holland's own classifier system this is done through the bucket-brigade algorithm, which rewards the classifier rules that lead to a good result. In neural networks it's done through a variety of learning algorithms, which present the network with a series of known inputs and then tweak the connection strengths up or down until it gives the right responses.

The second, more radical way of adjusting the connections is to change the network's whole wiring diagram. Rip out some of the old connections and put in new ones. This corresponds to what Holland calls *exploration* learning: taking the risk of screwing up big in return for the chance of winning big. In Holland's classifier systems, for example, this is exactly what happens when the genetic algorithm mixes rules together through its inimitable version of sexual recombination; the new rules that result will often link messages that have never been linked before. This is also what happens in the autocatalytic set model when occasional new polymers are allowed to form spontaneously—as they do in the real world. The resulting chemical connections can give the autocatalytic set an opening to explore a whole new realm of polymer space. This is *not* what usually happens in neural networks, since the connections were originally supposed to model synapses that can't be moved. But recently, says Farmer, a number of neural network aficionados have been experimenting with networks that do rewire themselves as they learn, on the grounds that any fixed wiring diagram is arbitrary and ought to be open to change.

So, in short, says Farmer, the connectionist idea shows how the capacity for learning and evolution can emerge even if the nodes, the individual agents, are brainless and dead. More generally, by putting the power in the connections and not the nodes, it points the way toward a very precise theory of what Langton and the artificial lifers mean when they say that the essence of life is in the organization and not the molecules. And it likewise points the way toward a deeper understanding of how life and mind could have gotten started in a universe that began with neither.

The Edge of Chaos

However, says Farmer, as beautiful as that prospect may be, connectionist models are a long way from telling you everything you'd like to know about the new second law. To begin with, they don't tell you much about how emergence works in economies, societies, or ecosystems, where the nodes are "smart" and constantly adapting to each other. To understand systems like that, you have to understand the coevolutionary dance of cooperation and competition. And that means studying them with coevolutionary models such as Holland's Echo, which have started to get popular only in the past few years.

More important, says Farmer, neither connectionist models nor coevolutionary models tell you what makes life and mind possible in the first place. What is it about the universe that allows these things to happen? It isn't enough to say "emergence"; the cosmos is full of emergent structures like galaxies and clouds and snowflakes that are still just physical objects; they have no independent life whatsoever. Something more is required. And this hypothetical new second law will have to tell us what that something is.

Clearly, this is a job for models that try to get at the basic physics and chemistry of the world, such as the cellular automata that Chris Langton is so fond of. And by no coincidence, says Farmer, Langton's discovery of this weird, edge-of-chaos phase transition in cellular automata seems to be a big part of the answer. Langton kept a discreet silence on this subject during the artificial life conference, given the state of his dissertation at the time. But from the start, says Farmer, a lot of people around Los Alamos and Santa Fe have found the edge-of-chaos idea awfully compelling. Langton is basically saying that the mysterious "something" that makes life and mind possible is a certain kind of balance between the forces of order and the forces of disorder. More precisely, he's saying that you should look at systems in terms of how they behave instead of how they're made. And when you do, he says, then what you find are the two extremes of *order* and *chaos*. It's a lot like the difference between solids, where the atoms are locked into place, and fluids, where the atoms tumble over one another at random. But right in between the two extremes, he says, at a kind of abstract phase transition called "the edge of chaos," you also find *complexity*: a class of behaviors in which the components of the system never quite lock into place, yet never quite dissolve into turbulence, either. These are the systems that are both stable enough to store information, and yet evanescent enough to transmit it. These are the systems that can be organized to perform complex computations, to react to the world, to be spontaneous, adaptive, and alive.

Strictly speaking, of course, Langton demonstrated the connection between complexity and phase transitions only in cellular automata. No one really knows if it holds true in other models—or in the real world, for that matter. On the other hand, says Farmer, there are some strong hints that it might. With 20/20 hindsight, for example, you can see that phase-transitionlike behavior has been cropping up in connectionist models for years. Back in the 1960s it was one of the first things that Stuart Kauffman discovered about his genetic networks. If the connections were too sparse, the networks would basically just freeze up and sit there. And if the connections were too dense, the networks would churn around in total chaos. Only right in between, when there were precisely two inputs per node, would the networks produce the stable state cycles Kauffman was looking for.

Then, in the mid-1980s, says Farmer, it was much the same story with the autocatalytic set model. The model had a number of parameters such as the catalytic strength of the reactions, and the rate at which "food" molecules are supplied. He, Packard, and Kauffman had to set all these parameters by hand, essentially by trial and error. And one of the first things they discovered was that nothing much happened in the model until they got those parameters into a certain range—whereupon the autocatalytic sets would take off and develop very quickly. Again, says Farmer, the behavior is strongly reminiscent of a phase transition—although it's still far from clear how it relates to phase transitions in the other models. "One senses the analogies, but it's more difficult to make them precise," he says. "That's another area where somebody needs to do some careful cross comparisons, analogous to the Rosetta Stone paper."

Meanwhile, says Farmer, it's even less clear whether the edge-of-chaos idea applies to coevolutionary systems. When you get to something like an ecosystem or an economy, he says, it's not obvious how concepts like order, chaos, and complexity can even be defined very precisely, much less a phase transition between them. Nonetheless, he says, there's something about the edge-of-chaos principle that still feels *right*. Take the former Soviet Union, he says: "It's now pretty clear that the totalitarian, centralized approach to the organization of society doesn't work very well." In the long run, the system that Stalin built was just too stagnant, too locked in, too rigidly controlled to survive. Or look at the Big Three automakers in Detroit in the 1970s. They had grown so big and so rigidly locked in to certain ways of doing things that they could barely recognize the growing challenge from Japan, much less respond to it.

On the other hand, says Farmer, anarchy doesn't work very well, either— as certain parts of the former Soviet Union seemed determined to prove in the aftermath of the breakup. Nor does an unfettered laissez-faire system: witness the Dickensian horrors of the Industrial Revolution in England or, more recently, the savings and loan debacle in the United States. Common sense, not to mention recent political experience, suggests that healthy economies and healthy societies alike have to keep order and chaos in balance—and not just a wishy-washy, average, middle-of-the road kind of balance, either. Like a living cell, they have to regulate themselves with a dense web of feedbacks and regulation, at the same time that they leave plenty of room for creativity, change, and response to new conditions. "Evolution thrives in systems with a bottom-up organization, which gives rise to flexibility," says Farmer. "But at the same time, evolution has to channel the bottom-up approach in a way that doesn't destroy the organization. There has to be a hierarchy of control—with information flowing from the bottom up as well as from the top down." The dynamics of complexity at the edge of chaos, he says, seems to be ideal for this kind of behavior.

The Growth of Complexity

In any case, says Farmer, "at a vague, heuristic level we think we know something about the domain where this interesting organizational phenomenon appears." However, this can't be the whole story, either. Even if you assume, for the sake of argument, that this special edge-of-chaos domain really exists, the hypothetical new second law will still have to explain how emergent systems *get* there, how they *keep* themselves there, and what they *do* there.

At that same vague, heuristic level, says Farmer, it's easy to persuade yourself that those first two questions have already been answered by Charles Darwin (as generalized by John Holland). Since the systems that are capable of the most complex, sophisticated responses will always have the edge in a competitive world, goes the argument, then frozen systems can always do better by loosening up a bit, and turbulent systems can always do better by getting themselves a little more organized. So if a system isn't on the edge of chaos already, you'd expect learning and evolution to push it in that direction. And if it is on the edge of chaos, then you'd expect learning and evolution to pull it back if it ever starts to drift away. In other words, you'd expect learning and evolution to make the edge of chaos stable, the natural place for complex, adaptive systems to be.

The third question, what systems do once they get to the edge of chaos, is a bit more subtle. In the space of all possible dynamical behaviors, the edge of chaos is like an infinitesimally thin membrane, a region of special, complex behaviors separating chaos from order. But then, the surface of the ocean is only one molecule thick, too; it's just a boundary separating water from air. And the edge of chaos region, like the surface of the ocean, is still vast beyond all imagining. It contains a near-infinity of ways for an agent to be both complex and adaptive. Indeed, when John Holland talks about "perpetual novelty," and adaptive agents exploring their way into an immense space of possibilities, he may not say it this way—but he's talking about adaptive agents moving around on this immense edge-of-chaos membrane.

So, what might the new second law have to say about that? Partly, of course, it could talk about building blocks, internal models, coevolution, and all the other adaptation mechanisms that Holland and others have studied. But Farmer, for one, suspects that at heart it will not be about mechanism so much as direction: the deceptively simple fact that evolution is constantly coming up with things that are more complicated, more sophisticated, more *structured* than the ones that came before. "A cloud is more structured than the initial miasma after the Big Bang," says Farmer, "and the prebiotic soup was more structured than a cloud." We, in turn, are more structured than the prebiotic soup. And, for that matter, a modern economy is more structured than those of the Mesopotamian city-states, just as modern technology is more sophisticated than that of Rome. It seems that learning and evolution don't just pull agents to the edge of chaos; slowly, haltingly, but inexorably, learning and evolution move agents *along* the edge of chaos in the direction of greater and greater complexity. Why?

"It's a thorny question," says Farmer. "It's very hard to articulate a notion of 'progress' in biology." What does it mean for one creature to be more advanced than another? Cockroaches, for example, have been around for several hundred million years longer than human beings, and they are very, very good at being cockroaches. Are we more advanced than they are, or just different? Were our mammalian ancestors of 65 million years ago more advanced than *Tyrannosaurus rex*, or just luckier in surviving the impact of a marauding comet? With no objective definition of fitness, says Farmer, "survival of the fittest" becomes a tautology: survival of the survivors.

"But I don't believe in nihilism, either—the idea that nothing is better than anything else," he says. "It isn't that evolution led inevitably toward us; that's silly. But if you stand back and take in the broad sweep of the entire evolutionary process, I do think you can talk meaningfully about progress. You see an overall trend toward increasing sophistication, complexity, and functionality; the difference between a model-T and a Ferrari is nothing compared with the difference between the earliest organisms and the latest organisms. As elusive as it is, this overall trend toward increasing 'quality' of evolutionary design is one of the most fascinating and profound clues as to what life is all about."

One of his favorite examples is the way evolution works in the autocatalytic set model that he did with Packard and Kauffman. One of the wonderful things about autocatalysis is that you can follow emergence from the ground up, he says. The concentration of a few chemicals gets spontaneously pumped up by orders of magnitude over their equilibrium concentration because they can collectively catalyze each other's formation. And that means that the set as a whole is now like a new, emergent individual sticking up from the equilibrium background—exactly what you want for explaining the origin of life. "If we knew how to do this in real chemical experiments, we'd have something poised between living and nonliving," he says. "These autocatalytic individuals don't have a genetic code. And yet on a crude level, they can maintain and propagate themselves—not nearly as well as seeds, for example, but much better than a pile of rocks."

In the original computer model, of course, there was no evolution of the sets because there was no interaction with any kind of outside environment. The model assumed that everything was happening in one well-stirred pot of chemicals, so once the sets emerged they were stable. In the real world of four billion years ago, however, the environment would have subjected these fuzzily defined autocatalytic individuals to all manner of buffeting and fluctuations. So to see what would happen in that kind of situation, Farmer and graduate student Rick Bagley subjected the model autocatalytic sets to fluctuations in their "food" supply: the stream of small molecules that served as raw material for the sets. "What was really cool was that some sets were like Panda bears, which can only digest bamboo," says Farmer. "If you changed their food supply they just collapsed. But others were like omnivores; they had lots of metabolic pathways that allowed them to substitute one food molecule for another. So when you played around with the food supply they were virtually unchanged." Such robust sets, presumably, would have been the kind that survived on the early Earth.

More recently, says Farmer, he, Bagley, and Los Alamos postdoc Walter Fontana made another modification in the autocatalytic model to allow for occasional spontaneous reactions, which are known to happen in real chemical systems. These spontanous reactions caused many of the autocatalytic sets to fall apart. But the ones that crashed paved the way for an evolutionary leap. "They triggered avalanches of novelty," he says. "Certain variations would get amplified, and then would stabilize again until the next crash. We saw a succession of autocatalytic metabolisms, each replacing the other."

Maybe that's a clue, says Farmer. "It will be interesting to see if we can articulate a notion of 'progress' that would involve emergent structures having certain feedback loops [for stability] that weren't present in what went before. The key is that there would be a sequence of evolutionary events structuring the matter in the universe in the Spencerian sense, in which each emergence sets the stage and makes it easier for the emergence of the next level."

"Actually," says Farmer, "I'm frustrated in talking about all this. There's a real language problem. People are thrashing around trying to define things like 'complexity' and 'tendency for emergent computation.' I can only evoke vague images in your brain with words that aren't precisely defined in mathematical terms. It's like the advent of thermodynamics—but we're where they were in about 1820. They knew there was something called 'heat,' but they were talking about it in terms that would later sound ridiculous." In fact, he says, they weren't even sure what heat was, much less how it worked. Most reputable scientists of the day were convinced that a red-hot poker, say, was densely laden with a weightless, invisible fluid known as *caloric*, which would flow out of the poker into cooler, less caloric-rich bodies at the slightest opportunity. Only a minority thought that heat might represent some kind of microscopic motion in the poker's atoms. (The minority was right.) Moreover, no one at the time seems to have imagined that messy, complicated things like steam engines, chemical reactions, and electric batteries could all be governed by simple, general laws. It was only in 1824 that a young French engineer named Sadi Carnot published the first statement of what would later be known as the second law of thermodynamics: the fact that heat will not spontaneously flow from cold objects to hot ones. (Carnot, who was writing a popular book about steam engines meant for his fellow engineers, quite correctly pointed out that this simple, everyday fact placed severe limits on how efficient a steam engine could be—not to mention internal combustion engines, power plant turbines, or any other engine that runs on heat. The statistical explanation for the second law, that atoms are constantly trying to randomize themselves, came some seventy years later.)

Likewise, says Farmer, it was only in the 1840s that the English brewer and amateur scientist James Joule laid the experimental foundations for the *first* law of thermodynamics, also known as the conservation of energy: the fact that energy can change from one form to another—thermal, mechanical, chemical, electrical—but can never be created or destroyed. And it was only in the 1850s that the two laws were stated in explicit, mathematical form.

"We're creeping toward that point in self-organization," says Farmer. "But organization turns out to be a lot harder to understand than disorganization. We're still missing the key idea—at least in a clear and quantitative form. We need something equivalent to the hydrogen atom, something we can pull apart to get a nice, clear description of what makes it tick. But we can't do that yet. We only understand little pieces of the puzzle, each in its own isolated context. For example, we now have a good understanding of chaos and fractals, which show us how simple systems with simple parts can generate very complex behaviors. We know quite a bit about gene regulation in the fruit fly, *Drosophila*. In a few very specific contexts we have some hints as to how self-organization is achieved in the brain. And in artificial life we are creating a new repertoire of 'toy universes.' Their behavior is a pale reflection of what actually goes on in natural systems. But we can simulate them completely, we can alter them at will, and we can understand exactly what makes them do what they do.

The hope is that we will eventually be able to stand back and assemble all these fragments into a comprehensive theory of evolution and self-organization.

"This not a field for people who like sharply defined problems," Farmer adds. "But what makes it exciting is the very fact that things *aren't* laid in stone. It's still happening. I don't see anybody with a clear path to an answer. But there are lots of little hints flying around. Lots of little toy systems and vague ideas. So it's conceivable to me that in twenty or thirty years we will have a real theory."

The Arc of a Howitzer Shell

Stuart Kauffman, for his part, is devoutly hoping that it will take far less time than that.

"I've heard Doyne say that this is like thermodynamics before Sadi Carnot came along," he says. "I think he's right. What we're really looking for in the science of complexity is the general law of pattern formation in non-equilibrium systems throughout the universe. And we need to invent the proper concepts to do it. But with all these hints, like the edge of chaos, I feel as though we're on the verge of a breakthrough, as though we're just a few years before Carnot."

Indeed, Kauffman clearly hopes that the new Carnot will be named—well, Kauffman. Like Farmer, he envisions a new second law that explains how emergent entities will do the most interesting things when they're at the edge of chaos, and how adaptation will inexorably build these entities up into higher and higher levels of complexity. But unlike Farmer, Kauffman hasn't been tied down and frustrated by the bureaucratic necessities of running a research group. He's been throwing himself headlong into the problem almost since the day he arrived at the Santa Fe Institute. He talks like someone who *needs* to find the answer—as though thirty years of trying to understand the meaning of order and self-organization has made the closeness of it like a physical ache.

"For me, this idea that there's an evolution to the edge of chaos is just the next step in an enormous struggle to understand the marriage of self-organization and selection," he says. "It's so annoying because I can almost taste it, almost see it. I'm not being a careful scientist. Nothing's finished. I've only had a first glance at a bunch of things. I feel more like a howitzer shell piercing through wall after wall, leaving a mess behind. I feel that I'm rushing through topic after topic, trying to see where the end of the arc of the howitzer shell is, without knowing how to clean up anything on the way back."

That arc began back in the 1960s, says Kauffman, when he first started playing around with autocatalytic sets and his network models of the genome. In those days he really wanted to believe that life was shaped almost entirely by self-organization, that natural selection was just a sideshow. And nothing showed that better than embryonic development, where interacting genes organized themselves into different configurations corresponding to different cell types, and where interacting cells organized themselves into various tissues and structures within a developing embryo. "I never doubted that natural selection worked," he says. "It just seemed to me that the deepest things had to do with self-organization.

"But then one day in the early 1980s," he says, "I was visiting John Maynard Smith," an old friend and an eminent population biologist at the University of Sussex. This was at a time when Kauffman had just started thinking seriously about self-organization again, after a ten-year hiatus when he worked on embryonic development in *Drosophila* fruit flies. "John and his wife Sheila and I went out for a walk on the Downs," he says, "and John pointed out that we weren't far from Darwin's home. Then he opined that, by and large, those who would take natural selection seriously were English country gentlemen—like Darwin. And then he looked at me and he gave me his little smile, and he said, 'Those who thought that natural selection didn't have much to do with biological evolution have been urban Jews!' Well, it cracked me up. I just sat down laughing in the hedgerow. But then he said, 'You really must think about selection, Stuart.' And I didn't want to. I wanted it all to be spontaneous."

But Kauffman had to admit that Maynard Smith was right. Self-organization couldn't do it all alone. After all, mutant genes can self-organize themselves just as easily as normal ones can. And when the result is, say, a fruit fly monstrosity with legs where its antennae should be, or no head, then you still need natural selection to sort out what's viable from what's hopeless.

"So I sat down in 1982 and I outlined my book." (*The Origins of Order*, a much revised summary of Kauffman's thinking over the past thirty years, was finally published in 1992.) "The book was to be about self-organization and selection: How do you put the two together? And the way I conceived of it at the beginning was that there is a struggle between them. Selection may want one thing, but there are limits to what the self-organizing behavior of the system will allow. So they tug on one another until they come to some equilibrium where selection can't budge things. That image stayed with me through the first two-thirds of the book"—or more precisely, until the mid-1980s, when Kauffman arrived in Santa Fe and started hearing about the edge of chaos.

Ultimately, says Kauffman, the edge-of-chaos concept transformed his point of view on the self-organization-versus-selection question yet again. At the time, however, he had decidedly mixed feelings about it. Not only had he been seeing phase-transitionlike behavior in his genetic networks since the 1960s, but in 1985 he had come very close to working out the edge-of-chaos idea himself.

"That," he says with the air of a man still kicking himself, "is one of those papers that I never wrote, and that I've always regretted." The idea had come to him in the summer of 1985, he says, when he was spending a sabbatical year at the Ecole Normale Supérieure in Paris. Along with physicist Gérard Weisbuch and graduate student Françoise Fogelman-Soule, who was doing her thesis on Kauffman's genetic networks, he had gone off from Paris to spend a few months at the Hadassah Hospital in Jerusalem. And one morning there, Kauffman got to thinking about what he called "frozen components" in his networks. He'd first noticed them back in 1971, he says. In his light bulb analogy, it was as if connected clusters of nodes here and there in the network would either all light up or all go dark, and then *stay* that way while the light bulbs elsewhere in the network continued to flicker on and off. The frozen components didn't appear at all in the densely connected networks, which were a solid mass of chaotic flickering. And yet they seemed to dominate the very sparsely connected networks, which is why those systems tended to freeze up entirely. But what happens in the middle, he wondered? That's where you find the more-or-less sparsely connected networks that seem to correspond most closely to real genetic systems. And that's where the networks are neither completely frozen nor completely chaotic. . . .

"I remember bursting in on Françoise and Gérard that morning," says Kauffman. "I said, 'Look, you guys, just where the frozen components are melting and are tenuously connected to one another, and the isolated unfrozen islands are just beginning to join tendrils, you ought to be able to get the most complex computations!' We talked about it extensively that morning, and we all agreed that was interesting. I noted it down as something to get to. But—we got off onto other things. Besides, it was still in that period of my own life when I thought, 'Ah, nobody's going to care about these things.' So I never focused on it again."

The upshot was that Kauffman listened to all this talk about the edge of chaos with an odd mix of déjà vu, regret, and excitement. Here was an idea he couldn't help feeling a bit proprietary about. And yet he had to admit that Langton was the one who made the connection between phase transitions, computation, and life into something far more than a morning's passing fancy. Langton had done the hard work of making the idea rigorous and precise. Moreover, Langton had recognized what Kauffman had not: that the edge of chaos was much more than just a simple boundary separating purely ordered systems from purely chaotic systems. Indeed, it was Langton who finally got Kauffman to understand the point after several long conversations: the edge of chaos was a special region unto itself, the place where you could find systems with lifelike, *complex* behaviors.

So Langton had clearly done an elegant and important piece of work, says Kauffman. Nonetheless, what with his own involvement in economics, autocatalysis, and all sorts of other projects at the Santa Fe Institute—not to mention the time he was putting in on his book about the tension between self-organization and selection—it was several years before the full implications of the edge of chaos finally hit home. It didn't really happen until the summer of 1988, in fact, when Norman Packard came through the institute on a visit from Illinois and gave a seminar about his own work regarding the edge of chaos.

Packard, who had independently hit upon the phase transition idea at about the same time Langton did, had also been giving a good deal of thought to adaptation. So he couldn't help but wonder: Are the systems that can adapt the best also the ones that can compute the best—the ones at this funny boundary? It was an appealing thought. So Packard had done a simple simulation. Starting with a lot of cellular automata rules, he had demanded that each of them perform a certain calculation. He had then applied a Holland-style genetic algorithm to evolve the rules according to how well they did. And he had discovered that the final rules, the ones that could do the calculation fairly well, did indeed end up clumped at the boundary. In 1988 Packard had published his results in a paper entitled "Adaptation to the Edge of Chaos"—which, as it happened, was the first time that anyone had actually used the phrase "edge of chaos" in print. (When Langton was being formal he still said "*onset* of chaos.")

Listening to this, Kauffman was thunderstruck. "It was one of those moments when I said, 'Of course!' It was almost a shock of recognition. It had crossed my mind that you could get complex computation at the phase transition. But the thought that I hadn't had, which was silly, was that selection would get you there. The thought just didn't cross my mind."

Now that it *had* crossed his mind, however, his old problem of self-organization versus natural selection took on a wonderful new clarity. Living systems are *not* deeply entrenched in the ordered regime, which was essentially what he'd been saying for the past twenty-five years with his claim that self-organization is the most powerful force in biology. Living systems are actually very close to this edge-of-chaos phase transition, where things are much looser and more fluid. And natural selection is *not* the antagonist of self-organization. It's more like a law of motion—a force that is constantly pushing emergent, self-organizing systems toward the edge of chaos.

"Let's talk about networks as a model of the genetic regulatory system," says Kauffman with the enthusiasm of a convert. "My claim is that sparsely connected networks in the ordered regime, but not too far from the edge, do a pretty good job of fitting lots of features about real embryonic development, and real cell types, and real cell differentiation. And insofar as that's true, then it's a good guess that a billion years of evolution has in fact tuned real cell types to be near the edge of chaos. So that's very powerful evidence that there must be something good about the edge of chaos.

"So let's say the phase transition is the place to be for complex computation," he says. "Then the second assertion is something like, 'Mutation and selection will get you there.'" Packard, of course, had already demonstrated this assertion in his simple cellular automata model. But that was just one model. And anyway, Kauffman wanted to see it happen in his own genetic networks, if only to bolster his argument about evolution bringing real cell types to the edge of chaos. So shortly after he heard Packard's talk, he worked up a simulation in collaboration with a young programmer, Sonke Johnsen, who had just graduated from the University of Pennsylvania. Following Packard's same basic strategy, Kauffman and Johnsen presented pairs of simulated networks with a challenge: the "mismatch" game. The idea was to wire up each network so that six of their simulated light bulbs were visible to its opponent, and then set them to flashing the light bulbs at each other in various patterns; the "fittest" network was the one that could flash a series of patterns that were as *different* as possible from its opponent's patterns. The mismatch game could be adjusted to make it more complicated or less complicated for the networks, says Kauffman. The question was whether selection pressure, coupled with the genetic algorithm, would be enough to lead the networks to the phase transition zone where they were just on the verge of going chaotic. And the answer was yes in every case, he says. In fact, the answer continued to be yes whether he and Johnsen started the networks from the ordered regime or the chaotic regime. Evolution always seemed to lead to the edge of chaos.

So does that prove the conjecture? Hardly, says Kauffman. A handful of simulations don't prove anything by themselves. "If it turned out to be true for a wide variety of complicated games that the edge of chaos was the best place to be, and that mutation and selection got you there, then maybe the whole loose, wonderful conjecture *could* be answered," he says. But that, he admits, is one of those piles of rubble he hasn't had time to clean up. He's felt too many other wonderful conjectures beckoning.

The Danish-born physicist Per Bak was something of a wild card in the edge-of-chaos game. He and his colleagues at Long Island's Brookhaven National Laboratory had first published their ideas on "self-organized criticality" in 1987, and Phil Anderson, for one, had been raving about their work ever since. When Bak finally came through Los Alamos and the Santa Fe Institute to talk about it in the fall of 1988, he proved to be a rotund young man in his mid-thirties, with cherubic face and a Teutonic manner that ranged from brusque to confrontational. "I know what I'm talking about," he replied when Langton asked a question at one seminar. "Do you know what *you're* talking about?" He was also undeniably brilliant. His formulation of the phase transition idea was at least as simple and as elegant as Langton's, and yet so utterly different that it was sometimes hard to see how they related at all.

Bak explains that he and his coworkers Chao Tang and Kurt Wiesenfeld discovered self-organized criticality in 1986 as they were studying an esoteric condensed-matter phenomenon known as charge-density waves. But they quickly recognized it as something much more general and far-reaching. For the best and most vivid metaphor, he says, imagine a pile of sand on a tabletop, with a steady drizzle of new sand grains raining down from above. (This experiment has actually been done, by the way, both in computer simulations and with real sand.) The pile grows higher and higher until it can't grow any more: old sand is cascading down the sides and off the edge of the table as fast as the new sand dribbles down. Conversely, says Bak, you could reach exactly the same state by starting with a huge pile of sand: the sides would just collapse until all the excess sand had fallen off.

Either way, he says, the resulting sand pile is *self-organized*, in the sense that it reaches the steady state all by itself without anyone explicitly shaping it. And it's in a state of *criticality*, in the sense that sand grains on the surface are just barely stable. In fact, the critical sand pile is very much like a critical mass of plutonium, in which the chain reaction is just barely on the verge of running away into a nuclear explosion—but doesn't. The microscopic surfaces and edges of the grains are interlocked in every conceivable combination, and are just ready to give way. So when a falling grain hits there's no telling what might happen. Maybe nothing. Maybe just a tiny shift in a few grains. Or maybe, if one tiny collision leads to another in just the right chain reaction, a catastrophic landslide will take off one whole face of the sand pile. In fact, says Bak, all these things do happen at one time or another. Big avalanches are rare, and small ones are frequent. But the steadily drizzling sand triggers cascades of all sizes—a fact that manifests itself mathematically as the avalanches' "power-law" behavior: the average frequency of a given size of avalanche is inversely proportional to some power of its size.

Now the point of all this, says Bak, is that power-law behavior is very common in nature. It's been seen in the activity of the sun, in the light from galaxies, in the flow of current through a resistor, and in the flow of water through a river. Large pulses are rare, small ones are common, but all sizes occur with this power-law relationship in frequency. The behavior is so common, in fact, that explaining its ubiquity has become one of the nagging mysteries of physics: Why?

The sand pile metaphor suggests an answer, he says. Just as a steady trickle of sand drives a sand pile to organize itself into a critical state, a steady input of energy or water or electrons drives a great many systems in nature to organize themselves the same way. They become a mass of intricately interlocking subsystems just barely on the edge of criticality—with breakdowns of all sizes ripping through and rearranging things just often enough to keep them poised on the edge.

A prime example is the distribution of earthquakes, says Bak. Anyone who lives in California knows that little earthquakes that rattle the dishes are far more common than the big earthquakes that make international headlines. In 1956, geologists Beno Gutenberg and Charles Richter (of the famous Richter scale) pointed out that these tremors in fact follow a power law: in any given area, the number of earthquakes each year that release a certain amount of energy is inversely proportional to a certain power of the energy. (Empirically, the power is about $3/2$.) This sounded like self-organized criticality to Bak. So he and Chao Tang did a computer simulation of a fault zone like the San Andreas, where the two sides of the fault are being pulled in opposite directions by the steady, inexorable motion of the earth's crust. The standard earthquake model says that the rocks on either side are locked together by enormous pressure and friction; they resist the motion until suddenly they slip catastrophically. In Bak and Tang's version, however, the rocks on either side bend and deform until they are just ready to slip past each other—whereupon the fault undergoes a steady cascade of little slips and bigger slips that are just sufficient to keep the tension at that critical point. So a power law for earthquakes is exactly what you would expect, they argued; it's just a statement that the earth has long since tortured all its fault zones into a state of self-organized criticality. And indeed, their simulated earthquakes follow a power law very similar to the one found by Gutenberg and Richter.

Soon after that paper was published, says Bak, people started finding evidence for self-organizing criticality in all sorts of areas. Fluctuations in stock prices, for example, or the vagaries of city traffic. (Stop-and-go traffic jams correspond to critical avalanches.) There is still no general theory specifying which systems will go to a critical state and which won't, he admits. But clearly a lot of systems do.

Unfortunately, he adds, self-organizing criticality only tells you about the overall statistics of avalanches; it tells you nothing about any particular avalanche. This is yet another case where understanding is not the same thing as prediction. The scientists who try to predict earthquakes may ultimately succeed, but not because of self-organized criticality. They're in the same position as an imaginary group of tiny scientists living on a critical sand pile. These microscopic researchers can certainly perform a lot of detailed measurements on the sand grains in their immediate neighborhood, and—with a tremendous effort—predict when those particular sand grains are going to collapse. But knowing the global power-law behavior of the avalanches doesn't help them a bit, because the global behavior doesn't depend on the local details. In fact, it doesn't even make any difference if the sand pile scientists try to prevent the collapse they've predicted. They can certainly do so by putting up braces and support structures and such. But they just end up shifting the avalanche somewhere else. The global power law stays the same.

"Absolutely wonderful stuff," declares Kauffman. "When Per came through the institute, I fell in love with his self-organized criticality." Langton, Farmer, and all the rest of the Santa Fe crew felt much the same way, despite the prickliness of the messenger. Here, clearly, was another crucial piece of the edge-of-chaos puzzle. The trick was to figure out exactly where it fit in.

Self-organized criticality was obviously on the edge of something. And in many ways, that something was very much like the phase transitions that Langton was trying so hard to write about in his dissertation. In the kind of "second-order" phase transitions that he thought were important for the edge of chaos, for example, a real substance shows microscopic density fluctuations on all size scales; right at the transition, in fact, the fluctuations follow a power law. In the more abstract second-order phase transitions that Langton had discovered in the von Neumann universe, moreover, Class IV cellular automata such as the Game of Life also show structures and fluctuations and "extended transients" on all size scales.

In fact, you could even make the analogy mathematically precise. Langton's ordered regime, where systems always converge to a stable state, was like a subcritical piece of plutonium where chain reactions always die out, or like a tiny sand pile where the avalanches never really get going. His chaotic regime, where systems always diverge into unpredictable thrashing, was like a supercritical piece of plutonium where the chain reaction runs away or like a huge sand pile that collapses because it can't support itself. The edge of chaos, like the state of self-organized criticality, lies right at the boundary.

However, there were also some puzzling differences. The whole point of Langton's edge of chaos was that systems at the edge had the potential to do complex computations and show lifelike behaviors. Bak's critical state didn't seem to have anything to do with life or computation. (Can earthquakes compute?) Furthermore, there was nothing in Langton's formulation that said systems *had* to be at the edge of chaos; as Packard had pointed out, they can get there only through some form of natural selection. Bak's systems moved to the critical state spontaneously, driven by the input of sand, or energy, or whatever. It was (and is) an unsolved problem to understand precisely how these two phase transition concepts fit together.

Kauffman, however, wasn't terribly worried about that. It was clear enough that the concepts did fit together; whatever the details, there was something about self-organized criticality that felt *right*. Better still, Bak's way of looking at things clarified something that had been bugging him for a while now. It was one thing to talk about individual agents being on the edge of chaos. That's precisely the dynamical region that allows them to think and be alive. But what about a collection of agents taken as a whole? The economy, for example: people talk as if it had moods and responses and passing fevers. Is *it* at the edge of chaos? Are ecosystems? Is the immune system? Is the global community of nations?

Intuitively you'd like to believe that they all are, says Kauffman, if only to make sense of emergence. Molecules collectively make a living cell, and the cell is presumably at the edge of chaos because it is alive. Cells collectively make an organism, and organisms collectively make ecosystems, et cetera. So arguing by analogy, it seems reasonable to think that each new level is "alive" in the same sense—by virtue of being at or very near the edge of chaos.

But that was just the problem: Reasonable or not, how could you even test such a notion? Langton had been able to recognize a phase transition by watching for cellular automata that showed manifestly complex behavior on a computer screen. Yet it was not at all obvious how to do that for economies or ecosystems out in the real world. How are you supposed to tell what's simple and what's complex when you're looking at the behavior of Wall Street? Precisely what does it *mean* to say that global politics or the Brazilian rain forest is on the edge of chaos?

Bak's self-organized criticality suggested an answer, Kauffman realized. You can tell that a system is at the critical state and/or the edge of chaos if it shows waves of change and upheaval on all scales and if the size of the changes follows a power law. Of course, that was just a mathematically more precise way of saying what Langton had been saying all along: that a system can exhibit complex, lifelike behavior only if it has just the right balance of stability and fluidity. But a power law was something you could hope to measure.

To see how it might work, says Kauffman, imagine a stable ecosystem or a mature industrial sector where all the agents have gotten themselves well adapted to each other. There is little or no evolutionary pressure to change. And yet the agents can't stay there forever, he says, because eventually one of the agents is going to suffer a mutation large enough to knock him out of equilibrium. Maybe the aging founder of a firm finally dies and a new generation takes over with new ideas. Or maybe a random genetic crossover gives a species the ability to run much faster than before. "So that agent starts changing," says Kauffman, "and then he induces changes in one of his neighbors, and you get an avalanche of changes until everything stops changing again." But then someone else mutates. Indeed, you can expect the population to be showered with a steady rain of random mutations, much as Bak's sand pile is showered with a steady drizzle of sand grains—which means that you can expect any closely interacting population of agents to get themselves into a state of self-organized criticality, with avalanches of change that follow a power law.

In the fossil record, says Kauffman, this process would show up as long periods of stasis followed by rapid bursts of evolutionary change—exactly the kind of "punctuated equilibrium" that many paleontologists, notably Stephen J. Gould and Niles Eldridge, claim that they do see in the record. Taking that idea to its logical conclusion, moreover, you can argue that these avalanches lie behind the great extinction events in the earth's past, where whole groups of species vanish from the fossil record and are replaced by totally new ones. A falling asteroid or comet may very well have killed off the dinosaurs some 65 million years ago; all the evidence points that way. But most or all of the other great extinctions may have been purely internal affairs—just bigger-than-usual avalanches in a global ecosystem at the edge of chaos. "There's not enough fossil data on extinction events to persuade yourself," says Kauffman. "But you can plot what there is to see if you find a power law, and you sort of roughly do." Indeed, he made such a plot not long after hearing Bak's talk. The graph wasn't a perfect power law by any means. It was bent over, so that there weren't enough big avalanches compared with little avalanches. On the other hand, the graph didn't have to resemble a power law at all. So the results may not have been compelling, he says, but given the uncertainties in the data, they were certainly suggestive.

This tentative success led Kauffman to wonder if power-law cascades of change would be a general feature of "living" systems on the edge of chaos— the stock market, interdependent webs of technology, rain forests, et cetera. And while the evidence isn't all in yet, by a long shot, he feels that that prediction is still quite plausible. In the meantime, however, thinking about ecosystems at the edge of chaos had deflected his attention to another issue: How do they get there?

Packard's original answer, and his own, had been that systems get to the edge of chaos through adaptation. And Kauffman still believed that this was basically the right answer. The problem, however, was that when he and Packard had actually done their models, they had both demanded that their systems adapt to some arbitrary definition of fitness that *they* had imposed from the outside. And yet in real ecosystems, fitness isn't decreed from the outside at all. It arises from the dance of coevolution, as each individual constantly tries to adapt to all the others. This is exactly the issue that had led John Holland to start his work with the Echo model: imposing a definition of fitness from the outside is cheating. So the real question wasn't whether adaptation per se could get you to the edge of chaos, Kauffman realized. The real question was whether *coevolution* could get you there.

To find out—or, at least, to clarify the issues in his own mind—Kauffman decided to do yet another computer simulation, again in collaboration with Sonke Johnsen. As ecosystem models go, he admits, the simulation was a pretty good connectionist network. (At the heart of the program was a variant of his "NK landscape" model, which he had been developing over the past few years to get a better understanding of natural selection, and what it really means for a species' fitness to depend upon many different genes. The name refers to the fact that each species has N genes, with the fitness of each one depending on K other genes.) The model was even more abstract that Holland's Echo, which was already pretty rarefied. But conceptually, says Kauffman, it was pretty straightforward. You start by imagining an ecosystem where species are free to mutate and evolve by natural selection, but where they can interact with each other only in certain specified ways. So the frog always tries to catch the fly with its sticky tongue, the fox always hunts the rabbit, and so on. Alternatively, you could think of the model as an economy where each firm is free to organize itself however it likes internally, but where its relationships to other firms are fixed by a network of contracts and regulations.

Either way, says Kauffman, there is still plenty of room within those constraints for coevolution. If the frog evolves a longer tongue, for example, the fly has to learn how to make a faster getaway. If the fly evolves a chemical to make itself taste ghastly, the frog has to learn how to tolerate that taste. So how can you visualize this? One way, he says, is to look at each species in turn, starting with, say, the frog. At any given instant, the frog will find that some strategies work better than others. So at any given instant, the set of all strategies available to the frog forms a kind of imaginary landscape of "fitness," with the most useful strategies being at the peaks and the least useful being somewhere down in the valleys. As the frog evolves, moreover, it moves around on this landscape. Every time it undergoes a mutation it takes a step from its current strategy to a new strategy. And natural selection, of course, ensures that the average motion is always uphill toward greater fitness: mutations that move the organism downhill tend to die out.

It's the same story with the fly, the fox, the rabbit, et cetera, says Kauffman. Everyone is moving around on his own landscape. However, the whole point of coevolution is that these landscapes are *not* independent. They are coupled. What's a good strategy for the frog depends on what the fly is doing, and vice versa. "So as each agent adapts, it changes the fitness landscape of all the other agents," says Kauffman. "You have to imagine this picture of a frog climbing up hills toward a peak in its strategy space, and a fly climbing up hills toward a peak in *its* space, but the landscapes are deforming as they go." It's as if everyone were walking on rubber.

Now, says Kauffman: What kind of dynamics do you get in such a system? What kind of global behaviors do you see, and how do those behaviors relate to one another? This is where the simulation came in, he says. When he and Johnsen got their NK ecosystem model up and running, they found exactly the same three regimes that Langton had: an ordered regime, a chaotic regime, and an edge-of-chaoslike phase transition.

That was gratifying, says Kauffman. "It didn't have to be, but it was." In retrospect, however, it's easy enough to see why. "Picture a big ecosystem, with all the landscapes coupled. Well, there are only two things that could happen. Either all the species keep walking uphill and the landscape keeps deforming under them, so that they keep mushing around and never stop moving. Or a group of neighboring species does stop, because they've reached what John Maynard-Smith would call an evolutionarily stable strategy." That is, each species in the group has gotten so well adapted to the others that there's no immediate incentive to change.

"Now both processes can occur in the same ecosystem at the same time, depending on the precise structure of landscapes and how they are coupled," says Kauffman. "So look at the set of players who've quit moving because they're at a local optimum. Paint those guys red. And paint the others green." He and Johnsen actually did that as a way of displaying the simulation on a computer screen, he says. When the system is deep in the chaotic regime, then almost nobody is standing still. So the display shows a sea of green, with just a few islands of red twinkling into existence here and there where a handful of species manage to find a temporary equilibrium. When the system is deep in the ordered regime, conversely, then almost everyone is locked into an equilibrium. So the display shows a field of red, with bits of green snaking around here and there where individual species can't quite manage to settle down.

When the system is at the phase transition, of course, order and chaos are in balance. And fittingly enough, the display seems to pulse with life. Red islands and green islands intertwine with each other, shooting out tendrils like random fractals. Parts of the ecosystem are forever hitting equilibrium and turning red, while other parts are forever twinkling and turning green as they find new ways to evolve. Waves of change wash across the screen on all size scales—including the occasional huge wave that spontaneously washes across the screen and transforms the ecosystem beyond recognition.

It looks like punctuated equilibrium in action, says Kauffman. But as much fun as it was to see the three dynamical regimes displayed in this way—and as gratifying as it was to see that the coevolutionary model really did have an edge-of-chaos phase transition—that was only half the story. It still didn't explain how ecosystems could get to this boundary region. On the other hand, he says, even with all this business about rubbery, deformable fitness landscapes, the only thing he's talked about so far has been the process of mutation in individual genes. What about changes to the *structure* of each species' genome—the internal organization chart that tells how one gene interacts with another? That structure is presumably just as much a product of evolution as the genes themselves are, he says. "So you can imagine an evolutionary metadynamic, a process that would tune the internal organization of each agent so that they all reside at the edge of chaos."

To test that idea, Kauffman and Johnsen allowed the agents in their simulation to change their internal organization. This was tantamount to what John Holland calls "exploratory learning." It was also very much like the radical rewiring that Farmer talked about in his Rosetta Stone paper on connectionist models. But the upshot was that when species were given the ability to evolve their internal organization, the ecosystem as a whole did indeed move toward the edge of chaos.

Once again, says Kauffman, it's easy enough in retrospect to see why. "If we're deep in the ordered regime," he says, "then everybody is at a peak in fitness and we're all mutually consistent—but these are lousy peaks." Everybody is trapped in the foothills, so to speak, with no way to break loose and head for the crest of the range. In terms of human organizations, it's as if the jobs are so subdivided that no one has any latitude; all they can do is learn how to perform the one job they've been hired for, and nothing else. Whatever the metaphor, however, it's clear that if each individual in the various organizations is allowed a little more freedom to march to a different drummer, then everyone will benefit. The deeply frozen system will become a little more fluid, says Kauffman, the aggregate fitness will go up, and the agents will collectively move a bit closer to the edge of chaos.

Conversely, says Kauffman, "If we're deep in the chaotic regime, then every time I change I screw you up, and vice versa. We never get to the peaks, because you keep kicking me off and I keep kicking you off, and it's like Sisyphus trying to roll the rock uphill. Therefore, my overall fitness tends to be pretty low, and so does yours." In organizational terms, it's as if the lines of command in each firm are so screwed up that nobody has the slightest idea what they're supposed to do—and half the time they are working at cross-purposes anyway. Either way, it obviously pays for individual agents to tighten up their couplings a bit, so that they can begin to adapt to what other agents are doing. The chaotic system will become a little more stable, says Kauffman, the aggregate fitness will go up, and once again, the ecosystem as a whole will move a bit closer to the edge of chaos.

Somewhere in between the ordered and chaotic regimes, of course, the aggregate fitness has to reach a maximum. "From the numerical simulations that we've done," says Kauffman, "it turns out that the maximum fitness is occurring right at the phase transition. So the crux is, as if by an invisible hand, all the players change their landscape, each to its own advantage, and the whole system coevolves to the edge of chaos."

So there it is, says Kauffman: the evidence consists of a sort of power law in the fossil record suggesting that the global biosphere is near the edge of chaos; a couple of computer models showing that systems can adapt their way to the edge of chaos through natural selection; and now one computer model showing that ecosystems may be able to get to the edge of chaos through coevolution. "So far," he says, "that's the only evidence that I know that the edge of chaos is actually where complex systems go in order to solve a complex task. It's pretty sketchy. So, while I'm absolutely in love with this hypothesis—I think it's absolutely plausible and credible and intriguing—I don't know if it's generally true.

"But if this hypothesis *is* generally true," he says, "Then it's really important. It would apply to economic systems and everything else." It would help us make sense of our world in a way we never were able to before. It would be a linchpin of this hypothetical new second law. And, not incidentally, it would go a long way toward Stuart Kauffman's thirty-year quest for a lawful marriage of self-organization and selection.

Finally, says Kauffman, there has to be at least one further aspect of the new second law: "There has to be something in it about the fundamental fact that organisms have gotten more complex since life began. We need to know *why* organisms have gotten more complex. What's the advantage?"

The only honest answer, of course, is that nobody knows—yet. "But that question is behind this whole other strand in my thinking," he says. "It starts with the origin-of-life–autocatalytic-polymer-set model, and goes on through a theory of complexity and organization that may follow from that." That theory is still nebulous and tentative in the extreme, he admits. He can't claim that he's satisfied with it. "But it's where my own deepest hopes for Carnot's whispers remain."

From his own point of view, ironically, the autocatalytic set idea sat around in limbo for quite a while. By the time that he, Farmer, and Packard had published the origin-of-life simulation in 1986, says Kauffman, Farmer had gone off to do some more work on prediction theory, Packard was helping Stephen Wolfram set up a complex systems institute at the University of Illinois, and he just didn't feel he could develop the model any further by himself. Quite aside from wanting to pursue about a dozen hot ideas per day at the Santa Fe Institute, he lacked either the patience or the programming skills to sit in front of a computer screen day after day hunting down bugs in a complicated piece of software. (Indeed, work on the origin-of-life model resumed again only in 1987, when Farmer found a graduate student, Richard Bagley, who was interested in taking it up as his dissertation topic. Bagley greatly improved the simulation by including a more realistic account of thermodynamics and several other refinements, not to mention speeding up the computer code about 1000-fold. He received his Ph.D. in 1991.)

The upshot was that he did relatively little on autocatalysis for about four years, says Kauffman—not until May 1990, in fact, when he heard a seminar given by Walter Fontana, a young German-Italian postdoc who had recently joined Farmer's complex systems group up at Los Alamos.

Fontana started with one of those cosmic observations that sound so deceptively simple. When we look at the universe on size scales ranging from quarks to galaxies, he pointed out, we find the complex phenomena associated with life only at the scale of molecules. Why?

Well, said Fontana, one answer is just to say "chemistry": life is clearly a chemical phenomenon, and only molecules can spontaneously undergo complex chemical reactions with one another. But again, Why? What is it that allows molecules to do what quarks and quasars can't?

Two things, he said. The first source of chemistry's power is simple variety: unlike quarks, which can only combine to make protons and neutrons in groups of three, atoms can be arranged and rearranged to form a huge number of structures. The space of molecular possibilities is effectively limitless. The second source of power is reactivity: structure A can manipulate structure B to form something new—structure C.

Of course, this definition left out a lot things like rate constants and temperature dependence, which are crucial to understanding real chemistry. But that was intentional, said Fontana. His contention was that "chemistry" is a concept that actually applies to a wide variety of complex systems, including economies, technologies, and even minds. (Goods and services interact with goods and services to produce new goods and services, ideas react with ideas to produce new ideas, et cetera.) And for that reason, he said, a computer model that distilled chemistry down to its purest essence— variety and reactivity—ought to give you a whole new way to study the growth of complexity in the world.

To accomplish that, Fontana went back to the essence of computer programming itself to define what he called algorithmic chemistry, or "Alchemy." As John von Neumann had pointed out long ago, he said, a piece of computer code leads a double life. On the one hand it's a program, a series of commands telling the computer what to do. But on the other hand it's just data, a string of symbols sitting somewhere inside the computer's memory. So let's use that fact to define a chemical reaction between two programs, said Fontana: program A simply reads in program B as input data, and then "executes" to produce a string of output data—which the computer now interprets as a new program, program C. (Since this obviously wouldn't work very well with a computer language such as FORTRAN or PASCAL, Fontana actually wrote his reactive programs in a variant of the computer language LISP, in which almost any string of symbols can represent a valid program.)

Next, said Fontana, take a few zillion of these symbol-string programs, put them in a simulated pot where they can interact at random, and watch what happens. In fact, the results are not unlike those of the autocatalytic model of Kauffman, Farmer, and Packard, he said, but with weird and wonderful variations. There are self-sustaining autocatalytic sets, of course. But there are also sets that can grow without bound. There are sets that can repair themselves if some of their component "chemicals" are deleted, and there are sets that adapt and change themselves if new components are injected. There are even pairs of sets that have no members in common, but that mutually catalyze each other's existence. In short, he said, the Alchemy program suggests that populations of pure processes—his symbol-string programs—are enough for the spontaneous emergence of some very lively structures indeed.

"Well, I was really very, very excited about what Walter had done," says Kauffman. "I had been thinking about my autocatalytic polymer stuff for a long time as a model of economic and technological webs, but I couldn't see my way past polymers. But as soon as I heard Walter, that was it. He had figured it out."

Kauffman immediately decided to follow Fontana's lead and get back into the autocatalysis game in a big way—but with his own twist. Fontana had identified abstract chemistry as a whole new way of thinking about emergence and complexity, he realized. But were his results a general property of abstract chemistry? Or just of the way he implemented his Alchemy program?

It was the same question Kauffman had asked about genomic regulatory systems in 1963, when he first devised his network models. "Just as I wanted to find the generic properties of genetic networks," he says, "I wanted to look at generic properties of abstract chemistries. As you tune the complexity of the chemistry, and other things, such as how much diversity you have in the initial set of molecules, what are the generic consequences for the unfolding behavior?" So instead of following Fontana's Alchemy approach directly—it was Fontana's, in any case—Kauffman abstracted the idea even further. He still used symbol strings to represent the "molecules" of the system. But he didn't even insist that they be programs. They could just be strings of symbols: 110100111, 10, 111111, et cetera. The "chemistry" of his model was then just a set of rules specifying how certain symbol strings could transform certain other symbol strings. And since strings of symbols are like words in a language, he called that set of rules—what else?—a "grammar." (In fact, such grammars of symbol-string transformations have been extensively studied in the context of computer languages, which is where Kauffman got the idea.) The upshot was that he could sample the kind of behaviors that result from various chemistries by generating a set of grammar rules at random, and then seeing what kinds of autocatalytic structures result.

"Here's the intuition," he says. "Start with a pot of symbol strings, and let them act on one another according to the grammar rules. It might be the case that the new strings are always longer than the old strings, so that you can never make a string you've made before." Call that a "jet": in the space of all possible strings, it's a structure that shoots farther and farther outward without ever looking back. "Or, when you make a cloud of strings," he says, "you might start making a string you've made before, but by a different route. Call that a 'mushroom.' Those are my autocatalytic sets; they're a model of how you can bootstrap yourself into existence. Then you might get a set of strings that collectively make themselves and nothing else, just hovering there in string space. Let me call that an 'egg.' It's a self-reproducing thing, but no single entity in there reproduces itself. Or you might make what I'll call a 'filigreed fog'—you make all kinds of strings all over the place, but there are certain strings you cannot make, like 110110110. So here's some new kinds of objects to play with."

And what does all this have to do with the mysterious, inexorable growth of complexity? Maybe a lot, says Kauffman. "The growth of complexity really does have something to do with far-from-equilibrium systems building themselves up, cascading to higher and higher levels of organization. Atoms, molecules, autocatalytic sets, et cetera. But the key thing is that once those higher level entities emerge, they can also interact among themselves." A molecule can connect to a molecule to make a new molecule. And much the same thing happens with these emergent objects in the string world, he says: the same chemistry that creates them allows them to have a rich set of interactions, simply by exchanging strings. "For example, here's an egg, you throw in a string from the outside, and it might turn into a jet, or change to another egg, or become a filigreed fog. And the same for any of these other objects."

In any case, says Kauffman, once you have the interactions, it ought to be true in general that autocatalysis occurs whenever the conditions are right—whether you're talking about molecules or economies. "Once you've accumulated a sufficient diversity of objects at the higher level, you go through a kind of autocatalytic phase transition—and get an enormous proliferation of things at that level." These proliferating entities then proceed to interact and produce autocatalytic sets at a still higher level. "So you get a hierarchical cascade from lower-order things to higher-order things— each going through something like this autocatalytic phase transition."

If that's really true, says Kauffman, then you can begin to see why the growth of complexity seems so inexorable: it's just a reflection of the same law of autocatalysis that (perhaps) was responsible for the origin of life. And surely that has to be a part of the putative new second law. That said, however, he's also convinced that it's not the whole story—for exactly the same reasons that he finally came to realize that self-organization isn't all there is to biology. When you think about it, in fact, this upward cascade of levels upon levels is just another kind of self-organization. So how is the cascade shaped by selection and adaptation?

This is where things really get tentative, says Kauffman. But he does have some ideas. "This is either a deep insight or stupid, but it dawned on me one day recently: if you start with some founder set of strings, they may give rise to autocatalytic sets of strings, they may give rise to autocatalytic sets squirting jets, they may give rise to mushrooms, eggs, or whatever. But they may also make dead strings. A 'dead' string means one that's inert. It can't be a catalyst, and it doesn't react with anything."

Now clearly, he says, if the system ends up making a lot of dead strings, then it's not going to expand very far—rather like an economy that diverts most of its output into knickknacks that no one wants to buy and that can't be made into anything else. "But if the 'live,' productive strings can some-how organize themselves so they don't make so many dead strings, then there are more live strings." So the net productivity goes up, and this group of live strings has a selective advantage over groups that don't organize themselves in this way. And, in fact, when you look at the computer models you find that the flow of strings into dead strings does decrease as the simulation proceeds.

"I also think that there are refinements of that idea," he says. "Suppose you have two jets coming out of a founder set. The two jets can compete with one another for strings. But if one jet can learn to help a second jet avoid making dead strings, and the second jet can learn to help the first jet avoid making dead strings, then you have mutualists." That pair of cooperative jets might then become the foundation for a new "multijet" structure that would emerge as a new and more complex individual at a still higher level. "I have a hunch that higher-order things emerge because they can suck more flow of stuff into themselves, faster," says Kauffman, "whether we're talking about *E. coli*, prebiotic evolution, or firms. So what I'd like to see all this lead to is a theory of coupled processes that build themselves up into things that compete for and win the flow—and that get themselves to the edge of chaos at the same time."

Now admittedly, says Kauffman, none of this is anything more than intuition so far. "But it feels right to me. Somehow, the next step in the new second law is to understand the natural unfolding of this upward, billowing cascade. If I can just show that those entities that happen fastest and suck the most flow through them are what you see, with some characteristic distribution, that will be it."

At Home in the Universe

Science is about a great many things, says Doyne Farmer. It's about the systematic accumulation of facts and data. It's about the construction of logically consistent theories to account for those facts. It's about the discovery of new materials, new pharmaceuticals, and new technologies.

But at heart, he says, science is about the telling of stories—stories that explain what the world is like, and how the world came to be as it is. And like older explanations, such as creation myths, epic legends, and fairy tales, the stories that science tells help us understand something about who we are as human beings, and how we relate to the universe. There is the story of how the universe exploded into existence some 15 billion years ago at the instant of the Big Bang; the story of how quarks, electrons, neutrinos, and all the rest came flying out of the Big Bang as an indescribably hot plasma; the story of how those particles gradually condensed into the matter we see around us today in the galaxies, the stars, and the planets; the story of how the sun is a star like other stars, and how Earth is a planet like other planets; the story of how life arose on this Earth and evolved over 4 billion years of geological time; the story of how the human species first arose on the African savannah some 3 million years ago and slowly acquired tools, culture, and language.

And now there is the story of complexity. "I almost view it as a religious issue," says Farmer. "For me as a physicist, as a scientist, my deep-down motivation has always been to understand the universe around me. For me as a pantheist, nature *is* God. So by understanding nature I get a little closer to God. Up until I was in my third year of grad school, in fact, I never even dreamed I would be able to get a job as a scientist. I just viewed it as what I was doing instead of joining a monastery.

"So when we ask questions like how life emerges, and why living systems are the way they are—these are the kind of questions that are really fundamental to understanding what we are and what makes us different from inanimate matter. The more we know about these things, the closer we're going to get to fundamental questions like, 'What is the purpose of life?' Now, in science we can never even attempt to make a frontal assault on questions like that. But by addressing a different question—like, Why is there an inexorable growth in complexity?—we may be able to learn something fundamental about life that suggests its purpose, in the same way that Einstein shed light on what space and time are by trying to understand gravity. The analogy I think of is averted vision in astronomy: if you want to see a very faint star, you should look a little to the side because your eye is more sensitive to faint light that way—and as soon as you look right at the star, it disappears."

Likewise, says Farmer, understanding the inexorable growth of complexity isn't going to give us a full scientific theory of morality. But if a new second law helps us understand who and what we are, and the processes that led to us having brains and a social structure, then it might tell us a lot more about morality than we know now.

"Religions try to impose rules of morality by writing them on stone tablets," he says. "We do have a real problem now, because when we abandon conventional religion, we don't know what rules to follow anymore. But when you peel it all back, religion and ethical rules provide a way of structuring human behavior in a way that allows a functioning society. My feeling is that all of morality operates at that level. It's an evolutionary process in which societies constantly perform experiments, and whether or not those experiments succeed determines which cultural ideas and moral precepts propagate into the future." If so, he says, then a theory that rigorously explains how coevolutionary systems are driven to the edge of chaos might tell us a lot about cultural dynamics, and how societies reach that elusive, ever-changing balance between freedom and control.

"I draw a lot of fairly speculative conclusions about the implications of all this," says Chris Langton. "It comes from viewing the world very much through these phase transition glasses: you can apply the idea to a lot of things and it kind of fits."

Witness the collapse of communism in the former Soviet Union and its Eastern European satellites, he says: the whole situation seems all too reminiscent of the power-law distribution of stability and upheaval at the edge of chaos. "When you think of it," he says, "the Cold War was one of these long periods where not much changed. And although we can find fault with the U.S. and Soviet governments for holding a gun to the world's head—the only thing that kept it from blowing up was Mutual Assured Destruction—there was a lot of stability. But now that period of stability is ending. We've seen upheaval in the Balkans and all over the place. I'm more scared about what's coming in the immediate future. Because in the models, once you get out of one of these metastable periods, you get into one of these chaotic periods where a lot of change happens. The possibilities for war are much higher—including the kind that could lead to a world war. It's much more sensitive now to initial conditions.

"So what's the right course of action?" he asks. "I don't know, except that this is like punctuated equilibrium in evolutionary history. It doesn't happen without a great deal of extinction. And it's not necessarily a step for the better. There are models where the species that dominate in the stable period after the upheaval may be less fit than the species that dominated beforehand. So these periods of evolutionary change can be pretty nasty times. This is the kind of era when the United States could disappear as a world power. Who knows what's going to come out the other end?

"The thing to do is to try to determine whether we can apply this sort of thing to history—and if so, whether we also see this kind of punctuated equilibrium. Things like the fall of Rome. Because in that case, we really are part of the evolutionary process. And if we really study that process, we may be able to incorporate this thinking into our political, social, and economic theories, where we realize that we have to be very careful and put some global agreements and treaties in place to get us through. But then the question is, do we want to gain control of our own evolution or not? If so, does that stop evolution? It's good to have evolution progress. If

single-celled things had found a way to stop evolution to maintain themselves as dominant life-forms, then we wouldn't be here. So you don't want
to stop it. On the other hand, maybe you want to understand how it can
keep going without the massacres and the extinctions.

"So maybe the lesson to be learned is that evolution hasn't stopped,"
says Langton. "It's still going on, exhibiting many of the same phenomena
it did in biological history—except that now it's taking place on the social-
cultural plane. And we may be seeing a lot of the same kinds of extinctions
and upheaval."

"I have partial answers to what it all means," says Stuart Kauffman, who
speaks as a man who's had reason to be reflective of late. Shortly before
Thanksgiving 1991, he and his wife Liz were passengers in a car crash that
left them both severely injured and could easily have been fatal; they spent
months recuperating.

"For example, suppose that these models about the origin of life are
correct. Then life doesn't hang in the balance. It doesn't depend on whether
some warm little pond just happens to produce template-replicating molecules like DNA or RNA. Life is the natural expression of complex matter.
It's a very deep property of chemistry and catalysis and being far from
equilibrium. And that means that we're at home in the universe. We're to
be expected. How welcoming that is! How far that is from the image of
organisms as tinkered-together contraptions, where everything is bits of
widgetry piled on top of bits of ad hocery, and it's all blind chance. In that
world there are no deep principles in biology, other than random variation
and natural selection; we're not at home in the universe in the same way.

"Next," says Kauffman, "suppose that you come back many years later,
after the autocatalytic sets have been coevolving with one another and
squirting strings at one another. The things that would still be around would
be those things that had come to evolve competitive interactions, food webs,
mutualism, symbiosis. The things that you would see would be those that
made the world they now mutually live in. And that reminds us that we
make the world we live in with one another. We're participants in the story
as it unfolds. We aren't victims and we aren't outsiders. We are part of the
universe, you and me, and the goldfish. We make our world with one
another.

"And now suppose it's really true that coevolving, complex systems get
themselves to the edge of chaos," he says. "Well, that's very Gaia-like. It
says that there's an attractor, a state that we collectively maintain ourselves

in, an ever-changing state where species are always going extinct and new ones are coming into existence. Or if we imagine that this really carries over into economic systems, then it's a state where technologies come into existence and replace others, et cetera. But if this is true, it means that the edge of chaos is, on average, the best that we can do. The ever-open and ever-changing world that we must make for ourselves is in some sense as good as it possibly can be.

"Well, that's a story about ourselves," says Kauffman. "Matter has managed to evolve as best it can. And we're at home in the universe. It's not Panglossian, because there's a lot of pain. You can go extinct, or broke. But here we are on the edge of chaos because that's where, on average, we all do the best."

Roasted

In late 1989 it finally happened, just as Doyne Farmer had always been afraid it would. Chris Langton applied for an internal grant from Los Alamos headquarters. And in the course of processing the paperwork, the laboratory higher-ups discovered that Langton had been a postdoc there for three full years and *still* didn't have the "doc." "It hit the fan," says Farmer. "I remember it because I was in Italy at the time on vacation. They somehow tracked me down in this small town on the Ligurian coast, and I had to make a series of phone calls where I was plunking down these thousand-lira coins into a phone that looked like it had been made by Alexander Graham Bell himself. And when I got back, I had to meet in front of the postdoc committee to defend Chris—and to defend myself as his supervisor. I really got roasted. 'How could this have happened?' et cetera. All I could do was point to the fact that Chris was the founder of a whole new field called artificial life. Of course, all that accomplished was to raise their suspicions even more. In the end, because he still hadn't finished, we even had to ask for a three-month extension of his postdoc appointment."

Farmer and David Campbell, director of the Center for Nonlinear Studies where Langton worked, continued to be supportive. But there was no doubt in their minds or in Langton's that the pressure was on. On top of that, a second artificial life conference had already been scheduled for February 1990. And while Langton had some organizational help this time around from Farmer and several others, the workshop was still his baby. He had to get this damned dissertation out of the way. So he worked like a fiend. And in November 1989 he flew to Ann Arbor, ready to defend the thing in front of his dissertation committee, which was cochaired by John Holland and Art Burks. If they found the work acceptable, they would award him his Ph.D. on the spot and the agony would be over.

The opinion of the committee, unfortunately, was unanimous: "Not yet." The basic edge-of-chaos idea is wonderful, they said, and you've done lots of computer experiments to back it up. But you've also made some pretty sweeping statements in here about the Wolfram classes, the emergence of computation, and such, and the link to the data is pretty slippery. So the thing to do is to tone down the statements, make them more supportable, and get them lined up better with your data.

But that means rewriting the whole thing! said a despairing Langton.

Then you'd better get started, said Holland, Burks, and the others.

"This was a very depressing time," says Langton. "Here I was thinking I was ready to defend. But I couldn't. And then A-Life II was coming up that February. So I had to put it aside again."

Shortly before Christmas of 1989, as Brian Arthur drove west from Santa Fe with a car packed full of books and clothes for his return home to Stanford, he found himself staring straight into a spectacular New Mexico sunset that bathed the desert in a vast red glow. "I thought, 'This is too bloody romantic to be true!' " he laughs.

But appropriate. "I had been at the institute just about eighteen months at that time," he says, "and I felt that I needed to go home—to write, and think, and get things clear in my mind. I was just loaded down with ideas. I'd felt that I was learning at Santa Fe more in a month than I would have in a year at Stanford. The experience had almost been too rich. And yet it was a wrench to leave. I felt very, very, very sad, in a good way, and very nostalgic. The whole scene—the desert, the light, the sunset—brought home to me that those eighteen months might well have been the high point of my scientific life, and they were over. That time would not be easily recaptured. I knew other people would come and follow up. I knew I could probably go back—even go back and run the economics program again in some future years. But I suspected that the institute might never be the same. I felt lucky to have been in on a golden time."

The Tao of Complexity

Three years later, sitting in his corner office overlooking the tree-shaded walkways of Stanford University, the Dean and Virginia Morrison Professor of Population Studies and Economics admits that he *still* hasn't gotten the Santa Fe experience completely clear in his mind. "I'm beginning to appreciate it more as time passes," says Arthur. "But I think the story of what's been accomplished in Santa Fe is still very much unfolding."

Fundamentally, he says, he's come to realize that the Santa Fe Institute was and is a catalyst for changes that would have taken place in any case—but much more slowly. Certainly that was the case for the economics program, which continued after his departure under the joint directorship of Minnesota's David Lane and Yale's John Geanakoplos. "By about 1985," says Arthur, "it seems to me that all sorts of economists were getting antsy, starting to look around and sniff the air. They sensed that the conventional neoclassical framework that had dominated over the past generation had reached a high water mark. It had allowed them to explore very thoroughly the domain of problems that are treatable by static equilibrium analysis. But it had virtually ignored the problems of process, evolution, and pattern formation—problems where things were not at equilibrium, where there's a lot of happenstance, where history matters a great deal, where adaptation and evolution might go on forever. Of course, the field had kind of gotten stymied by that time, because theories were not held to be theories in economics unless they could be fully mathematized, and people only knew how to do that under conditions of equilibrium. And yet some of the very best economists were sensing that there had to be other things going on and other directions that the subject could go in.

"What Santa Fe did was to act as a gigantic catalyst for all that. It was a place where very good people—people of the caliber of Frank Hahn and Ken Arrow—could come and interact with people like John Holland and Phil Anderson, and over a period of several visits there realize, Yes! We can deal with inductive learning rather than deductive logic, we can cut the Gordian knot of equilibrium and deal with open-ended evolution, because many of these problems have been dealt with by other disciplines. Santa Fe provided the jargon, the metaphors, and the expertise that you needed in order to get the techniques started in economics. But more than that, Santa Fe legitimized this different vision of economics. Because when word got around that people like Arrow and Hahn and Sargent and others were writing papers of this sort, then it became perfectly reasonable and perfectly kosher for others to do so."

Arthur sees evidence for that development every time he goes to an economics meeting these days. "The people who were interested in process and change in the economy were there all along," he says. Indeed, many of the essential ideas were championed by the great Austrian economist

Joseph Schumpeter as far back as the 1920s and 1930s. "But my sense is that in the past four or five years, the people who think this way have gotten much more confident. They aren't apologetic any more about just being able to give wordy, qualitative descriptions of economic change. Now they're armed. They have technique. They form a growing movement that is becoming part of the neoclassical mainstream everywhere."

That movement has certainly made his own life easier, notes Arthur. His ideas on increasing returns, once virtually unpublishable, now have a following. He finds himself getting invitations to give this or that distinguished lecture in far-off places. In 1989 he was invited to write a feature article on increasing-returns economics for *Scientific American*. "That was one of the biggest thrills," he says. And that article, published in February 1990, helped him become a co-winner of the International Schumpeter Society's 1990 Schumpeter Prize for the best research on evolutionary economics.

For Arthur, however, the most gratifying assessment of the Santa Fe approach came in September 1989, as Ken Arrow was summarizing a big, week-long workshop that had reviewed the program's progress to date. At the time, ironically, Arthur barely heard what Arrow was saying. That noontime, he says, as he'd headed out the front door of the convent on his way to lunch, he'd managed to trip and sprain his ankle terribly. He'd spent that whole afternoon in the convent's chapel-turned-conference room listening to the closing session of the workshop through a haze of pain, with his foot carefully wrapped by Dr. Kauffman and propped up with a bag of ice on the chair in front of him. In fact, the full impact of Arrow's words only hit him a few days later, after he'd defied all advice of doctors, colleagues, and wife and hobbled off to a long-planned conference in Irkutsk, on the shores of Lake Baikal in Siberia.

"It was one of these flashes of extreme clarity you get at three in the morning," he says. "The Aeroflot jet was just coming into Irkutsk, and there was this guy riding a bicycle down the runway, waving a light stick to show us where to taxi. And when I thought about what Arrow had said in his closing summary, it finally struck home. He said, 'I think we can safely say we have another type of economics here. One type is the standard stuff that we're all familiar with'—he was too modest to call it the Arrow-Debreu system, but he basically meant the neoclassical, general equilibrium theory—'and then this other type, the Santa Fe–style evolutionary economics.' He made it clear that, to his mind, what the program had demonstrated in a year was that this was another valid way to do economics, equal in status to the traditional theory. It wasn't that the standard formulation was wrong, he said, but that we were exploring into a new way of looking at parts of the economy that are not amenable to conventional methods. So this new approach was complementary to the standard ones. He also said that we didn't know where this new sort of economics was taking us. It was the beginnings of a research program. But he found it very interesting and exciting.

"That pleased me enormously," says Arthur. "But Arrow said a second thing also. He compared the Santa Fe program of research with the Cowles Foundation program that he had been associated with in the early 1950s. And he said that the Santa Fe approach seemed to be much more accepted at this stage, given that it's now at most two years old, than the Cowles Foundation group had been at the same point. Well, I was amazed to hear that, and tremendously flattered. Because the Cowles Foundation people were the Young Turks of their day—Arrow, Koopmans, Debreu, Klein, Hurwicz, et cetera. Four of them got Nobel Prizes, with maybe a few more to come. They were the people who mathematized economics. They were the people who had set the agenda for the following generations. They were the people who had actually revolutionized the field."

From the Santa Fe Institute's point of view, of course, this effort to catalyze a sea change in economics is only a part of its effort to catalyze the complexity revolution in science as a whole. That quest may yet prove quixotic, says Arthur. But nonetheless, he's convinced that George Cowan, Murray Gell-Mann, and the others have gotten hold of exactly the right set of issues.

"Nonscientists tend to think that science works by deduction," he says. "But actually science works mainly by metaphor. And what's happening is that the kinds of metaphor people have in mind are changing." To put it in perspective, he says, think of what happened to our view of the world with the advent of Sir Isaac Newton. "Before the seventeenth century," he says, "it was a world of trees, disease, human psyche, and human behavior. It was messy and organic. The heavens were also complex. The trajectories of the planets seemed arbitrary. Trying to figure out what was going on in the world was a matter of art. But then along comes Newton in the 1660s. He devises a few laws, he devises the differential calculus—and suddenly the planets are seen to be moving in simple, predictable orbits!

"This had an incredibly profound effect on people's psyche, right up to the present," says Arthur. "The heavens—the habitat of God—had been explained, and you didn't need angels to push things around anymore. You didn't need God to hold things in place. So in the absence of God, the age became more secular. And yet, in the face of snakes and earthquakes, storms and plagues, there was still a profound need to know that something had it all under control. So in the Enlightenment, which lasted from about 1680 all through the 1700s, the era shifted to a belief in the primacy of nature: if you just left things alone, nature would see to it that everything worked out for the common good."

The metaphor of the age, says Arthur, became the clockwork motion of the planets: a simple, regular, predictable Newtonian machine that would run of itself. And the model for the next two and a half centuries of reductionist science became Newtonian physics. "Reductionist science tends to say, 'Hey, the world out there is complicated and a mess—but look! Two or three laws reduce it all to an incredibly simple system!'

"So all that remained was for Adam Smith, at the height of the Scottish Enlightenment around Edinburgh, to understand the machine behind the economy," says Arthur. "In 1776, in *The Wealth of Nations*, he made the case that if you left people alone to pursue their individual interests, the 'Invisible Hand' of supply and demand would see to it that everything worked out for the common good." Obviously, this was not the whole story: Smith himself pointed to such nagging problems as worker alienation and exploitation. But there was so much about his Newtonian view of the economy that was simple and powerful and *right* that it has dominated Western economic thought ever since. "Smith's idea was so brilliant that it just dazzled us," says Arthur. "Once, long ago, the economist Kenneth Boulding asked me, 'What would you like to do in economics?' Being young and brash, I said very immodestly, 'I want to bring economics into the twentieth century.' He looked at me and said, 'Don't you think you should bring it into the eighteenth century first?' "

In fact, says Arthur, he feels that economics in the twentieth century has lagged about a generation behind a certain loss of innocence in all the sciences. As the century began, for example, philosophers such as Russell, Whitehead, Frege, and Wittgenstein set out to demonstrate that all of mathematics could be founded on simple logic. They were partly right. Much of it can be. But not all: in the 1930s, the mathematician Kurt Gödel showed that even some very simple mathematical systems—arithmetic, for example—are inherently incomplete. They always contain statements that cannot be proved true or false within the system, even in principle. At about the same time (and by using essentially the same argument), the logician Alan Turing showed that even very simple computer programs can be undecidable: you can't tell in advance whether the computer will reach an answer or not. In the 1960s and 1970s, physicists got much the same message from chaos theory: even very simple equations can produce results that are surprising and essentially unpredictable. Indeed, says Arthur, that message has been repeated in field after field. "People realized that logic and philosophy are messy, that language is messy, that chemical kinetics is messy, that physics is messy, and finally that the economy is naturally messy. And it's not that this is a mess created by the dirt that's on the microscope glass. It's that this mess is inherent in the systems themselves. You can't capture any of them and confine them to a neat box of logic."

The result, says Arthur, has been the revolution in complexity. "In a sense it's the opposite of reductionism. The complexity revolution began the first time someone said, 'Hey, I can start with this amazingly simple system, and look—it gives rise to these immensely complicated and unpredictable consequences.'" Instead of relying on the Newtonian metaphor of clockwork predictability, complexity seems to be based on metaphors more closely akin to the growth of a plant from a tiny seed, or the unfolding of a computer program from a few lines of code, or perhaps even the organic, self-organized flocking of simpleminded birds. That's certainly the kind of metaphor that Chris Langton has in mind with artificial life: his whole point is that complex, lifelike behavior is the result of simple rules unfolding from the bottom up. And it's likewise the kind of metaphor that influenced Arthur in the Santa Fe economics program: "If I had a purpose, or a vision, it was to show that the messiness and the liveliness in the economy can grow out of an incredibly simple, even elegant theory. That's why we created these simple models of the stock market where the market appears moody, shows crashes, takes off in unexpected directions, and acquires something that you could describe as a personality."

While he was actually at the institute, ironically, Arthur had almost no time at all for Chris Langton's artificial life, or the edge of chaos, or the hypothetical new second law. The economics program was taking up 110 percent of his workday as it was. But what he did hear he found fascinating. It seemed to him that artificial life and the rest captured something essential about the spirit of the institute. "Martin Heidegger once said that the fundamental philosophical question is *being*," notes Arthur. "What are we doing here as conscious entities? Why isn't the universe just a turbulent mess of particles tumbling around each other? Why are there structure, form, and pattern? Why is consciousness possible at all?" Very few people at the institute were grappling with that problem quite as directly as Langton, Kauffman, and Farmer were. But in one way or another, says Arthur, he sensed that everyone was working on a piece of it.

Furthermore, he felt that the ideas resonated strongly with what he and his coconspirators were trying to accomplish in economics. When you look at the subject through Chris Langton's phase transition glasses, for example, all of neoclassical economics is suddenly transformed into a simple assertion that the economy is deep in the ordered regime, where the market is always in equilibrium and things change slowly if at all. The Santa Fe approach is likewise transformed into a simple assertion that the economy is at the edge of chaos, where agents are constantly adapting to each other and things are always in flux. Arthur always knew which assertion *he* thought was more realistic.

Like other Santa Fe folk, Arthur is hesitant when it comes to speculating about the larger meaning of all this. The results are still so—embryonic. And it's entirely too easy to come off sounding New Age and flaky. But like everyone else, he can't help thinking about the larger meaning.

You can look at the complexity revolution in almost theological terms, he says. "The Newtonian clockwork metaphor is akin to standard Protestantism. Basically there's order in the universe. It's not that we rely on God for order. That's a little too Catholic. It's that God has arranged the world so that the order is naturally there if we behave ourselves. If we act as individuals in our own right, if we pursue our own righteous self-interest and work hard, and don't bother other people, then the natural equilibrium of the world will assert itself. Then we get the best of all possible worlds—the one we deserve. That's probably not quite theological, but it's the impression I have of one brand of Christianity.

"The alternative—the complex approach—is total Taoist. In Taoism there is no inherent order. 'The world started with one, and the one became two, and the two became many, and the many led to myriad things.' The universe in Taoism is perceived as vast, amorphous, and ever-changing. You can never nail it down. The elements always stay the same, yet they're always rearranging themselves. So it's like a kaleidoscope: the world is a matter of patterns that change, that partly repeat, but never quite repeat, that are always new and different.

"What is our relation to a world like that? Well, we are made of the same elemental compositions. So we are a part of this thing that is never changing and always changing. If you think that you're a steamboat and can go up the river, you're kidding yourself. Actually, you're just the captain of a paper boat drifting down the river. If you try to resist, you're not going to get anywhere. On the other hand, if you quietly observe the flow, realizing that you're part of it, realizing that the flow is ever-changing and always leading to new complexities, then every so often you can stick an oar into the river and punt yourself from one eddy to another.

"So what's the connection with economic and political policy? Well, in a policy context, it means that you observe, and observe, and observe, and occasionally stick your oar in and improve something for the better. It means that you try to see reality for what it is, and realize that the game you are in keeps changing, so that it's up to you to figure out the current rules of the game as it's being played. It means that you observe the Japanese like hawks, you stop being naive, you stop appealing for them to play fair, you stop adhering to standard theories that are built on outmoded assumptions about the rules of play, you stop saying, 'Well, if only we could reach this equilibrium we'd be in fat city.' You just observe. And where you can make an effective move, you make a move."

Notice that this is *not* a recipe for passivity, or fatalism, says Arthur. "This is a powerful approach that makes use of the natural nonlinear dynamics of the system. You apply available force to the maximum effect. You don't waste it. This is exactly the difference between Westmoreland's approach in South Vietnam versus the North Vietnamese approach. Westmoreland would go in with heavy forces and artillery and barbed wire and burn the villages. And the North Vietnamese would just recede like a tide. Then three days later they'd be back, and no one knew where they came from. It's also the principle that lies behind all of Oriental martial arts. You don't try to stop your opponent, you let him come at you—and then give him a tap in just the right direction as he rushes by. The idea is to observe, to act courageously, and to pick your timing extremely well."

Arthur is reluctant to get into the implications of all this for policy issues. But he does remember one small workshop that Murray Gell-Mann persuaded him to cochair in the fall of 1989, shortly before he left the institute. The purpose of the workshop was to look at what complexity might have to say about the interplay of economics, environmental values, and public policy in a region such as Amazonia, where the rain forest is being cleared for roads and farms at an alarming rate. The answer Arthur gave during his own talk was that you can approach policy-making for the rain forest (or for any other subject) on three different levels.

The first level, he says, is the conventional cost-benefit approach: What are the costs of each specific course of action, what are the benefits, and how do you achieve the optimum balance between the two? "There is a place for that kind of science," says Arthur. "It does force you to think through the implications of the alternatives. And certainly at that meeting we had a number of people arguing the costs and benefits of rain forests. The trouble is that this approach generally assumes that the problems are well defined, that the options are well defined, and that the political wherewithal is there, so that the analyst's job is simply to put numbers on the costs and benefits of each alternative. It's as though the world were a railroad switch yard: We're going down this one track, and we have switches we can turn to guide the train onto other tracks." Unfortunately for the standard theory, however, the real world is almost never that well defined—particularly when it comes to environmental issues. All too often, the apparent objectivity of cost-benefit analyses is the result of slapping arbitrary numbers on subjective judgments, and then assigning the value of zero to the things that nobody knows how to evaluate. "I ridicule some of these cost-benefit analyses in my classes," he says. "The 'benefit' of having spotted owls is defined in terms of how many people visit the forest, how many will see a spotted owl, and what's it worth to them to see a spotted owl, et cetera. It's all the greatest rubbish. This type of environmental cost-benefit analysis makes it seem as though we're in front of the shop window of nature looking in, and saying, 'Yes, we want this, or this, or this'—but we're not inside, we're not part of it. So these studies have never appealed to me. By asking only what is good for human beings, they are being presumptuous and arrogant."

The second level of policy-making is a full institutional-political analysis, says Arthur: figuring out who's doing what, and why. "Once you start to do that for, say, the Brazilian rain forest, you find that there are various players: landowners, settlers, squatters, politicians, rural police, road builders, indigenous peoples. They aren't out to get the environment, but they are all playing this elaborate, interactive Monopoly game, in which the environment is being deeply affected. Moreover, the political system isn't some exogenous thing that stands outside the game. The political system is actually an outcome of the game—the alliances and coalitions that form as a result of it."

In short, says Arthur, you look at the system as a *system*, the way a Taoist in his paper boat would observe the complex, ever-changing river. Of course, a historian or a political scientist would look at the situation this way instinctively. And some beautiful studies in economics have recently started to take this approach. But at the time of the workshop in 1989, he says, the idea still seemed to be a revelation to many economists. "In my talk I put in a strong plea for this kind of analysis," he says. "If you really want to get deeply into an environmental issue, I told them, you have to ask these questions of who has what at stake, what alliances are likely to form, and basically understand the situation. Then you might find certain points at which intervention may be possible.

"So all of that is leading up to the third level of analysis," says Arthur. "At this level we might look at what two different world views have to say about environmental issues. One of these is the standard equilibrium viewpoint that we've inherited from the Enlightenment—the idea that there's a duality between man and nature, and that there's a natural equilibrium between them that's optimal for man. And if you believe this view, then you *can* talk about 'the optimization of policy decisions concerning environmental resources,' which was a phrase I got from one of the earlier speakers at the workshop.

"The other viewpoint is complexity, in which there is basically no duality between man and nature," says Arthur. "We are part of nature ourselves. We're in the middle of it. There's no division between doers and done-to because we are all part of this interlocking network. If we, as humans, try to take action in our favor without knowing how the overall system will adapt—like chopping down the rain forest—we set in motion a train of events that will likely come back and form a different pattern for us to adjust to, like global climate change.

"So once you drop the duality," he says, "then the questions change. You can't then talk about optimization, because it becomes meaningless. It would be like parents trying to optimize their behavior in terms of 'us versus the kids,' which is a strange point of view if you see yourself as a family. You have to talk about accommodation and coadaptation—what would be good for the family as a whole.

"Basically, what I'm saying is not at all new to Eastern philosophy. It's never seen the world as anything else but a complex system. But it's a world view that, decade by decade, is becoming more important in the West—both in science and in the culture at large. Very, very slowly, there's been a gradual shift from an exploitative view of nature—man versus nature—to an approach that stresses the mutual accommodation of man and nature. What has happened is that we're beginning to lose our innocence, or naiveté, about how the world works. As we begin to understand complex systems, we begin to understand that we're part of an ever-changing, interlocking, nonlinear, kaleidoscopic world.

"So the question is how you maneuver in a world like that. And the answer is that you want to keep as many options open as possible. You go for viability, something that's workable, rather than what's 'optimal.' A lot of people say to that, 'Aren't you then accepting second best?' No, you're not, because optimization isn't well defined anymore. What you're trying to do is maximize robustness, or survivability, in the face of an ill-defined future. And that, in turn, puts a premium on becoming aware of nonlinear relationships and causal pathways as best we can. You observe the world very, very carefully, and you don't expect circumstances to last."

So what is the role of the Santa Fe Institute in all this? Certainly not to become another policy think tank, says Arthur, although there always seem to be a few people who expect it to. No, he says, the institute's role is to help us look at this ever-changing river and understand what we're seeing.

"If you have a truly complex system," he says, "then the exact patterns are not repeatable. And yet there are themes that are recognizable. In history, for example, you can talk about 'revolutions,' even though one revolution might be quite different from another. So we assign metaphors. It turns out that an awful lot of policy-making has to do with finding the appropriate metaphor. Conversely, bad policy-making almost always involves finding inappropriate metaphors. For example, it may not be appropriate to think about a drug 'war,' with guns and assaults.

"So from this point of view, the purpose of having a Santa Fe Institute is that it, and places like it, are where the metaphors and a vocabulary are being created in complex systems. So if somebody comes along with a beautiful study on the computer, then you can say 'Here's a new metaphor. Let's call this one *the edge of chaos*,' or whatever. So what the SFI will do, if it studies enough complex systems, is to show us the kinds of patterns we might observe, and the kinds of metaphor that might be appropriate for systems that are moving and in process and complicated, rather than the metaphor of clockwork.

"So I would argue that a wise use of the SFI is to let it do science," he says. "To make it into a policy shop would be a great mistake. It would cheapen the whole affair. And in the end it would be counterproductive, because what we're missing at the moment is any precise understanding of how complex systems operate. This is the next major task in science for the next 50 to 100 years."

"I think there's a personality that goes with this kind of thing," Arthur says. "It's people who like process and pattern, as opposed to people who are comfortable with stasis and order. I know that every time in my life that I've run across simple rules giving rise to emergent, complex messiness, I've just said, 'Ah, isn't that lovely!' And I think that sometimes, when other people run across it, they recoil."

In about 1980, he says, at a time when he was still struggling to articulate his own vision of a dynamic, evolving economy, he happened to read a book by the geneticist Richard Lewontin. And he was struck by a passage in which Lewontin said that scientists come in two types. Scientists of the first type see the world as being basically in equilibrium. And if untidy forces sometimes push a system slightly out of equilibrium, then they feel the whole trick is to push it back again. Lewontin called these scientists "Platonists," after the renowned Athenian philosopher who declared that the messy, imperfect objects we see around us are merely the reflections of perfect "archetypes."

Scientists of the second type, however, see the world as a process of flow and change, with the same material constantly going around and around in endless combinations. Lewontin called these scientists "Heraclitians," after the Ionian philosopher who passionately and poetically argued that the world is in a constant state of flux. Heraclitus, who lived nearly a century before Plato, is famous for observing that "Upon those who step into the same rivers flow other and yet other waters," a statement that Plato himself paraphrased as "You can never step into the same river twice."

"When I read what Lewontin said," says Arthur, "it was a moment of revelation. That's when it finally became clear to me what was going on. I thought to myself, 'Yes! We're finally beginning to recover from Newton.'"

The Hair Shirt

Meanwhile, at about the same time that Brian Arthur was driving off into the sunset, the Heraclitian-in-chief back in Santa Fe was getting ready to call it quits. For all the undeniable success of the economics program, and for all the intellectual ferment over the edge of chaos, artificial life, and the rest, George Cowan was acutely aware that the institute's permanent endowment fund still stood at zero. And after six years, he was tired of constantly begging people for operating cash. He was tired of fretting over the economics program, lest it become the 800-pound gorilla that took over the institute. And speaking of 800-pound gorillas, he was tired of the endless contest of wills with Murray Gell-Mann to define what the Santa Fe Institute was all about—including, not incidentally, what the complexity revolution could tell us about building a more sustainable future for the human race. Cowan was just—tired. Now that he'd gotten the Santa Fe Institute up and running, he wanted to spend the time he had left in life working on the *science* of the institute, this strange new science of complexity. So at the first opportunity—the annual meeting of the institute's board of trustees in

March 1990—Cowan submitted his formal letter of resignation. One more year, he told the board members. He'd give them one more year to pick a successor, while he did his best to get the institute's funding stabilized. But that was it.

"I just felt it was time for a fresh face at the helm," he says. "The board meeting was the week after my seventieth birthday. I suppose that I promised myself as a much younger person that I wouldn't think I was essential to anything when I was seventy. I've seen too many old farts just getting in the way. There were lots of other people with their own ideas. It was time for them to have a chance."

Cowan's announcement didn't exactly come as a surprise to anyone who'd spent any time around the convent. He'd been looking so beaten down of late that his colleagues had started to worry about his health. His temper was erratic; he could be all smiles one day, then thundery and doleful the next. He frequently told people that he'd announced his resignation from the presidency the day he took the job in 1984, that he'd only taken it to keep the chair warm for a younger person. He'd already threatened more than once to resign, and had been talked out of it. At the previous board meeting in March 1989, in fact, he had hinted broadly that the time had come and he had appointed a search committee to find his replacement—a committee that now had to shift into high gear and actually do something.

But that was just the problem, for the search committee and for everyone else. Cowan was the one who had conceived the institute in the first place. He was the one who had envisioned a science of complexity before anyone had even known what to call it. He was the one who had done more than anyone else to make the Santa Fe Institute happen, to make it the most intellectually exciting place that any of them had ever been in. As Chris Langton said, when you saw George sitting there in the mother superior's office, you somehow knew that everything was okay. And it wasn't at all clear whether anyone else could carry that off.

So if not George Cowan, then who?

Cowan himself had not the slightest idea. And for the moment, at least, he didn't have much time to worry about it; for the next twelve months, the pressure was only going to get more relentless. "Before I could, in all good conscience, step down," he says, "I wanted the funding for the next three years to be reasonably well in hand, so that my successor wouldn't be immediately impoverished." That meant that his most urgent priority had to be the completion of a pair of massive proposals to the National Science Foundation (NSF) and the Department of Energy (DOE). The original three-year grants from those agencies—a total of some $2 million—had been awarded in 1987, and were coming up for renewal; if they *weren't* renewed, there wouldn't be much of an institute left for anyone to be president of.

To Cowan, however, there was a lot more at stake in those proposals than money per se. If it had just been a matter of money, in fact, his life would have been a good deal easier. The institute could have simply done what many university science and engineering departments do, which is to insist that individual researchers hustle up their own grants from the funding agencies. It wouldn't have been hard; the place was full of smart and experienced academics who'd been hustling grants all their adult lives. They knew how the game was played. But Cowan was convinced that such an approach would have ended up destroying the very thing that had made the institute so special.

"To me," says Cowan, "the overriding issue was that we were inventing a new kind of scientific community—one that was more or less ecumenical, covering all of the aspects of the hard sciences, the mathematical arts, and the social sciences. We started with the very best people we could find, together with some black magic that I can only define as matters of taste, in that we went out of our way to bring together the kinds of people who would inevitably produce a kind of intellectual donnybrook. I think that the community we've built is unique in it's breadth and quality. I have not seen a similar roster in any other scientific institution in history—and I've looked for them, in an effort to emulate the successes.

"But if we'd gotten strictly bits and pieces of funding," he continues, "we would have fragmented right away." Quite aside from the fact that the funding agencies generally restrict their individual grants to cover one specific piece of research in one specific, recognized discipline—exactly the opposite of the Santa Fe approach—individual grants tend to create individual baronies. "You see, when somebody writes in for a grant, he spends a lot of time working on it," says Cowan. "Then he gets his $50,000 or $100,000, and he becomes, in effect, an entrepreneur who owns that money. If you try in any way to infringe on his autonomy, you've committed a mortal sin." So even with the best will in the world, he says, even with everybody trying hard to stay loose, collegial, and interdisciplinary, the individual investigators inexorably end up spending more and more time doing their own thing on their own projects, and less and less time communicating with each other. "There's no central coordination. You become academic all over again."

In practice, of course, the Santa Fe community went out hustling for specialized grants anyway. Finances being what they were, the institute didn't have the luxury of foregoing them entirely. Indeed, Citicorp's funding for the economics program was just one very large example of a specialized grant—and Brian Arthur had spent a good bit of his time as director writing proposals to this foundation or that foundation, asking for even more money. So, to counteract the centrifugal forces, Cowan had very badly wanted to get what he called an "umbrella grant": money that could cover anyone who seemed to have a good idea in complexity, whether or not that idea fit into a predefined slot. A Chris Langton, for example, or a John Holland, or a Stuart Kauffman. "If you want a coherent program on complexity," says Cowan, "then you've got to invent a community in which that coherence emerges from the bottom up—without trying to tell people what to do. The umbrella grant was an essential part of that."

That's why he had gone to the NSF and the DOE in the first place. Until such time as an angel appeared and gave the institute an endowment, those agencies were the only places he could hope to get umbrella money without forcing the program into a disciplinary mold. And that's why he felt it was so crucial to get the renewals: if the umbrella folded, the creative ferment that Arthur, Kauffman, Holland, and the others had found so incredibly exciting would curdle and sour in very short order.

So Cowan spent endless hours working on the new proposal that spring, along with his executive vice president, Mike Simmons, and the various science board members. They all knew that this had to be one hell of a persuasive document. It had been tough enough to talk the two agencies into funding the institute on the first go-around in 1987, when all the Santa Fe team had to do was prove that they had some very good people and a very good idea. It was going to be a far tougher sell on this second go-around, when they were asking the NSF and the DOE to jointly up the ante by a factor of 10—from $2 million over three years to some $20 million over five years. Moreover, they were proposing this increase at a time when federal science budgets were inexorably getting tighter; researchers in the conventional disciplines were screaming for cash more fiercely than ever, and midlevel managers in both the NSF and the DOE had been heard to wonder aloud why money was going to this speculative, interdisciplinary stuff while perfectly solid projects out in the universities went begging.

So Cowan, Simmons, and company obviously couldn't make their case on promises anymore. They had to show that they had actually accomplished something in the past three years, and that they had the ability to do something worth $20 million over the next five years. This was a bit tricky, of course, since they couldn't honestly claim to have solved the whole mystery of complexity. At best they'd made a start. But what they could and did claim was that in three years of full-time operation, they had created a viable institution devoted to *attacking* the problem of complexity. As promised in the original 1987 proposal, they wrote, the Santa Fe Institute "has developed a comprehensive program, an innovative system of governance, a cadre of remarkably qualified investigators, and a support base that has only begun to match the large overall need."

And, in fact, making due allowance for the proposalese in that statement, Cowan and Simmons could make a pretty strong argument. In three years, they pointed out, the institute had sponsored 36 interdisciplinary workshops attended by more than 700 people. It had also hosted more extended visits by another 100 researchers, who in turn had published some 60 papers on complexity in established scientific journals. It had launched an annual Complex Systems Summer School, with month-long sessions designed to teach the mathematical and computational techniques of complexity research to some 150 scientists at a time. It had started publishing a series of volumes known as the *Santa Fe Institute Studies in the Sciences of Complexity*. And at the time the proposal was being written, it was negotiating with several academic publishers who wanted to launch a new research journal on complexity.

Then there was the research itself. "It is particularly noteworthy," Cowan and Simmons wrote, "that the commitment to SFI programs of gifted associates, ranging from brilliant graduate students to Nobel laureates, senior corporate executives, and prominent public officials, has grown and is no longer an untested aspect of the SFI approach. The formation and support of interactive groups and networks representative of the highest levels of ability in the many disciplines of interest rank among the most significant contributions made to date by the institute."

Once again, they could back up the proposalese with a long list of specific accomplishments. In fact, the bulk of the proposal was devoted to doing just that, with extended discussion of programs ranging from artificial life to economics. "The most mature of the Santa Fe Institute's programs," Cowan and Simmons said of economics, "this initiative is viewed as a paradigm, in terms of both substance and organization, for the development of other institute endeavors."

. . .

Of course, like any reasonably happy family putting on their best face for company, the Santa Fe team had a few intimate details that they *didn't* spell out in the proposal—such as the fact that the economics program had done as much as anything else to drive George Cowan nuts.

Part of it was just the same old problem of money: in his less charitable moments, Cowan sometimes felt as though the economists wanted the institute to raise all the cash while they went off and had all the fun. And even when he wasn't being quite that dyspeptic, he was acutely aware that the economics program had been far more successful intellectually than it had ever been financially. Citicorp was happy enough, and had continued to renew its $125,000-per-year funding for the program. But that hadn't covered the full cost by any means. And Arthur's efforts to get more money out of the big foundations—Russell Sage, Sloan, Mellon—had all fallen flat. The brutal fact was that there was very little research funding out there even for mainstream economics, much less for this speculative Santa Fe stuff.

"It turns out that economics is very poorly supported in this country," says Cowan. "Individual economists are paid very well, but they don't get paid for doing basic research. They get paid from corporate sources, for doing programmatic things. At the same time, the field gets remarkably little in the way of research money from the National Science Foundation and other government agencies because it's a social science, and the government is not a big patron of the social sciences. It smacks of 'planning,' which is a bad word." As a result, he says, many of the economists seemed to look upon the Santa Fe Institute as just another source of support, and brought very little additional support to it. So the institute had had to supplement the Citicorp funding with chunks of its own federal grants— money that Cowan had hoped to use for other projects.

And then on top of all that, says Cowan, Ken Arrow had been trying to recruit a top-notch economist to replace Brian Arthur as resident director after he left at the end of 1989. "Well, we were funding ourselves from year to year, and could hardly look past the current year's budget," he says. "But when you're trying to attract people who can go anywhere and do anything, you have to start making promises about what resources are going to be available down the road. And although the chancy nature of the institute was very obvious when the economics program started, it became less obvious after a year or two. It started to look more substantial than it was. So the people we talked to began to treat us as though we were Stanford or Yale. And since there was no endowment we either had to disillusion them or else act as though they were absolutely right and create some resources. It was a different kind of pressure. The nature of the game changed."

Once again, however, Cowan's real concern wasn't the money per se, but the fragile Santa Fe community. The very success of the program threatened to turn the place into a full-time economics institute, which was not the idea at all. "Creating an institution without departments, and then just pursuing one discipline, is a contradiction in terms," says Cowan. "You might as well set up a department in the first place. We had to start somewhere, but we also had to make sure from the beginning that economics didn't become the one interest of the institute."

Perhaps not surprisingly, this had led to more than one dustup with Arthur over the funding and pace of the economics program. "On the science board," says Cowan, "Brian took a partisan economist's position, which was that the program was a great success—and that as long as this program was proceeding as successfully as it was, we shouldn't divert support from it for anything else. Don't stop betting on a winning horse. Now, Brian is a fierce defender of his views. That's great. But the entire philosophy of the institute was that complex systems consist of many aspects. These include—especially if you're going to talk about complex systems that involve people—neural behavior, human behavior, societal behavior, and many other things that economics does not specifically deal with. So I pushed hard to support at least one other program that would be equal in size to the economics program. We needed to broaden our academic agenda, and spread our bets. And the science board as a whole was quite supportive of that—though with a lot of discussion about it."

The particular program Cowan had in mind was "adaptive computation": an effort to develop a set of mathematical and computational tools that could be applied to all the sciences of complexity—including economics. "If there's a common conceptual framework," he says, "there ought to be a common analytic framework." In part, he adds, starting such a program would just be a matter of recognizing what was already there and giving it broader support. John Holland's ideas about genetic algorithms and classifier systems had long since permeated the institute, and would presumably form the backbone of adaptive computation. But there were also all these similar ideas growing out of Stuart Kauffman's Boolean networks and autocatalytic sets, Chris Langton's artificial life, and the various economy-under-glass models that Brian Arthur and the economists were building. A lively cross-fertilization was well under way—witness Doyne Farmer's "Rosetta Stone for Connectionism" paper, in which he pointed out that neural networks, the immune system model, autocatalytic sets, and classifier systems were essentially just variations on the same underlying theme. Indeed, Mike Simmons had invented the phrase "adaptive computation" one day in 1989, as he and Cowan had been sitting in Cowan's office kicking around names that would be broad enough to cover *all* these ideas—but that wouldn't carry the intellectual baggage of a phrase like "artificial intelligence."

So at one level, says Cowan, an adaptive computation program would simply give this ferment some formal recognition and coordination, not to mention some extra money for graduate students, visitors, and workshops. In the long run, however, he was also hoping that the program would give economists, sociologists, political scientists, and even historians some of the same precision and rigor that Newton brought to physics when he invented calculus. "What we're still waiting for—it may take ten or fifteen years—is a really rich, vigorous, general set of algorithmic approaches for quantifying the way complex adaptive agents interact with one another," he says. "The usual way debates are conducted now in the social sciences is that each person takes a two-dimensional slice through the problem, and then argues that theirs is the most important slice. 'My slice is more important than your slice, because I can demonstrate that fiscal policy is much more important than monetary policy,' and so forth. But you can't demonstrate that, because in the end it's all words, whereas a computer simulation provides a catalog of explicitly identified parameters and variables, so that people at least talk about the same things. And a computer lets you handle many more variables. So if a simulation has both fiscal policy and monetary policy in it, then you can start to say *why* one turns out to be more important than the other. The results may be right or they may be wrong. But it's a much more structured debate. Even when the models are wrong, they have an enormous advantage of structuring the discussions."

Whether or not the simulations ever got that good, however, starting an adaptive computation program would certainly have at least one happy side effect: it would give Cowan and company an excuse to hire John Holland away from Michigan to be their first full-time faculty member. Not only was he the natural and unanimous choice to be the program's resident director, but he was a nonstop font of energy and ideas. People just liked having him around.

Cowan and Simmons accordingly gave adaptive computation its own special 10-page section in the NSF-DOE proposal—much of it written by an enthusiastic John Holland himself—and shipped the whole 150-page package off to Washington on July 13, 1990. From there on out, about all they could do was wait with their fingers well crossed, and hope that the reviewers would be kind.

There was a certain irony in the institute's courting of Holland. Back in the early days of the institute, Cowan and the other founders had had every intention of hiring a permanent faculty and making the place into a full-fledged research institution along the lines of Rockefeller University in New York. But fiscal reality had intervened. And by 1990, Cowan, Simmons, and quite a number of the other Santa Fe regulars had begun to suspect that this particular restriction had at least one large virtue: the institute might actually be much better off *without* a permanent faculty.

"The virtue was that we were more flexible than we would have been," says Cowan. After all, he'd realized, once you hire a bunch of people full time, your research program is pretty well cast in concrete until those people leave or die. So why not just keep the institute going in its catalyst role? It had certainly worked beautifully so far. Keep going with a rotating cast of visiting academics who would stay for a while, mix it up in the intellectual donnybrook, and then go back to their home universities to continue their collaborations long distance—and, not incidentally, to spread the revolution among their stay-at-home colleagues.

That said, however, everyone was more than willing to make an exception in Holland's case. And best of all, a source of funds to support him had already presented itself, in the flamboyant form of one Robert Maxwell: former Czech resistance fighter, self-made billionaire press baron in London, and—it turned out—a man given to quirky enthusiasms about things like complexity.

In retrospect, of course, Robert Maxwell is also famous for his mysterious death by drowning in late 1991, and the spectacular collapse of his debt-ridden media empire immediately thereafter. But at the time, he looked like a fairy godmother. The institute's contact with Maxwell had begun more than a year earlier, when Murray Gell-Mann had happened to meet Maxwell's daughter, Christine. Christine Maxwell, in turn, had arranged for Gell-Mann to have lunch with her father in May 1989. And when Gell-Mann had reported back to Cowan that the elder Maxwell seemed intrigued by what the institute was doing, the Santa Fe team had immediately gone into fund-raising mode. Nobody had the slightest idea what Maxwell was worth, but it had to be—zillions.

Many faxes and phone calls later, in February 1990, there arrived one particular fax from London making two key points. First, said Maxwell, he wanted to begin his association with the institute with a contribution of $100,000 to be used for the study of adaptive complex systems. Second, he liked the institute's idea of founding a new scientific journal on complexity, and would be interested in publishing that journal through his subsidiary, the academic publishing house Pergamon Press.

Wanted to *begin* his association!? Cowan and Simmons mulled over that little gem for a while. Finally, Cowan decided to take a gamble and up the ante: "I want to ask him for more." In his reply he enclosed a draft of the work of the institute's journal committee, outlining what they had in mind for the journal, and added a proposal that the publisher establish "The Robert Maxwell Professorship" at the institute, funded at the level of $300,000 per year. That sum would cover not just the salary for the Maxwell professor alone, Cowan explained, but would also pay for postdocs, graduate students, travel money, a secretary, and assorted other expenses.

The response from London took some time in coming. Maxwell, as Cowan and Simmons had long since learned, delegated almost nothing. All they could do was keep the fax lines hot with reminders, along with letters, telephone calls, and contacts through Gell-Mann, Christine Maxwell, and her brothers. The answer—"Accept in principle"—finally came just in time for the board meeting in March 1990 to formally offer the Maxwell professorship to John Holland for five years.

Up in Michigan, Holland proceeded to parlay that offer for all it was worth. By that point, still bitter over the merger of his old Computer and Communications Science department into the engineering school, and hating the kind of short-term, applications-oriented mindset that prevailed there, Holland had already leveraged himself halfway out. A few years earlier, UCLA had started to hint that it might offer him an endowed chair. So Holland, showing a previously unsuspected talent for academic gamesmanship, had immediately gone to the university's provost. "To stay here," he said, "I need at least a half-time appointment in psychology"—a large, nationally ranked department where he had extensive contacts from his days of working on the book *Induction*. The provost, Edie Goldenberg, being both sympathetic and eager to keep him at Michigan, had made the necessary arrangements.

Now, with the offer from Santa Fe in his pocket, Holland went to Goldenberg again. "This Maxwell professorship is almost ideal in terms of doing research," he told her, "and I'm very inclined to take it—*unless* I can spend more time on research here at Michigan." Once again, Goldenberg was ready to listen. She found money, made arrangements, and helped him work out a quid pro quo: Holland would get a full-time appointment in psychology, plus a reduced teaching load to give him more time for research. And in return, he would set up a permanent link between the Santa Fe Institute and the university—an arrangement wherein professors, postdocs, and graduate students from Michigan would regularly spend time in Santa Fe, and the two institutions would regularly sponsor joint conferences. It would be a kind of Santa Fe outpost in the snows of Ann Arbor.

The deal was consummated by the summer of 1990. To inaugurate the outpost, Holland organized a two-week seminar in the fall of 1990, with a special kickoff symposium starring Brian Arthur, Stanford's Marc Feldman, and Murray Gell-Mann. Holland had a great time, and from all accounts, so did everyone else. "[University president James] Duderstadt came to the kickoff symposium and stayed the whole time!" says Holland. "He even took notes. It was a lot of fun, and everybody was pleased." From then until now, moreover, with the exception of forays to Santa Fe and to various conferences, Holland has spent the majority of his time happily ensconced with his Macintosh II computer in the study of his home, a striking hilltop chateau overlooking the rolling woodlands west of Ann Arbor. Lately, in fact, he's even begun to talk seriously about retiring from the university altogether, so that he can have even more time for research. "It's the finite horizon effect," he says. "I'm getting old enough [he's sixty-three], and I have so many ideas in my folder that I want to work on more. . . ."

Back in Santa Fe, Cowan was sorry to hear Holland say no to the professorship. But he had to admit he was impressed by the way Holland had finessed his way out of a bad situation at home. And he was even more impressed by the fact that Holland had played "bet your job" to secure the ongoing link with Michigan—something the institute was overjoyed to have, and probably never would have had otherwise.

In the meantime, however, Cowan had to deal with Maxwell. He and Simmons spent the early summer of 1990 keeping the fax lines hot with very polite reminders to London: please don't forget to send the money. Maxwell's personal check for $150,000—the first installment of the first year's $300,000—finally arrived in August. And it was only then that they told him that Holland wouldn't be able to accept. "Do you think it would help if I went to Michigan and talked to him?" Maxwell responded.

Well, no. But Santa Fe was able to offer a compromise: Holland and Gell-Mann would share the professorship for the fall 1990 semester that was just starting, during which time Holland would lay the groundwork for this new adaptive computation program. In 1991 the position would rotate between Stuart Kauffman and David Pines. And in the meantime, the institute would use the flexibility to bring in some first-class younger people, such as Seth Lloyd, James Crutchfield, and Alfred Hubler.

That, said the fax machine, was quite acceptable to Maxwell. It also proved to be quite acceptable to everyone that the new complexity journal be published through Maxwell's Pergamon Press. The details were worked out by Cowan and Maxwell during a long transatlantic telephone conversation—shortly before Maxwell suddenly decided to sell Pergamon to help finance his other acquisitions. And in late February 1991, after a series of increasingly urgent transatlantic reminders, Maxwell even remembered to send along another $150,000 to pay for the second half of the professorship's academic year.

All through the summer and fall of 1990, whenever the subject of Cowan's successor came up, Murray Gell-Mann could be heard to sigh in tones of resignation, "I guess I'll have to do it."

Gell-Mann certainly didn't *want* to be president of the institute, one was given to understand. He loathed bureaucratic busywork. He'd been turning down jobs like this all his life—the chairmanship of Caltech's Division of Physics, Mathematics, and Astronomy, for example. But the Santa Fe Institute and the sciences of complexity were so incredibly important that—well, who else had such a clear vision of what needed to be done? Who else could articulate the sciences of complexity so well? Who else had the prestige and the network of contacts to give the institute the clout it needed?

Who else indeed? The institute's search committee immediately went into paralysis. No one was fooled: Murray Gell-Mann wanted very badly to be president of the Santa Fe Institute. The question was whether they dared let him do it. Some people felt that they ought to seriously consider the possibility. After all, they said, what we have here is a seminal figure in the history of science—with a Nobel Prize to boot. If he really wants the job, why not give him a shot?

Others, who knew him better, were appalled at the thought of Murray Gell-Mann actually trying to run anything. No one doubted his intellectual vision, his energy, or his fund-raising power. He was a nonstop source of ideas about what scientific questions would be interesting to tackle. And he seemed to know everybody; he had an incredible knack for getting together groups of people who were absolutely tops in their fields. The Santa Fe Institute wouldn't be what it was without him. But—*president?* They had visions of his desk accumulating geological strata of unsigned papers and unreturned phone calls while Gell-Mann was off saving the rain forest. Worse, they had visions of the Santa Fe Institute becoming, de facto, "The Gell-Mann Institute."

"Murray has more of a purely intellectual approach to life than anyone I've ever known," says one physicist who has known Gell-Mann for many years. "All of his conversations and everything else about his life is driven by his intellectual concerns. He cares deeply about the intellectual agenda of the Santa Fe Institute. He sees the direction that he wants it to go. He's thought about it very deeply, and wants to be sure we move in that direction.

"Now, that's both good and bad. I think it's good for the institute to have a strong intellect like Murray driving it in productive directions. But the flip side is that when Murray is around, it's hard for anyone else to get a word in edgewise. Once he analyzes a problem, he feels that it has been fully analyzed. If someone disagrees, he tends to think that they must not have heard him or must not have understood. And if he doesn't write them off entirely, he tends to repeat his argument for greater clarity. So by sheer intellectual power and force of personality, he tends to displace every other point of view. The danger that everyone saw was that the Santa Fe Institute would just become a vehicle for Gell-Mann's personal enthusiasms."

That was certainly the danger that Cowan saw. In fairness, he'd heard Gell-Mann say all the right words about the need for diversity and multiple points of view at the institute. But he was also convinced that Gell-Mann as president would wreck the institute's tumultuous, multifaceted community without even meaning to, as all the truly original thinkers left to preserve their sanity. "Murray would be the Herr Professor who was running things," says Cowan. "He always feels that his point of view is the only possible point of view. He's always straightening people out."

Cowan had reason to know. In one way or another, he'd been fighting this battle with Gell-Mann since the institute was founded. He did his best to keep a lid on it, of course. Cowan was acutely aware of how much he and the institute needed Gell-Mann; he'd felt compelled to defer to the man so often that many people wondered if he were simply intimidated by Nobel Prize winners. But there were days when Cowan just couldn't stand it anymore.

Take their long-running debate over the proper subject matter of the institute. "I think of the subject as the study of simplicity *and* complexity," says Gell-Mann. "The simple law of the universe and it's probabilistic character seem to me to underlie the whole subject—that, and the nature of information and quantum mechanics. Well, we have had information and the universe discussed twice at Santa Fe. And in the early days we had a wonderful workshop on superstrings, with an overview of mathematics, cosmology, and particle physics. But there was all this pressure against studying simplicity, and we've never done superstrings again. The president of the institute, George Cowan, hated these things bitterly. I don't know why."

Actually, Cowan didn't hate these things. Superstring theory—a hypothetical "Theory of Everything" that attempts to describe all elementary particles as infinitesimally tiny, furiously vibrating threads of pure energy—was wonderful stuff. It's just that there were plenty of other places where people could work on strings, quarks and cosmology, and he didn't think that the institute had any time or money to waste on duplicating them. (Nor was Cowan the only one who thought that: a majority of the science board looked at that superstring workshop and said, "Never again.") But for Cowan, the really annoying thing about Gell-Mann's "simplicity" was that it sounded to him like reductionism in disguise. He found it telling that Gell-Mann still took such obvious relish in clever put-downs of anything he wasn't personally interested in, such as chemistry or solid-state physics. (He would call the latter "*squalid*-state physics" to Phil Anderson's face, apparently just to irritate Anderson.) Maybe Gell-Mann was simply trying to be funny, says Cowan. But the not-very-veiled message was that the study of collective behavior was somehow pragmatic, messy, and not "intellectual."

To outsiders, the testiness over Gell-Mann's notion of simplicity sounded a bit like one of those arcane medieval debates over the finer points of theology. But Cowan and Gell-Mann got quite angry about it, with the subject leading to any number of arguments and abruptly terminated telephone calls. The one occasion that Cowan particularly remembers was about 1987, when five or six of the senior Santa Fe regulars were sitting around the table in a small private meeting and discussing how the Santa Fe Institute should describe itself. "Every time we would say we were interested in the sciences of complexity," says Cowan, "Murray would add, 'and the fundamental principles of which it was composed'—meaning quarks. The implication was that social organizations were made up of a lot of quarks, and you could follow the quarks through to the various aggregations.

"Now, this is what I would call the religion of theoretical physics," says Cowan: "this belief in symmetry and total reductionism. I saw no reason to adopt that statement, and I said we weren't going to do that." Cowan's argument, supported by most of the others around the table, was that emergent, complex systems represented something new—that the fundamental concepts needed to understand their macroscopic behavior go well beyond the fundamental laws of force.

"Murray said flatly that he wouldn't go along with it," says Cowan. "Well, this was the first time I realized that Murray, merely by asserting that this was how he wanted it to be, expected other people to do things his way. And I felt that this was so monumentally egocentric that I lost my temper."

Indeed, Cowan was so furious that he threw what was, for him, a tantrum: he picked up his papers, said "I quit," and walked out of the room—with Ed Knapp and Pete Carruthers in hot pursuit, shouting, "George, come back!"

He did, eventually. And after that incident, Gell-Mann rarely mentioned "simplicity" again.

However, Cowan's annoyance over simplicity was nothing compared with what he felt about the institute's "Global Sustainability" program. To begin with, it had started out as *his* program, a small effort reflecting his most deeply held concerns about humanity's future on this planet. Moreover, he hadn't even called it "sustainability" then. His original concept was "Global Stability" or "Global Security"—the latter being the title of the first small workshop he organized in December 1988. "The subject started out as something like national security, but it rapidly became much broader," says Cowan: "How do we survive the next hundred years without a 'Class A' catastrophe? That is, something that can't be set straight in a generation." In edge-of-chaos terms, avoiding such cataclysms would mean finding some way to damp out the very largest, most destructive avalanches of change. "Originally, number one on my list of Class A catastrophes was nuclear war," says Cowan, "with a Class B catastrophe being something like World War II. But by the time of the first meeting, rapprochement between Russia and the United States was such that the nuclear war problem was down around number five on the list. And what emerged very quickly instead was the population explosion, the [Paul] Ehrlich-type catastrophe. Then came possible environmental catastrophe, such as the greenhouse warming, which I myself didn't think of as a Class A catastrophe, but which others have seized upon."

This effort perked along for a while in a low-key way, largely because Cowan kept organizing small meetings on his own whenever he could find the time. But then Gell-Mann began to get more interested. The idea of taking a global, integrated look at humanity's long-term viability was something that resonated strongly with him. After all, Gell-Mann's first introduction to science had been his nature walks in Central Park at age five. *His* most deeply held concern was the preservation of the global environment in general, and the biological diversity of the rain forest in particular. So he waded in, inexorably pushing Cowan's global stability program in the direction he wanted. And by 1990, he had effectively redefined the agenda and made it his own.

It was a far more activist agenda than Cowan's. Gell-Mann wasn't interested simply in avoiding catastrophe. He wanted to achieve a state of global "sustainability"—whatever that notoriously slippery word might mean.

Speaking at a Santa Fe workshop in May 1990—by now he was a co-chairman with Cowan—Gell-Mann pointed out that "sustainability" has in truth become a trendy cliché of late, the source of endless platitudes. For most people it seems to mean something like business as usual—but, you know, *sustainable*. And yet business as usual is precisely the problem, he said. At the World Resources Institute, a Washington-based environmental think tank that Gell-Mann had helped set up in his capacity as a director of the MacArthur Foundation, founding director Gus Speth and others have argued that global sustainability is possible only if human society undergoes at least six fundamental transitions within a very few decades:

1. A *demographic* transition to a roughly stable world population.
2. A *technological* transition to a minimal environmental impact per person.
3. An *economic* transition to a world in which serious attempts are made to charge the real costs of goods and services—including environmental costs—so that there are incentives for the world economy to live off nature's "income" rather than depleting its "capital."
4. A *social* transition to a broader sharing of that income, along with increased opportunities for nondestructive employment for the poor families of the world.
5. An *institutional* transition to a set of supranational alliances that facilitate a global attack on global problems and allow various aspects of policy to be integrated with one another.
6. An *informational* transition to a world in which scientific research, education, and global monitoring allow large numbers of people to understand the nature of the challenges they face.

The trick, of course, is to get from here to there without one of Cowan's Class A global catastrophes. And if we're to have any hope of doing that, said Gell-Mann, the study of complex adaptive systems is clearly critical. Understanding these six fundamental transitions means understanding economic, social, and political forces that are deeply intertwined and mutually dependent upon one another. You can't just look at each piece of the problem individually, as has been done in the past, and hope to describe the behavior of the system as a whole. The only way to do it is to look at the world as a strongly interconnected system—even if the models are crude.

But more than that, said Gell-Mann, the trick in getting from here to there is to make sure that "there" is a world worth living in. A sustainable human society could easily be some Orwellian dystopia characterized by rigid control and narrow, confined lives for almost everyone in it. What it *should* be is a society that is adaptable, robust, and resilient to lesser disasters, that can learn from mistakes, that isn't static, but that allows for growth in the quality of human life instead of just the quantity of it.

Achieving this will clearly be an uphill battle, he said. In the West, intellectuals and managers alike tend to be highly rationalistic, emphasizing the means by which undesirable effects occur and looking for technical fixes that will block those effects. Thus, we have contraceptives, emission controls, arms control agreements, and so forth. And those things are certainly important. But the real solution will require much more, he said. It will require the renunciation or sublimation or transformation of our traditional appetites: to outbreed, outconsume, and conquer our rivals, especially our rivals in other tribes. These impulses may once have been adaptive. Indeed, they may even be hard-wired into our brains. But we no longer have the luxury of tolerating them.

And yet therein lies one of the crucial problems, said Gell-Mann. On the one hand, humanity is gravely threatened by superstition and myth, the stubborn refusal to recognize the urgent planetary problems, and generalized tribalism in all its forms. To achieve those six fundamental transitions will require some kind of broad-scale agreement on principles and a more rational way of thinking about the future of the planet, not to mention a more rational way of governing ourselves on a global scale.

But on the other hand, he said, "How do you reconcile the identification and labeling of error with the tolerance—not only tolerance, but celebration and preservation—of cultural diversity?" This isn't a matter of political correctness, but of hard-edged practicality. Cultures can't be eradicated by fiat; witness the violent reaction to the Shah's efforts to westernize Iran. The world will have to be governed pluralistically or not at all. Moreover, cultural diversity will be just as important in a sustainable world as genetic diversity is in biology. We *need* cross-cultural ferment, said Gell-Mann. "Of particular importance may be discoveries about how [our own culture can] restrain the appetite for material goods and substitute more spiritual appetites." In the long run, he said, solving this dilemma may require much more than sensitivity. It may require profound new developments in the behavioral sciences. After all, the cure of individual neuroses is not easy; the same is true of social neuroses.

Of course, said Gell-Mann, looking at such multifaceted, densely interconnected systems is exactly what the Santa Fe Institute was set up to do. But he argued that the institute was far too small to undertake a study of global sustainability by itself. It needed partners such as the World Resources Institute, the Brookings Institution, and the MacArthur Foundation (by no coincidence, the cosponsors of that particular workshop). With them to look at the policy aspects, he said, and the Santa Fe Institute to focus on the basic research, they could begin to attack the problem of sustainability as a whole.

By the time of the May 1990 workshop, of course, what was now called the "Global Sustainability" program had long since slipped from Cowan's control. And about the only thing he could do about it was watch in quiet fury. After all, Gell-Mann was cochairman of the institute's science board, which gave him far more say in the direction of any given program than Cowan had. Gell-Mann could and did define the program the way he wanted it, while Cowan, as president, had the responsibility to go out and raise the money.

And as if that weren't infuriating enough, there was the actual content of Gell-Mann's agenda. Cowan didn't think it was wrong, exactly. Cowan was the first to agree that the world is far from sustainable now, and that some fundamental changes are sorely needed. No, what Cowan found enraging was that Gell-Mann and his buddies from Brookings and MacArthur and the World Resources Institute were so—*sure*. Despite all of Gell-Mann's protestations to the contrary, when you actually listened to them you couldn't help feeling that they knew the problems, they knew the solutions, and all they really wanted to do was get on with preserving the rain forest.

Cowan was hardly alone in that feeling. Then and now, many people at the institute were deeply suspicious that the Global Sustainability project would turn into some kind of global environmental activism. "If you already know what to do, then it's not a research program," says one Santa Fe regular. "Its a policy implementation program, and that's not our role."

Yet the fact was that Cowan, for one, just didn't have the energy to fight with Gell-Mann anymore. Let him have the damn Global Sustainability program. Cowan would get back to his own vision of global stability after he stepped down as president, if then. "I have a feeling that Murray and I don't have really deep intellectual differences," says Cowan. "We're both more similar than we ought to be. Maybe that's the problem. His social skills are such that I find myself easily offended by Murray. I'm not alone in that. But I have no reason to put up with it, so I probably lose patience more easily. If I were a more perfect person, there wouldn't be any problems. I've just reached the age when I don't bother with people I have to make allowances for."

Toward the end of 1990, at a time when Gell-Mann was still the only serious candidate for the Santa Fe Institute presidency, Cowan happened to be chatting with Ed Knapp, who was now back at Los Alamos heading up the laboratory's meson physics facility. Knapp, a tall, easygoing physicist with a distinguished crop of wavy silver hair, mentioned that Los Alamos was offering a very attractive early retirement package, at least partly to ease the pinch of post-Cold War defense cutbacks. In fact, said the fifty-eight-year-old Knapp, he was thinking of taking advantage of it.

Neither man remembers exactly who said what at that point. But very quickly, the obvious question was in the air: Would Knapp be interested in becoming president of the Santa Fe Institute?

It made a lot of sense to Cowan. Knapp had been present at the creation, back when the institute was still just an idea being kicked around among the laboratory's senior fellows. He'd always been willing to help out when he could—even agreeing to serve as chairman of the board of trustees for two years. He'd been head of the National Science Foundation back in Washington, then head of the Universities Research Association, the seventy-two-member university consortium that runs the Fermi National Accelerator Laboratory outside of Chicago and the DOE's new superconducting supercollider project. He clearly cared about the institute and what it stood for. And yet, unlike certain other candidates, Knapp had no personal axes to grind regarding what the institute should or shouldn't do.

"George," protested Knapp, "remember that I'm not a theoretical scientist, I'm an administrator."

"That's great," said Cowan.

The discussion went from there. Knapp agreed that if offered the job by the institute's board of trustees, he would take it. And once Cowan had passed that word along, the sense of relief among the board members was tangible. The question had always been whether Gell-Mann would or could remake himself into an administrator, and whether he would be willing to take enough time away from his many other interests to do a decent job in Santa Fe. By late 1990, the general consensus was that he wouldn't. And now that an acceptable alternative candidate was available, it very quickly became clear to everyone, including Gell-Mann, that if he forced the issue to a vote of the board, he would lose.

Meanwhile, Gell-Mann himself had begun to get a sense of what he'd been asking for. David Pines, among others, had spent quite a lot of time trying to explain what it meant to be an administrator—the budgets, the meetings, the endless hassling with personnel. "Murray," Pines kept saying, "this isn't the job you want at the Santa Fe Institute; you want to be a *professor*."

So in the end it was all very gentlemanly. A special board of trustees meeting was called for December 1990. Gell-Mann himself put Knapp's name in nomination. And Knapp was the unanimous choice.

"I was a little disappointed," says Gell-Mann. "I would have liked the job. This was the first time in my life I'd ever expressed interest in such a job. But I was quite pleased with the choice of Ed Knapp. I was happy that the person we chose was a good one and easy to work with."

As he'd promised a year earlier, George Cowan stepped down from the presidency of the Santa Fe Institute at the board of trustees meeting in March 1991. And as he'd hoped, he was able to do so in good conscience. The NSF and the DOE had renewed their grants for three years instead of five years, and for an unchanged $2 million instead of $20 million. But the grants had indeed been renewed. Meanwhile, the MacArthur Foundation had decided to boost its annual contribution from $350,000 to $500,000. Increases had been promised by several individual donors, including Gordon Getty and William Keck, Jr. And Robert Maxwell was committed to funding his professorship at the rate of $300,000 per year— although he was still doling out the money a semester at a time. So Cowan was indeed leaving the institute in sound financial shape for the near term; his successor, Ed Knapp, would have the luxury of pursuing an endowment without having to constantly scrounge for day-to-day operating expenses. (Life has not been quite that rosy in practice: the evaporation of the professorship after Maxwell's death in late 1991 left a rather large hole in Knapp's 1992 budget, and forced the institute to cut back on the number of visitors and postdocs. Fortunately, the shortfall was temporary and reparable.)

Once the mantle had safely passed, however, Cowan was out of there. He was now seventy-one years old, and after seven years of anxiety and administrative hassle, he badly needed a rest—which in his case meant reimmersing himself in the double beta decay experiment that he and several Los Alamos colleagues had been planning for the better part of a decade, and which was now nearing completion. For months he rarely showed his face around the institute. (The double beta decay experiment had been one of a long list of Cowan's research projects cited by the DOE the previous October when the department named him as the corecipient of its prestigious Fermi Award, which honors outstanding scientific achievement in the development, use, or control of atomic energy. Previous recipients had included such figures as John von Neumann and J. Robert Oppenheimer. Double beta decay is an exotic and exceedingly rare form of radioactivity that provides a sensitive experimental test of the standard elementary particle theories. Much to Cowan's delight, he and his colleagues were able to detect the decay and show that it was completely consistent with the standard theories.)

For Cowan, however, the break apparently had its healing effect. By the fall of 1991, he had once again become a regular at the institute, where he shared a small office with Chris Langton. And more than one person remarked how healthy and enthusiastic he was looking.

"I don't know quite how to explain how I felt about stepping down," Cowan says now. "One way is to repeat the story of the guy who sat in the presence of this continuously loud sound, and when it stopped said, 'What was that!?' Or, if you constantly wear a hair shirt, you feel a little funny when you take it off. If you've got a puritan streak, you even feel a little guilty when you take it off. But now I've put on a modified version of the hair shirt, and I feel a lot better."

In particular, he says, now that he has so much more time to think about the new sciences of complexity, he finds himself more enthralled that ever. "Talk about the coercive power of an intellectual idea! I feel as though I've been coerced more than anyone else. These things have grabbed me and kept me in a state of perpetual excitement. I feel as though I've taken a new lease on life, at the cranium part of my body. And that, to me, is a major accomplishment. It makes everything I've ever done here worthwhile."

The issue that grabs him the hardest, he says, is adaptation—or, more precisely, adaptation under conditions of constant change and unpredictability. Certainly he considers it one of the central issues in the elusive quest for global sustainability. And, not incidentally, he finds it to be an issue that's consistently slighted in all this talk about "transitions" to a sustainable world. "Somehow," he says, "the agenda has been put into the form of talking about a set of transitions from state A, the present, to a state B that's sustainable. The problem is that there is no such state. You have to assume that the transitions are going to continue forever and ever and ever. You have to talk about systems that remain continuously dynamic, and that are embedded in environments that themselves are continuously dynamic." Stability, as John Holland says, is death; somehow, the world has to adapt itself to a condition of perpetual novelty, at the edge of chaos. "I still haven't found the right words for that," says Cowan. "Just recently I was toying with the title of Havelock Ellis's book, *The Dance of Life*. But that isn't quite right. It isn't a dance. There's not even a given tempo. So if anything we're getting back to Heraclitus: 'Everything Moves.' A term like 'sustainable' doesn't really capture that."

Of course, adds Cowan, it may be that concepts such as the edge of chaos and self-organized criticality are telling us that Class A catastrophes are inevitable no matter what we do. "Per Bak has shown that it's a fairly fundamental phenomenon to have upheavals and avalanches on all scales, including the largest," he says. "And I'm prepared to believe that." But he also finds reason for optimism in this mysterious, seemingly inexorable increase in complexity over time. "The systems that Per Bak looks at don't have memory or culture," he says. "And for me it's an article of faith that if you can add memory and accurate information from generation to generation—in some better way than we have in the past—then you can accumulate wisdom. I doubt very much whether the world is going to be transformed into a wonderful paradise free of trauma and tragedy. But I think it's a necessary part of a human vision to believe we can shape the future. Even if we can't shape it totally, I think that we can exercise some kind of damage control. Perhaps we can get the probability of catastrophe to decrease in each generation. For example, ten years ago, the probability of nuclear war was maybe a few percent. Now it's way down. Now we're more concerned with environmental and population catastrophes. So I suspect that if we can iterate day to day, and constantly do some course corrections, then we will help provide a somewhat better future for society than if we say, 'Well, it's all the hand of God.' "

On another front, Cowan is characteristically cautious when it comes to evaluating what he's accomplished as a founding father of the Santa Fe Institute. "I feel very good about having attempted it," he says. "The jury's still out about how successful it will be. One sign that it wasn't time wasted is that many people feel that we have legitimized physical scientists' getting into what they think of as 'soft' science—whether we call it economics, or social sciences, or something else. In effect, these people are giving up one of the things to which they've held very strongly in their professional careers, which is to deal only with phenomena that can be handled analytically and rigorously, and they've gone into fields that they've always criticized as being 'fuzzy.' That opens them to criticism from some of their more conservative colleagues for having become fuzzy themselves. But the notion that there is a discipline coming along called the science of complexity has made it more respectable to do this—to become concerned with questions that are central to the nation's welfare and, for that matter, the welfare of the world. And I think that this represents a trend that can only pay off to both the country and to the academic community. Because if it works, something very important has happened. It represents, to me, a reintegration of a scientific enterprise that has become almost totally fragmented over the past few centuries—a recombining of the analysis and rigor of the physical sciences with the vision of the social scientists and the humanists."

So far, he adds, this effort has been a remarkable success at Santa Fe, particularly in the economics program. But who knows how long it can last. One day, despite the best efforts of everyone involved, even the Santa Fe Institute could grow settled, conventional and old. Institutions do. "It may be like a floating crap game," says Cowan. "You may have to shut it down in one place and start it up in another. I think it's a necessary enterprise. And whether it's sustained or not here, I think it has to go on."

A Moment in the Sunlight

Shortly after lunchtime on a Friday afternoon in late May 1991, as the clear New Mexico sunshine flooded the tiny courtyard of the Cristo Rey Convent, Christopher G. Langton, Ph.D., sat at one of the blindingly white patio tables and did his best to answer the questions of a particularly persistent reporter.

Dr. Langton was looking markedly more relaxed and confident these

days. Having finally and successfully defended his dissertation on the edge
of chaos some six months earlier, in November 1990, he had removed an
enormous black cloud from his life—and, not incidentally, had earned a
scientist's essential union card. The Santa Fe Institute had immediately
made him a member of its "external faculty"—the list of researchers whose
association with the institute is considered quasi-permanent, and who have
a strong voice in its scientific direction. Indeed, with Los Alamos' budgets
growing increasingly constrained and survival-oriented in the aftermath of
the Cold War, the Santa Fe Institute had become *the* major support for
artificial life. Langton could feel at home at the institute in a way he never
had before.

He clearly wasn't the only one who felt at home. In the early afternoon
sunshine the courtyard was crowded with visitors and residents alike. At
one table, Stuart Kauffman was holding forth with Walter Fontana and
several others about his latest ideas on autocatalysis and the evolution of
complexity. At another, economics codirector David Lane was talking with
his graduate student, Francesca Chiaromonte, about the economics program's newest effort: a computer study that was trying to explore the dynamics of multiple adaptive firms engaged in technological innovation.
And at still another table, Doyne Farmer was talking with several other
Young Turks about his start-up company, Prediction, Incorporated. Having
reached the end of his limited patience with the tight budgets and the
bureaucratic pettifoggery up at Los Alamos, Farmer had recently decided
that the only sane way to pursue his real research interests was to go off for
a few years and use his forecasting algorithms to make so much money that
he would never have to write a grant proposal again. He felt so strongly
about it, in fact, that he'd even trimmed off his ponytail, the better to deal
with the business types.

Of course, there was a certain wistful sense of being at the end of an era
that Friday afternoon. For more than four years, the Cristo Rey Convent
had been small, primitive, overcrowded, and somehow perfect. But the
institute was continuing to grow, and the fact was that the staff just couldn't
keep putting more desks in the hallway. And in any case, the lease was up
and the Catholic Church needed its convent back. So within a month, the
Santa Fe Institute was scheduled to move to larger rented quarters in the
Land of the Lawyers—a new office complex out on the Old Pecos Trail.
So far as anyone could tell, it was a perfectly fine space, but—well, there
wouldn't be many more lunchtimes on the sun-drenched patio.

As Langton continued his attempts to educate the reporter on the nuances
of artificial life and the edge of chaos, several of the institute's younger

postdocs began to pull up chairs around the table, not realizing that this was supposed to be an interview. The architect of artificial life was something of a celebrity in their circle, and always worth listening to. The interview quickly turned into a general bull session. How do you recognize emergence when you see it? What makes a collection of entities an individual? Everyone had opinions, and no one seemed to be particularly shy about offering them.

Melanie Mitchell, a computer science postdoc from Michigan, where she is the newest member of the BACH group, asked, "Are there degrees of being an individual?" Langton had no idea. "I can't think of evolution acting on individuals any more," he said. "It's always acting on an ecosystem, a population, with one part producing something another part needs."

That sparked off other questions: Is evolution a matter of survival of the fittest or survival of the most stable? Or is it just survival of the survivors? And what exactly is adaptation, anyway? The Santa Fe line is that adaptation requires changing an internal model, à la John Holland. But is that the only way to look at it?

And speaking of emergence, someone asked, is there more than one *kind* of emergence? And if so, how many different kinds are there? Langton started to answer, ground to a halt, and ended up just laughing. "I'm going to have to punt on that one," he said. "I just don't have a good answer. All these terms like emergence, life, adaptation, complexity—these are the things we're still trying to figure out."

Bibliography

For those who want to learn a bit more about complexity, the books and articles listed below will be a start. Few of them are intended for casual readers; the field is still so new that most of the written accounts are in the form of conference proceedings and journal articles. Nonetheless, many of the references here should be reasonably accessible to nonexperts. And virtually all of them contain further references to the technical literature.

The Santa Fe Institute
and the Sciences of Complexity in General

Davies, Paul C. W., ed. *The New Physics*. New York: Cambridge University Press (1989). Contains a number of survey articles on condensed-matter physics, collective phenomena, nonlinear dynamics, and self-organization, all written by leading scientists in the field.

Jen, Erica, ed. *1989 Lectures in Complex Systems*. Santa Fe Institute Studies in the Sciences of Complexity, Lectures vol. 2. Redwood City, CA: Addison-Wesley (1990). This volume, along with the two edited by Daniel Stein listed below, is based on lectures given at the Sante Fe Institute's annual summer school on complexity. They provide a broad overview of mathematical and computational techniques.

Nicolis, Grégoire, and Ilya Prigogine. *Exploring Complexity*. New York: W. H. Freeman (1989).

Perelson, Alan S., ed. *Theoretical Immunology, Part One* and *Theoretical Immunology, Part Two*. Sante Fe Institute Studies in the Sciences of Complexity, Proceedings vols. 2 and 3. Redwood City, CA: Addison-Wesley (1988).

Perelson, Alan S., and Stuart A. Kauffman, eds. *Molecular Evolution on Rugged Landscapes: Proteins, RNA, and the Immune System.* Santa Fe Institute Studies in the Sciences of Complexity, Proceedings vol. 9. Redwood City, CA: Addison-Wesley (1990).

Pines, David, ed. *Emerging Syntheses in Science.* Santa Fe Institute Studies in the Sciences of Complexity, Proceedings vol. 1. Redwood City, CA: Addison-Wesley (1986). This is the proceedings volume of the institute's founding workshops in the fall of 1984.

Prigogine, Ilya. *From Being to Becoming.* San Francisco: W. H. Freeman (1980).

Santa Fe Institute, *Bulletin of the Santa Fe Institute* (1987–present). Published two or three times per year, the institute's bulletin contains extended interviews with some of the major figures there, as well as summaries of its various workshops and meetings.

Stein, Daniel L., ed. *Lectures in the Sciences of Complexity.* Santa Fe Institute Studies in the Sciences of Complexity, Lectures vol. 1. Redwood City, CA: Addison-Wesley (1989).

Stein, Daniel L., and Lynn Nadel, eds. *1990 Lectures in Complex Systems.* Santa Fe Institute Studies in the Sciences of Complexity, Lectures vol. 3. Redwood City, CA: Addison-Wesley (1991).

Zurek, Wojciech H., ed. *Complexity, Entropy, and the Physics of Information.* Santa Fe Institute Studies in the Sciences of Complexity, Proceedings vol. 8. Redwood City, CA: Addison-Wesley (1990).

Economics and the Santa Fe Economics Program

Anderson, Philip W., Kenneth J. Arrow, and David Pines, eds. *The Economy as an Evolving Complex System.* Santa Fe Institute Studies in the Sciences of Complexity, vol. 5. Redwood City, CA: Addison-Wesley (1988). This is the proceedings volume of the institute's first big economics meeting in September 1987.

Arthur, W. Brian. "Positive Feedbacks in the Economy." *Scientific American* (February 1990): 92–99.

Arthur, W. Brian, et al. *Emergent Structures: A Newsletter of the Economic Research Program* (March 1989 and August 1990). Santa Fe: The Santa Fe Institute. Detailed accounts of the various projects undertaken during the program's first 18 months.

Evolution and Order

Judson, Horace Freeland. *The Eighth Day of Creation*. New York: Simon & Schuster (1979).

Kauffman, Stuart A. "Antichaos and Adaptation." *Scientific American* (August 1991): 78–84.

Kauffman, Stuart A. *Origins of Order: Self-Organization and Selection in Evolution*. Oxford: Oxford University Press (1992).

Neural Networks, Genetic Algorithms,
Classifier Systems, and Coevolution

Anderson, James A., and Edward Rosenfeld, eds. *Neurocomputing: Foundations of Research*. Cambridge, MA: MIT Press (1988). Excerpts from many of the basic books and papers in the field of neural networks. Among them are some of the original works of McCulloch and Pitts, Hebb, and von Neumann. Also included is the first neural network paper by Rochester, Holland, and their colleagues.

Axelrod, Robert. *The Evolution of Cooperation*. New York: Basic Books (1984).

Forrest, Stephanie, ed. *Emergent Computation: Self-Organizing, Collective, and Cooperative Phenomena in Natural and Artificial Computing Networks*. Cambridge, MA: MIT Press (1991). The proceedings of a conference sponsored by Los Alamos' Center for Nonlinear Studies. Contains papers by Chris Langton on computation at the edge of chaos, Doyne Farmer on the "Rosetta Stone for Connectionism," and much more.

Goldberg, David E. *Genetic Algorithms in Search, Optimization, and Machine Learning*. Reading, MA: Addison-Wesley (1989).

Holland, John H. *Adaptation in Natural and Artificial Systems*. Ann Arbor: University of Michigan Press (1975).

Holland, John H., Keith J. Holyoak, Richard E. Nisbett, and Paul R. Thagard. *Induction: Processes of Inference, Learning, and Discovery*. Cambridge, MA: MIT Press (1986).

Cellular Automata, Artificial Life, the Edge of Chaos,
and Self-Organized Criticality

Bak, Per, and Kan Chen. "Self-Organized Criticality." *Scientific American* (January 1991): 46–53.

Burks, Arthur W., ed. *Essays on Cellular Automata*. Champaign-Urbana: University of Illinois Press (1970).

Dewdney, A. K., "Computer Recreations." *Scientific American* (May 1985). A basic discussion of cellular automata and computation.

Farmer, Doyne, Alan Lapedes, Norman Packard, and Burton Wendroff, eds. *Evolution, Games, and Learning.* Amsterdam: North-Holland (1986). [Reprinted from *Physica D* 22D (1986) Nos. 1–3.] The proceedings of a conference sponsored by Los Alamos' Center for Nonlinear Studies. Includes one of the first public discussions of Farmer, Kauffman, and Packard's autocatalytic origin-of-life model, as well as one of the first presentations of Chris Langton's discovery of a phase transition in cellular automata.

Langton, Christopher G., ed. *Artificial Life.* Santa Fe Institute Studies in the Sciences of Complexity, Proceedings vol. 6. Redwood City, CA: Addison-Wesley (1989). Proceedings of the first artificial life workshop in September 1987. Includes Chris Langton's introduction and overview of the artificial life concept, as well as extensive bibliographies of the field.

Langton, Christopher G., Charles Taylor, J. Doyne Farmer, and Steen Rassmussen, eds. *Artificial Life II.* Santa Fe Institute Studies in the Sciences of Complexity, Proceedings vol. 10. Redwood City, CA: Addison-Wesley (1992). Proceedings of the second artificial life workshop in 1990. Includes papers by Chris Langton on the edge of chaos, Stuart Kauffman and Sonke Johnsen on coevolution to the edge of chaos, Walter Fontana on "algorithmic chemistry," and Rick Bagley, Doyne Farmer, and Walter Fontana on further development of the autocatalytic origin-of-life model.

von Neumann, John. *Theory of Self-Reproducing Automata.* Completed and edited by Arthur W. Burks. Champaign-Urbana: University of Illinois Press (1966).

Wolfram, Stephen. "Computer Software in Science and Mathematics." *Scientific American* (September 1984). Includes a basic discussion of cellular automata.

Wolfram, Stephen, ed. *Theory and Applications of Cellular Automata.* Singapore: World Scientific (1986). Reprints of many of the cellular automata papers published by Wolfram and his colleagues in the early 1980s.

Acknowledgments

To the dozens of people who have given so generously of their time and patience to help make this book what it is: a heartfelt thank-you. All books are a team effort, and this one is more so than most. Many of you have received only a passing mention in the text, or none at all. Believe me, that does not diminish either the importance of your contribution or the gratitude I feel.

To Brian Arthur, George Cowan, John Holland, Stuart Kauffman, Chris Langton, Doyne Farmer, Murray Gell-Mann, Kenneth Arrow, Phil Anderson, and David Pines: a special thank-you for enduring endless interviews and telephone calls, for sharing your own struggles, for educating me in the ways of complexity and, not least, for reading various drafts of the manuscript in whole or in part. For me, at least, the process has been a joy. I hope it has been for you as well.

To Ed Knapp, Mike Simmons, and the staff of the Sante Fe Institute: thank you for hospitality and assistance above and beyond the call of duty. The Santa Fe Institute is truly a place to come home to.

To my agent, Peter Matson: thank you for your guidance, advice, and reassurance, which always had a calming influence on a frequently nervous author.

To my editor, Gary Luke, and the production staff at Simon & Schuster: thank you for your enthusiastic support—and your yeoman efforts on behalf of a book that was very late.

And to Amy: thank you. For everything.

Index

Academy of Sciences, Budapest, 48
Adams, Robert McCormack, 86, 88, 91
Adaptation, 146. *See also* Complex
    adaptive systems
  arriving at edge of chaos through, 309
  coevolution and, 309
  evolution and, 179
  Holland's analysis of, 144–49, 166
  in human mind, 176
Adaptive agents, 145, 176
  in evolution, 179
  computer simulation of, 181–93
  Hebb's search for theory of, 180
  Holland on, 176–77
  in inductive mode, 253, 254
  simulated, 181–93
Adaptive computation, 341–43
  origin of term, 342
Adaptive systems. *See* Complex adaptive
    systems
Agents, in adaptive systems. *See* Adap-
    tive agents
Agnew, Harold, 62
AIDS, 60, 248
Alchemy model, 315–16
Anderson, Herb, 84
Anderson, Joyce, 92
Anderson, Philip, 12, 79–83, 95–96,
    136, 137, 139, 140, 143, 149, 195,
    244, 304, 348
  economics workshop organized by, 97

on emergent properties, 82–83
meeting with Reed, 92–93
on reductionism, 81
Anthropology, 86
Arrow, Kenneth, 12, 24, 52, 97, 136,
    137–38, 140, 142, 143, 195, 244,
    245, 246, 250, 325, 326, 340
  Arthur and, 52–53
  on evolutionary economics, 326–27
  on increasing returns economics,
    195–96, 327
  personality characteristics, 52
  on Santa Fe Institute versus Cowles
    Foundation research programs, 327
Arthur, Susan Peterson, 15, 29, 117
Arthur, William Brian, 15–54, 98, 240
  academic background, 15–16, 19–28
  at American Economics Association
    meeting, 50
  Arrow and, 52–53
  in Bangladesh, 26–27
  codirectorship of Santa Fe Institute of-
    fered to, 199
  "Competing Technologies, Increasing
    Returns, and Lock-In by Historical
    Events," 49–50
  on complexity as scientific revolution,
    327–30
  as director of Santa Fe Institute, 244–
    245, 247–48, 324–35
  dissertation, 24, 25–26

Arthur, William Brian (*Cont.*)
  Stewart Dreyfus and, 24
  in Germany, 21, 25
  Holland and, 144, 147–48, 194, 242–
    243
  ideas for Santa Fe Institute, 245–46
  on increasing returns economics, 29,
    34–46
  at International Institute for Applied
    Systems Analysis, 28–47
  Kauffman and, 100–101, 117–20,
    133–34, 137
  personality characteristics, 49
  point of view, 38–46
  on Santa Fe Institute approach, 326–
    335
  at Santa Fe Institute economics work-
    shops, 99–101, 117–20, 133, 136–
    143, 197–99
  speculations of, 327–35
  Stanford University professorship, 46–
    52
  as visiting professor at Santa Fe Insti-
    tute, 53–54
Artificial intelligence, 71, 157, 180,
    185. *See also* Neural network simu-
    lations
Artificial life, 199–240, 277–84, 299
  Farmer on, 284–99
  Langton and, 215–40
  Langton's use of term, 198, 237
  metaphor of, 329
  possibility of creation of, 282–84
  "real," 283–84
  rules of, 281–82
  science of, Langton as founder of,
    322, 358–59
  value of, 289
Artificial Life Workshop, Los Alamos,
    199–240
Asimov, Isaac, "The Final Question,"
    286
Aspen Center for Physics, 71, 83
Atomic Energy Commission, 58
Augusztinovics, Maria, 48
Autocatalytic sets/autocatalytic models,
    124–28, 130, 132–33, 148, 235–
    236, 291
  Alchemy program and, 315
  analogy to economics, 125–27
  emergence in, 289, 296–97

Kauffman on, 313, 315–18
  node-and-connection structure of, 290
  selection and adaptation in, 317–18
  simulating, 132–33
Automata theory, 109
Axelrod, Robert, 176, 263, 273
  Evolution of Cooperation, The, 264

Bagley, Richard, 133, 297, 314
Bak, Per, 304–6
Balance, between order and chaos, 294
Bangladesh population and develop-
    ment, 26–27
Behavior
  collective, emergent, 95
  patterns of, 66
Belin, Alletta, 284
Bethe, Hans, 57, 68
Big Bang theory, 62, 64, 76, 81, 295, 318
Biology, 16, 29–30
  developmental, 106
  molecular, 62, 76
  physicists and, 62–63
Bloch, Eric, 90
Bohr, Niels, 68
Boldrin, Michele, 196
Booker, Lashon, 191
Bottom-up organization, and evolution,
    294
Branscomb, Louis, 87
Brock, William, 246
Brookings Institute, 352
Bucket-brigade algorithm, 188–89, 191,
    272, 291
Bulletin of the Atomic Scientists, The,
    founding of, 58
Burks, Arthur, 160–61, 162, 166, 223,
    224

Cairns-Smith, Graham, 239
Calvin, Melvin, 128
Campbell, David, 322
Campbell, Jack, 79
Carruthers, Peter, 69–70, 84, 87, 349
  support of Santa Fe Institute, 69–70
Cell, as self-organizing system, 34
Cell assemblies, 182
Cellular automata, 87, 217, 219
  Class I rules, 225, 228, 231, 233
  Class II rules, 225–26, 228, 231, 233

Class III rules, 226, 227, 228, 232, 233
Class IV rules, 226, 227, 228–30, 234
  edge-of-chaos phase transition in, 292–93
  evolution and, 222
  Game of Life as special case of, 219
  Langton and, 221–22
  nonlinear systems and, 225
  rules followed by, 225–30
  self-reproduction and, 221–22, 279, 281
Cellular Automata (Codd, editor), 217
Cellular differentiation, 106
Change, power-law cascades of, 308–9
Chaos, 65–66, 311. *See also* Chaos theory
  classifier system and, 193
  evolution and, 303
  transition to, 230 (*see also* Edge of chaos)
Chaos (Gleick), 131, 287
Chaos theory, 131–32, 329. *See also* Edge of chaos
  as emerging science, 287
  Farmer on, 287–88
  Los Alamos as world center for, 70
  and self-organizing complex systems, 12
  use of, 95
Chemical Evolution (Calvin), 128
Chemistry model, 314–15
Chenery, Hollis, 137, 195–96
Chess analogy, 150–51
Chiaromonte, Francesca, 358
Church, Alonzo, 278
Citicorp, 91, 92, 94–95, 96, 184–94, 242–43, 244, 252, 258, 338, 340
Classifier systems, 188–89, 272, 289, 290
  Arthur on, 272–73
  bucket-brigade algorithm, 188–90, 191, 291
  cognitive theory and, 192–93
  deus ex machina, 258–59
  emergent system and, 193
  genetic algorithm, 188–90, 191, 199, 242–43, 252, 272, 303
  node-and-connection structure of, 290
  as self-adapting, 273

Codd, Ted, 217, 220, 221
Coevolution, 309
  edge-of-chaos principle and, 294
  as force for emergence, 259–60
Coevolutionary models, 292, 309–13. *See also* Echo model
  edge-of-chaos phase transition, 311
Cognition
  computer model of, 181–82
  learning and, 180
Cognitive science, 71, 76
  as interdisciplinary, 71
Cognitive theory, 192–203
  classifier system and, 190, 193
Cohen, Michael, 175, 176, 189
Colgate, Stirling, 70
Combat, as metaphor for interaction, 260
Competition, in adaptive systems, 185
  and cooperation, 185, 262–65
  in evolution, 262
Complex adaptive systems, 294–99
  building block structure and, 170
  chemistry and, 314–15
  economy as example of, 145 (*see also* Economics; Economy)
  emergent behavior in (*see* Emergence)
  generalized genotype versus phenotype of, 281–82
  Holland's work and, 166–94
  levels of organization, 145–46
  niches of, 147
  prediction and, 146, 176–77
  realignment of, 146
  rules of, 280–81
Complex systems. *See also* Complex adaptive systems; Complexity; Edge of chaos
  characteristics of, 11–12
  computers and, 63
  defined, 11
  spontaneous self-organizing, 11
Complexity, 294. *See also* Complexity research; Complexity theory
  in economics, 38
  at edge of chaos, 293
  growth of, 65, 165, 294–99, 316–18, 319
  phase transitions and, 232, 283
  Santa Fe Institute approaches to, 248
  scientists' fascination with, 63–64

Complexity (*Cont.*)
  Wolfram on, 86–87
Complexity research
  computers in, 87
  example of questions dealt with by, 9–11
  trouble in defining, 9
Complexity theory
  metaphors and, 334
  as scientific revolution, 327–30
  as world view, 333
Computation(s), of complex adaptive systems, 106, 232, 280–81, 341–43
  as essence of life, 232, 280
Computer graphics, 95
Computer simulation, 63–64, 93, 177, 251. *See also* Echo model; Modeling
  of adaptive agents, 181–93
  of autocatalysis, 132–33 (*see also* Autocatalytic sets/autocatalytic models)
  of complex adaptive systems, need for, 147
  connectionist, 289
  of cooperation, 262–65
  of earthquakes, 305–6
  of economics, 93, 95, 243, 251, 252–255, 268–74
  of ecosystems, 309–13
  emergent behavior in, 242, 270
  of evolution, 256–58
  of flocking behavior of birds, 241–43, 279, 329
  of genetic networks, 111–14, 116
  of immune system, 266
  lesson learned about complex adaptive systems from, 279
  local control in, 280
  of neural networks, 113, 159–60, 181, 289–90
  origin-of-life, 124–25, 313, 314
  of plant development, 279–80
  of policy options, 266–67
  rule-based systems, 181–82
  of stock market, 266, 273–74, 329
  value of, 289
Computer viruses, 238, 239, 283
  properties of, 283
Computers
  in complexity research, 63, 87
  conflict resolution strategies, 184–85
  development of, at IBM, 155–57

  first lectures on, 153
  new kinds of, 71, 76
  rule-based systems, Newell-Simon, 181–85
  software packages, debugging, 282
Connectionism, 289–92
  defined, 289
  models, 289
Connectionist theory of memory, 158–159, 160, 181
Connections, 289–90
  adjusting/changing, 291
  power of, in learning and evolution, 291, 292
  sparse, 303
Consistency, in adaptive systems, 185
Conway, John, 201, 219
Cooperation
  and competition, 262–65
  in evolution, 262–64
  in real world, 263
Cost-benefit analyses, 331–32
Cowan, George A., 13, 53–69, 143, 244, 276–77, 335–57
  academic background, 55–56, 57
  dissatisfaction with academic approach to science, 60–61
  Gell-Mann and, 73–74, 335, 347–54
  ideas for interdisciplinary institute for study of complexity, 68–69
  at Los Alamos, 57–59, 66–69
  as Manhattan Project scientist, 56–58
  personality characteristics, 54–55
  at Santa Fe Institute, after presidency, 355–57
  as Santa Fe Institute chairman of board and president, 72–73, 77–79, 84–85, 335–54
  White House Science Council and, 59–61
Cowan, Jack, 85, 87, 134
  Kauffman and, 116
Cowles Foundation, 327
Crick, Francis, 29, 62, 219
Critical states, self-organized, 304–7
Crutchfield, James, 287, 346

Dance of Life, The (Ellis), 356
Darwin, Charles, 75, 102, 179, 287, 295
David, Paul, 50
Dawkins, Richard, 232, 238, 261–62
Debreu, Gerard, 24

Default-hierarchy models, 192, 193
Detectors, 182–83
Dewdney, A. K., 239
DNA, 121, 122
  as molecular-scale computer, 31
  structure of, 29–30, 62, 219
  Von Neumann's analysis of self-repro-
    duction and, 219
Doolen, Gary, 274
Double Oral Auction Tournament, 273
Dreyfus, Stuart, 24
Dynamical behavior, Wolfram's classes
  of, 225–35
Dynamical systems, 287
  nonlinear, and cellular automata,
    225–30 (see also Cellular automata;
    Nonlinear dynamics)

Earthquakes
  computer models, 305–6
  distribution of, 305
  power law for, 305–6
  self-organized criticality and, 303–4
East-West Population Institute (Honolulu), 25, 29
Echo model, 260–62, 266, 292, 309,
  310
Econometric models, 93
Economics
  analogy to autocatalysis models, 125–
    127
  analogy to chess, 150–51
  computer simulations of, 93, 95, 243,
    251, 252–55, 268–74
  evolutionary approach to, 252
  "increasing returns" concept in (see
    Increasing returns economics)
  instability in, 17
  neoclassical view of, 17, 22–24, 27,
    34–35, 37–38, 43, 140–41, 328,
    330
  new (see Increasing returns economics)
  as nonlinear, 65
  Santa Fe Institute approach to, 250,
    252, 254–55, 330 (see also Increas-
    ing returns economics)
  Santa Fe Institute program in, 95–98
  as self-organizing system, 34
  spin glass metaphor, 138–39
Economists
  attitudes of, 142

historians, on increasing returns economics, 50
lack of attention to empirical fact, 141
openness to new view of economics,
  325–26
theoretical, on increasing returns economics, 49–50
Economy
  as complex adaptive system, 145
  computer-simulation of, with adaptive
    agents, 243, 251, 252–55
  Third World, patterns of develop-
    ment, 195–96
  world, computer simulations of, 93
Ecosystems
  computer simulation of, 309–13
  at edge of chaos, 308–13
Edge of chaos, 230–31, 292–94, 311,
  356
  as balance point, 12
  cellular automata at, 234
  complexity and, 295
  defined, 12
  ecosystems at, 308–13
  Farmer on, 293
  as Langton's phrase, 230
  Langton on, 302
  origin of life and, 235
  as region rather than boundary, 302
  self-organized criticality and, 307
  theory of, difficulty of testing in real
    world, 307–8
EDVAC, 161
Effectors, 182–83
Ehrlich, Paul, 25
Eighth Day of Creation, The (Judson),
  29, 34
Einstein, Albert, 103
Ellis, Havelock, 356
Emergence, 152, 200, 218
  and adaptation, 149, 183–84
  Anderson on, 82–83
  bottom-up, 278–79
  coevolution and, 259–60
  in computer simulations, 242, 270
  Farmer on, 288–92
  hierarchical building-block structure
    and, 169–70
  meaning of, 242
Emergent models, 270
  default-hierarchies models as, 193
ENIAC, 153, 161

Entropy, 286, 288
Equilibrium, 165, 167
  complex adaptive systems and, 147
  in economics, 255
Ermoliev, Yuri, 46, 246
Ervin, Frank, 200, 205
Evolution. *See also* Coevolution
  cellular automata and, 222
  competition and cooperation in, 262–265
  computer simulation of, 256–58
  development of theory of, 299
  edge of chaos and, 303
  Fisher versus Hebb on, 164–65
  hierarchy of control in, 294
  as longest term adaptation, Hebb's study of, 166–74
  movement of, toward greater complexity, 296
  paradox in, 262
  self-organized criticality and, 308–9
Evolution of Cooperation, The, 264
"Evolution, Games, and Learning" conference, Los Alamos, 148
"Evolutionary arms race," 261–62
Expectations
  in economics, 141
  rational, 270–71
Expert systems, 180, 181
Exploitation learning, 291
Exploration learning, 291, 312
"Extended transients," 229, 230, 233

Farmer, Doyne, 131–32, 134, 148, 235, 246, 258, 276, 284–99, 313, 314, 329, 358
  "Artificial Life: The Coming Evolution," 284, 285
  on chaos theory, 287–88
  on emergence, 288–92
  Langton and, 235, 322
  personality characteristics, 285
  "Rosetta Stone for Connectionism," 290, 294, 312, 341–42
  on science, 318–19
  on self-organization, 286–88
Feedback, 29
  Holland on, 180
  internal, 181
  natural selection and, 179
  negative, 139
  patterns and, 36

  positive, 138, 139, 196
  rule-based systems and, 183
  stock market and, 196
Feldman, Marc, 87, 134
Fermi, Enrico, 56, 68
Feynman, Richard, 68
Fisher, R. A., 163–64, 166–67, 259
Fishlow, 50
Flocking-of-birds computer simulation, 241–43, 279, 329
Fogelman-Soule, Françoise, 301
Fontana, Walter, 297, 314–15, 358
Forrest, Stephanie, 272
Free-market ideal, 47–48
  increasing returns economics and, 48
Freeman, Christopher, 43
Friedez, Gideon, 223

Game of Life, 201–3, 209, 218, 229, 231, 234
  artificial life and, 202–3
  as cellular automaton, 219
Games theory, 150, 262
Gamow, George, 62
Geanakoplos, John, 325
Gell-Mann, Murray, 12, 53–54, 73–77, 80, 81, 83, 84, 136, 149, 245, 258, 261, 284, 331, 346
  Cowan and, 73–74, 335, 347–51
  as head of Santa Fe Institute board of trustees, 89–90
  as head of Santa Fe Institute Science Board, 90
  on Holland, 148–49
  interests of, 74–75, 256–58
  personality characteristics, 74
Generalized genotype, 281–82
Generalized phenotype, 281–82
Genes, regulatory, 106, 107, 108
Genetic algorithms, 188–90, 191, 242–243, 252, 272, 303
  international conferences on, 199
Genetic circuits, 106
Genetic code, 29–30
  structure of, 62
Genetic network models, 111–14, 116, 289
  neural networks and, 113
  orderly states of, 111–13
  phase-transitionlike behavior in, 301
  regulatory, 110–13, 120–21
Genetic order, Kauffman on, 108–13

Genetics
  Fisher on, 163–64
  molecular, 106–7
Genomes, 106–8, 290
  regulatory systems, 315–16
Getty, Gordon, 354
Gleick, James, 131, 287
Global behavior, 306
Global sustainability program, 349–52
Gödel, Kurt, 278, 328
Goldberg, David, 191–92, 193
Goldenberg, Edie, 344–45
Goldstine, Herman, 161
Goodwin, Brian, 116
Gould, Stephen J., 308
Gutenberg, Beno, 305

Hahn, Frank, 247, 250, 255, 325
Haken, Hermann, 287
Hamilton, William, 175, 176
Hebb, Donald O., 157–60, 169, 176,
    182, 183–84, 188
  theory of learning and memory, 164–
    165
Hecker, Siegfred, 275
Heidegger, Martin, 329
"Heraclitians," 335
Heraclitus, 38
Hewlett-Packard, 69
Hicks, John R., 47
High technology industries
  government policy and, 41–42
  standards and, 42–43
Historical accidents, 36–37, 39, 41
Holistic approach, versus reductionism,
    60–61
Holland, John H., 144–94, 197, 220,
    246, 252, 253–54, 278–79, 289,
    290, 295, 309, 341, 342, 346
  academic background, 153–55, 161–
    162, 166, 175
  Adaptation in Natural and Artificial
    Systems, 174
  on adaptive systems, 144–47, 166–94
  at Artificial Life Workshop, Los Ala-
    mos, 198, 199–200, 238–40
  bucket-brigade algorithm, 188–90,
    191, 272, 291
  on classifier systems (see Classifier sys-
    tems)
  Gell-Mann and, 256–57
  at Santa Fe Institute, 244

Arthur and, 144, 147–48, 194, 242–
    243
  Burks and, 161–62
  on emergence and adaptation, 148,
    149–50
  on equilibrium, 167
  genetic algorithm, 188–90, 191, 242–
    243, 252, 272, 303
  and Hebb's theory, 158–60
  at IBM, 155–60
  Langton and, 224
  MIT senior thesis, 153–55
  on personal computers, 189–90
  programming knowledge of, 154–55
  Santa Fe Institute Maxwell professor-
    ship offered to, 343, 344
Holyoak, Keith, 192–93
Hubler, Alfred, 346

IBM, 156–57, 159
  computer development, 155–56
  Holland at, 155–60
IIASA. See International Institute for
    Applied Systems Analysis
Immune system model, 289
Increasing returns economics, 17–18,
    29, 34–38, 42
  acceptance of, 195–96, 325–26
  Arthur on, 38–46 (see also Arthur,
    William Brian)
  free-market ideal and, 48
  Kauffman on, 118
  "legitimization" of, by physicists, 139–
    140
  Marshall on, 44–45
  mathematical analysis of, 44–46
  properties of, 36–37
  technological change and, 119–20
Induction, 253, 254
Induction (Holland, Holyoak, Nisbett,
    and Thagard), 193, 272, 344
Information processing, 153
Insight, 193
Interdisciplinary research
  cognitive science as, 71
  universities and, 67
Internal combustion engine, 40–41
International Institute for Applied Sys-
    tems Analysis, 28, 32, 39–40
  Arthur at, 28–47
International Schumpeter Society, 326

Jacob, François, 31
Japan, economic development of, 43
Johnsen, Sonke, 303, 309–12
Joule, James, 298
Judson, Horace Freeland, 29–30, 31, 34

Kaniovski, Yuri, 46, 246
Kauffman, Stuart, 162, 194, 235, 246, 249, 257, 258, 259, 289, 299–304, 329, 341, 346, 358
    Arthur and, 100, 117–20, 133–34, 137
    on autocatalysis, 121–31, 315–18
    automobile accident, 321
    background, 103–13
    Jack Cowan and, 116
    death of daughter, 117, 134–35
    ecosystem model of, 309–13
    Farmer and, 288
    fruit fly development work, 129–30
    Goodwin and, 116
    Horsetail Falls home, 134
    marriage, 114
    McCullough and, 113–16
    on origin of life, 121–25
    Origins of Order, The, 300
    personality characteristics of, 101–2
    on self-organized criticality, 307–8
    speculations of, 321–22
Kaysen, Carl, 92
Keck, William, Jr., 354
Keynes, John Maynard, 23
Keyworth, George (Jay), II, 60, 69
Knapp, Ed, 88, 349, 353–54
    as Santa Fe Institute president, 354–355
Knief, Byron, 92
Koopmans, Tjalling, 24
Kopal, Zednek, 154

Lane, David, 272, 325, 358
Langton, Christopher G., 199–240, 257, 258, 274–84, 291, 311, 320
    and artificial life, science of, 215–40, 322, 358–59
    background, 200–38
    Farmer and, 235
    hang-gliding accident and recovery, 208–11
    Holland and, 224

intuition of, 202–3
    at Los Alamos Center for Nonlinear Studies, 236–40, 322
    marriage, 216
    speculations of, 320–21, 329, 330, 358–59
Learning, 291
    cognition and, 180
    computer modeling of, 185–89
    in economics, modeling effects of, 252
    Hebb's neurophysiological theory of, 155–56
    lack of interest in, 183–84
Levins, Dick, 116
Lewontin, Richard, 116, 335
Licklider, J. C. R., 157, 158, 159
Life
    artificial (see Artificial life)
    behavior of, 280
    as computation, 232, 280
    computer simulation of (see Artificial life)
    edge-of-chaos phase transition and, 302–3
    as emergent property, 82, 278
    organization as essence of, 288, 292
    origin of, 235, 257–58
Lindenmeyer, Aristid, 238–39, 279
Linear systems, versus nonlinear systems, 64–65. See also Reductionism
Lloyd, Seth, 346
Lock-in, of economic advantage, 36, 37, 39–40, 50, 139
    examples of, 40–42
    Japanese economy and, 43–44
    Uccello clock and, 40
Los Alamos, 54, 60, 61, 62
    Artificial Life Workshop, 199–240
    Carruthers at, 69–70
    Center for Nonlinear Studies, 236, 274
    chaos theory research, 70
    Complex Systems group, 285–86
    George Cowan at, 57–59, 66–69
    "Evolution, Games, and Learning" conference, 235
    mission of, 68
    nonlinear dynamics research at, 66–70

MacArthur Institute, 350, 352
McCarthy, John, 157
McCullough, Warren, 113–16
   death, 116
   Kauffman and, 113–16
McKenzie, Lionel, 24
McKinsey and Company, 20–21, 22, 24, 25
McMahon bill (1946), 58
McNicoll, Geoffrey, 26–27
Manhattan Project, 67–68
MANIAC, 70
Marimon, Ramon, 246, 270, 271
Marshall, Alfred, 44
Mathematics, ferment in, 71
Maxwell, Robert, 343–46, 355
Maynard Smith, John, 116, 120, 300, 311
Memory, Hebb's neurophysiological theory of, 155–56
Menezes, Victor, 92
Metaphors, and complex systems, 329, 334
Metropolis, Nick, 71, 72, 76, 78–79
   support of Santa Fe Institute, 70–71
Miller, Stanley, 121, 128, 132
Miller, John, 246–47, 266, 273
Mind, study of. See Cognitive science
Minsky, Marvin, 115
Mitchell, Melanie, 359
Modeling/models, 76. See also Computer simulations
   of complex adaptive systems, need for, 147
   econometric, 93
   emergent, 252, 253
   internal, 146, 147
   Kauffman and, 105
   mental, 177, 278
   prediction and, 177
Monod, Jacques, 31
Moravec, Hans, 239
Morris, Robert, 238
Mountain Bell, 78

Nagle, Darragh, 72
Natural law, belief in, 81
Nelson, Richard, 252
Networks
   "frozen components" in, 301
   genetic network models, 110–14, 289, 301, 303

neural network models, 113, 159–60, 181, 289–90
   node and connection structure of, 289 (see also Connectionism)
   random, of reactions, 132
   statistical properties of, 120–21
Neural network simulations, 113, 159–160, 181, 289–90
   McCullough-Pitts model, 113
Neuroscience, 16
Newell, Allen, 168, 181
Nisbett, Richard, 192–93
Nonlinear dynamics, 64, 65–66, 76, 121
   Carruthers and, 69–70
   cellular automata and, 225
   Eastern philosophy and, 331
   formation of patterns and, 65
   Los Alamos research on, 66–70
   self-organizing systems as, 65
Novelty, perpetual, 147
Nuclear power, 41–43, 60
   waste disposal, 60
   weapon research, 55–59

Operations research
   defined, 20
   Williams as student of, 20
Oppenheimer, J. Robert, 58, 68
Order, 293, 294, 311
   Christianity and, 330
   genetic, 108–13
   Kauffman on, 102
   from molecular chaos, 124
   Prigogine on, 32–33
Organization. See also Self-organization/ self-organizing systems
   as essence of life, 288, 292
   levels of, in complex adaptive systems, 145–46
   properties of, 288
Origin of life, 121–25, 319
   autocatalytic set and, 124–25
   computer simulation of, 124–25, 313, 314

Packard, David, 69
Packard, Norman, 131, 148, 235, 287, 288, 302–3, 313
Palmer, Richard, 246, 252, 272, 273
Papert, Seymour, 115
Path dependence, 50, 139

Patterns
  of active genes, 31, 107
  Arthur's awareness of, 27–28, 31–32
  of behavior, 66
  computer simulation of, 63–64
  of connections, 289–90
  of economic development, 195–96
  increasing complexity of, 66
  nonlinear dynamics and, 65
  reasons for, 36–37
Perelson, Alan, 248, 289
Pergamon Press, 344, 346
Periodic attractors, 226
Perpetual novelty, 147
Phase transitions, 228–30, 233–35,
    292–94
  to chaos, 230–31
  complexity and, 232, 293
  edge of chaos, 292–93, 302–3
  origin of life and, 235
  self-organized criticality and, 307
Physicists, 62, 287
  attitudes of, 142–43
  economists and, 139–40, 141–43
  and empirical fact, 140
Physics, 16–17
  Arthur's study of, 32
Pines, David, 71–72, 75, 79, 80, 83,
    92, 96–97, 196–97, 246, 276–77,
    346
  academic background, 71
  at Los Alamos, 71
Pipeline simulation, 191–93
Pitts, Walter, 113
"Platonists," 335
Policy-making, Arthur on, 330–34
Political policy, Taoist approach to, 331–
    334
Polynomial-time algorithms, 233
Population Council (New York), 26, 27,
    28
Population growth, Third World, 25–27
Power-law behavior, 305
  global, 306
  possibility of testing, 308–9
Predictability, 39, 142, 329
Prediction
  complex adaptive systems and, 146,
    176–77
  implicit, 146
  models and, 176–79
  in relation to science, 255

  understanding versus, 306
Prigogine, Ilya, 32, 287
Principles of Economics (Marshall), 44
Prisoners' Dilemma, 262, 263–65
  iterated, 263
  TIT FOR TAT strategy, 264–65
Process control, 278
Programming languages, 154
Prusinkiewcz, Przemyslaw, 279
Psychologists, 71, 195
Purmort, Wally, 165

Quarks, 53–54, 74, 314, 348
QWERTY keyboard, as example of lock-
    in, 35, 40, 42, 50, 139

Rational expectations, Sargents's work
    on, 270–71
Rationality, 141–42
  bounded, 250–51
  perfect, 250–51
Reagan administration, 46, 47
  "Star Wars" Strategic Defense Initia-
    tive, 60
Red Queen hypothesis, 261
Reductionism. See also Linear systems
  Anderson on, 81
  versus complexity, 329
  as dead end approach, 60, 61
  versus holism, 60–61
Reed, John, 91, 93, 136, 143, 149, 196,
    197, 244, 245
  meeting with Santa Fe Institute scien-
    tists, 92–96
  personality characteristics, 92
Reichenbach, Hans, 213
Reitman, Judy, 190–91
Religion, 330. See also Taoism
  Farmer on, 319
Research style, Arthur on, 29
Reynolds, Craig, 241–43, 279
Rice, Stuart, 127, 128
Richardson, Ginger, 100
Richter, Charles, 305
Rio Grande Institute, 79, 89. See also
    Santa Fe Institute
Riolo, Rick, 189–90
Risking. See Exploration learning
RNA, 121
Robotics, 283
Rochester, Nathaniel, 159, 160
Rockefeller, John D., III, 26

Roessler, Otto, 128
Rota, Gian-Carlo, 70–71, 72, 75
Rothenberg, 50
Ruelle, David, 137, 195
Russell Sage Foundation, 91, 92
Rust, John, 273

Salmon, Wesley, 213
Samuel, Arthur, 157, 165, 180, 194
Samuelson, Paul, 24
Sand pile metaphor, 304–5
Santa Fe Institute, 12, 13
  accomplishments of, 339
  approach of, 247–55, 326–35
  approach to complexity issue, 248
  approach to economics, 250–55
  Arthur as visiting fellow at, 53–54
  Arthur's introduction to, 52–54
  Artificial Life Workshop, 279
  Artifical Life II Workshop, 322, 323
  atmosphere of, 100, 249–50, 329
  attainment of name, 89
  building, 99–100, 358
  as catalyst, 325
  Complex Systems Summer School,
    339
  contributions to complexity science,
    324–27
  Cowan and, 53–56, 66–69, 78–79
    (see also Cowan, George A.)
  Double Oral Auction Tournament,
    273
  early plans for, 68–73
  economics program, 244–55, 339,
    341
  economics workshops, 95–98, 133–
    143, 149–51, 194–97, 247–55
  focus of study at, 53, 54–55
  founding workshops, 79, 84–89
  funding of, 77–78, 89, 90, 91, 96,
    99, 244, 247–48, 335–39, 354–55
  Global Sustainability program, 349–
    352
  Holland and, 225, 345 (see also Hol-
    land, John H.)
  impromptu seminar/bull sessions at,
    249–50
  incorporation of, 79
  mission, 72–73, 79
  as new kind of scientific community,
    337, 338
  organization of, 87

  senior fellows at, 69–73
  staffing of, 99
  University of Michigan, Ann Arbor,
    link with, 345
Sargeant, Tom, 136–37, 246, 266, 271,
  274
Scheinkman, José, 137, 196
Scholes, Christopher, 35
Schrödinger, Erwin, 62
Schumpeter, Joseph, 46, 119, 252, 326
Schwartz, Douglas, 85–86
Science, 318–19
  and prediction, 255
Scientists
  types of, 335
  world views, 334–35
Segura-Langton, Elvita, 216, 217, 221,
  223–24, 236
Selection, versus self-organization, 300–
  301, 313
Self-assembly, 218
Self-fulfilling prophecy, 274
Self-organization/self-organizing systems,
  317
  adaptive, 11
  autocatalytic sets and, 133
  defined, 102
  development of theory of, 299
  economics as, 34
  Farmer on, 286–88
  in genetic self-regulatory networks,
    121
  Kauffman on, 300
  knowledge of, 298–99
  living systems and, 102
  as nonlinear, 65
  Prigogine on, 32–34, 65
  rise and fall of civilizations and, 86
  versus selection, 300–301, 313
  spontaneous, 11
  Woolfram on, 86–87
Self-organized criticality, 304–5, 356
  of earthquake fault zones, 306
  Langton's formulations in relation to,
    306–8
Self-reinforcing mechanisms, Arthur on,
  137–39
Self-reproducing systems, 217, 237. See
    also Computer viruses
  cellular automata and, 221–22
  computers and, 237
Selfish Gene, The (Dawkins), 238

Shannon, Claude, 151
Shaw, Robert, 287
SimCity, 267
Simmons, Michael, 89, 338–39, 342,
    344
Simon, Herbert, 168, 181
Simulation Laboratories, 267
Singer, Eugenia, 92, 93, 95, 143, 195,
    244
Skills, as implicit models, 178–79
Smith, Adam, 279, 328
Smith, Stephen, 191
Society, Farmer on, 319–20
Sociologists, 195
Spencer, Herbert, 287
Spiegel, Arthur, 78
Spin glass metaphor, for economy, 138–
    139
Standards, economic rewards of, 42
Stanford University, Arthur as professor
    at, 46–52
Stock market, 269–70
    computer simulation of, 266, 273–74,
        329
    positive feedback mechanism and, 196
    self-fulfilling prophecy in, 274
Strange attractors, 226. See also Chaos
    theory
Strategies, in economics, 141
Strauss, Lewis, 58
Structure. See also Order; Patterns
    hierarchical building-block, 169–70
    increase in, 295
Summers, Larry, 137, 143
Superstring theory, 348
Symbolics Corporation, 240
Symbols, 183, 184
Syntheses in science, 75–76, 87–88

Tang, Chao, 304, 305–6
Taoism
    policy-making and, 330–34
    world view of, 330, 333–34
Tate, Lawson, 210
Taucher, George, 21
Technological change, 118–20
Thagard, Paul, 192–93

Theory of Self-Reproducing Automata,
    The (Von Neumann), 217
Thom, Rene, 116
Time-delayed control theory, 24, 25
Tobin, James, 97
Trivelpiece, Alvin, 90
Turing, Alan, 234, 278, 328–29

Uccello clock, 40
Ulam, Stanislas, 219
Uncertainties, and chaos, 66
Universal constructor, 218
Unpredictability, 36–37
Urey, Harold, 121, 128, 132

Vaupel, James, 40
VHS, lock-in to, 35–36, 40
Volker, Paul, 94, 95
Von Neumann, John, 68, 70, 161, 217
    analysis of self-reproduction, 217–19
    work on cellular automata, 218–20

Wada, Eiiti, 240
Waddington, Conrad, 116
Watson, James, 29, 62, 219
Weisbuch, Gérard, 301
Weisskopf, Victor, 81
What Is Life? (Schrödinger), 62
Whirlwind Project, at MIT, 153–55
White House Science Council, 59–61
Wicksell's triangle, 271
Wiener, Norbert, 287
Wiesenfeld, Kurt, 304
Wigner, Eugene, 68
Wilson, Robert, 87
Wilson, Stewart, 191
Winter, Sidney, 252
Wolfram, Stephen, 86–87, 225, 313–14
Wolpert, Lewis, 116
Wootters, William, 286–87
World Resources Institute, 75, 350, 352
World views, 333
    new, 249, 255
    of scientists, 334–35
Wriston, Walter, 91

Zegura, Steve, 214–16
Zurek, Wojciech, 210

READ MORE IN PENGUIN

In every corner of the world, on every subject under the sun, Penguin represents quality and variety – the very best in publishing today.

For complete information about books available from Penguin – including Puffins, Penguin Classics and Arkana – and how to order them, write to us at the appropriate address below. Please note that for copyright reasons the selection of books varies from country to country.

In the United Kingdom: Please write to *Dept. EP, Penguin Books Ltd, Bath Road, Harmondsworth, West Drayton, Middlesex UB7 0DA*

In the United States: Please write to *Consumer Sales, Penguin Putnam Inc., P.O. Box 12289 Dept. B, Newark, New Jersey 07101-5289.* VISA and MasterCard holders call 1-800-788-6262 to order Penguin titles

In Canada: Please write to *Penguin Books Canada Ltd, 10 Alcorn Avenue, Suite 300, Toronto, Ontario M4V 3B2*

In Australia: Please write to *Penguin Books Australia Ltd, P.O. Box 257, Ringwood, Victoria 3134*

In New Zealand: Please write to *Penguin Books (NZ) Ltd, Private Bag 102902, North Shore Mail Centre, Auckland 10*

In India: Please write to *Penguin Books India Pvt Ltd, 11 Community Centre, Panchsheel Park, New Delhi 110017*

In the Netherlands: Please write to *Penguin Books Netherlands bv, Postbus 3507, NL-1001 AH Amsterdam*

In Germany: Please write to *Penguin Books Deutschland GmbH, Metzlerstrasse 26, 60594 Frankfurt am Main*

In Spain: Please write to *Penguin Books S. A., Bravo Murillo 19, 1° B, 28015 Madrid*

In Italy: Please write to *Penguin Italia s.r.l., Via Benedetto Croce 2, 20094 Corsico, Milano*

In France: Please write to *Penguin France, Le Carré Wilson, 62 rue Benjamin Baillaud, 31500 Toulouse*

In Japan: Please write to *Penguin Books Japan Ltd, Kaneko Building, 2-3-25 Koraku, Bunkyo-Ku, Tokyo 112*

In South Africa: Please write to *Penguin Books South Africa (Pty) Ltd, Private Bag X14, Parkview, 2122 Johannesburg*

READ MORE IN PENGUIN

SCIENCE AND MATHEMATICS

Six Easy Pieces Richard P. Feynman

Drawn from his celebrated and landmark text *Lectures on Physics*, this collection of essays introduces the essentials of physics to the general reader. 'If one book was all that could be passed on to the next generation of scientists it would undoubtedly have to be *Six Easy Pieces*' John Gribbin, *New Scientist*

A Mathematician Reads the Newspapers John Allen Paulos

In this book, John Allen Paulos continues his liberating campaign against mathematical illiteracy. 'Mathematics is all around you. And it's a great defence against the sharks, cowboys and liars who want your vote, your money or your life' Ian Stewart

Dinosaur in a Haystack Stephen Jay Gould

'Today we have many outstanding science writers . . . but, whether he is writing about pandas or Jurassic Park, none grabs you so powerfully and personally as Stephen Jay Gould . . . he is not merely a pleasure but an education and a chronicler of the times' *Observer*

Does God Play Dice? Ian Stewart

As Ian Stewart shows in this stimulating and accessible account, the key to this unpredictable world can be found in the concept of chaos, one of the most exciting breakthroughs in recent decades. 'A fine introduction to a complex subject' *Daily Telegraph*

About Time Paul Davies

'With his usual clarity and flair, Davies argues that time in the twentieth century is Einstein's time and sets out on a fascinating discussion of why Einstein's can't be the last word on the subject' *Independent on Sunday*